BICONTINUOUS
LIQUID CRYSTALS

SURFACTANT SCIENCE SERIES

BICONTINUOUS LIQUID CRYSTALS

Edited by
Matthew L. Lynch
Proctor and Gamble Company
Cincinnati, Ohio, U.S.A.

Patrick T. Spicer
Proctor and Gamble Company
Cincinnati, Ohio, U.S.A.

CRC Press
Taylor & Francis Group
Boca Raton London New York

CRC Press is an imprint of the
Taylor & Francis Group, an **informa** business
A TAYLOR & FRANCIS BOOK

CRC Press
Taylor & Francis Group
6000 Broken Sound Parkway NW, Suite 300
Boca Raton, FL 33487-2742

First issued in paperback 2019

© 2005 by Taylor & Francis Group, LLC
CRC Press is an imprint of Taylor & Francis Group, an Informa business

No claim to original U.S. Government works

ISBN-13: 978-1-57444-449-0 (hbk)
ISBN-13: 978-0-367-39287-1 (pbk)
Library of Congress Card Number 2005041835

Library of Congress Cataloging-in-Publication Data

Bicontinuous liquid crystals / edited by Matthew L. Lynch, Patrick T. Spicer.
        p. cm. -- (Surfactant science series ; 127)
    Includes bibliographical references and index.
    ISBN 1-57444-449-2 (alk. paper)
    1. Liquid crystals. I. Lynch, Matthew L. II. Spicer, Patrick T. III. Series.

QD923.B53 2005
530.4'29--dc22   2005041835

**Visit the Taylor & Francis Web site at
http://www.taylorandfrancis.com**

**and the CRC Press Web site at
http://www.crcpress.com**

Preface

Bicontinuous structures are uniquely composed of two interpenetrating, yet non-intersecting, domains separated by a surface. In the case of liquid crystalline systems, one of the more common bicontinuous structures is based on a bilayer surface that separates two continuous but distinct aqueous domains. Some of the earliest dealings with bicontinuous liquid crystalline phases were reported by Luzzati and co-workers in soap/water systems; the subsequent observation of these phases in other systems (including those containing biological lipids) fascinated researchers and stimulated further investigation. In industrial practice, formation of a "viscous isotropic phase" was a significant nuisance worthy of avoidance at all costs. In academic circles, the first observations of the sponge phase so confounded researchers that it was termed the "anomalous" phase. As if indifferent to their size, polymeric surfactants and amphiphilic polymers assemble into comparably structured bicontinuous liquid crystalline phases. Finally, studies of lung tissue surfactant reveal the unmistakable architecture of nature demonstrating the presence of bicontinuous liquid crystalline phases. It is evident that, in addition to being intriguing, these phases are also quite ubiquitous! As developments have progressed in techniques as diverse as crystallography, differential geometry, and colloid science, rich and beautiful dynamics have been found lurking in the stark complexity of the bicontinuous liquid crystalline phases. Their structures, once mysterious and annoying, have been found to possess numerous properties beneficial to work in medicine, consumer products, material science, and biotechnology.

This volume considers four different equilibrium phases: cubic, mesh, ribbon, and sponge. The first three have intriguing bicontinuous microstructures and while the sponge is not strictly bicontinuous it exhibits many similar microstructure features justifying its inclusion in this volume. Bicontinuous cubic phases are labeled by their crystallographic nomenclature: I_{a3d}, P_{n3m}, and I_{m3m}, although they are more euphemistically referred to as the Gyroid, Double Diamond, and Plumber's Nightmare, respectively. Their structure is unusually symmetric, containing essentially all combinations of symmetry elements common in crystallography, resulting in a three-dimensional periodic network of oil and water channels. The stability of these microstructures is often described in terms of differential geometry and minimal surfaces; bending bilayers costs energy but bending the bilayers into the shape of minimal surfaces actually leads to a negligible net bending energy and stable structures. Mesh and ribbon phases, essentially two-dimensional analogues of the cubic phases, are often referred to as intermediate phases as they are sandwiched on the phase diagram between more common phases such as lamellar and hexagonal phases. Finally, sponge phase has a somewhat different set of structural characteristics. It resembles a "disordered" cubic phase or perhaps a "leaky" lamellar phase, in which holes are randomly punched through adjacent lamellar sheets. The structure contains very few symmetry elements, reflecting the lack of periodic structure, and resembles a common kitchen sponge in electron micrographs. While sponge is clearly a thermodynamic phase, its lack of long-range order excludes it from formal definitions of liquid crystals. Despite this, the bicontinuous microstructure of the sponge phase and its myriad applications warrant inclusion in this volume.

This volume attempts to present a comprehensive overview of the latest understanding of bicontinuous liquid crystals as well as a practical approach to applying these materials to manufacturing and laboratory processes. It was impossible, however, to secure authors on all the topics. In an attempt to compensate for this omission, the editors have put together a brief description of the most important topics along with pivotal references to help round out the volume. The volume is broken into three sections:

Section 1: Theoretical Aspects and Modeling of Bicontinuous Phases—provides a theoretical platform on which the concepts of bicontinuous liquid crystalline phases can be described and understood including differential geometry, conformational entropy, local and global packing. Thermodynamic considerations will be analyzed, beginning with a historical perspective

Section 2: Physical Chemistry and Characterization—provides a detailed discussion of the microscopic and macroscopic properties of these phases including aqueous systems and those involving biological lipids, pharmaceutical and biomolecular interactions, polymers and block copolymers, charged surfactants, high pressure environments, sponge phases, flow-induced effects, high viscosities, and unusual transport characteristics.

Section 3: Applications—presents an overview of unique applications of these phases including controlled release, materials development, fabrication, processing, polymerization, protein crystallization, membrane fusion, and treatment of human skin.

It is hoped that this work will provide a common ground to link researchers in varied fields and attract new ones to an area of common interest.

Introduction

Our goal for this book was to assemble a volume that represented current trends and leading-edge thoughts in the field of bicontinuous liquid crystals. We chose to include sections on theoretical and modeling aspects, physical chemistry and characterization, and applications of these fascinating materials that represent the cutting edge of research and applications of bicontinuous phases. Although the volume contains an impressive array of researchers and topics, it is inevitable that some of the relevant areas are not represented. For completeness, we compile below key references and reviews on important aspects of bicontinuous liquid crystals that could not be included in this volume.

One of the most intriguing (and challenging) aspects of working with bicontinuous systems is mathematically visualizing their structures. This is easily understood given that some bicontinuous structures have nicknames like "plumber's nightmare." A substantive barrier to visualizing bicontinuous liquid crystalline structures is that they cannot be analytically described. A significant advance came in the discovery that nodal surfaces, based on linear combinations of sine and cosine functions, are closely related to minimal surfaces and offer simple analytical expressions to approximate and plot bicontinuous phase structures (1). Such expressions are now routinely used for visualization in software such as *Mathematica* (2). A definitive reference is *The Language of Shape* (3), a compilation of research prior to 1997 that stimulated much of the work in this volume. Mesoscale modeling offers a numeric alternative to plotting analytic expressions by modeling bicontinuous phase self-assembly from a random state (4). Wënnerstrom provides a comprehensive discussion of related treatments for sponge phases (5–7).

Significant advances in the study of bicontinuous liquid crystals have come with the ability to characterize their structures. Polarized light microscopy traditionally characterizes anisotropic liquid crystals such as lamellar or hexagonal phases (8). However, it is also valuable for the study of bicontinuous isotropic liquid crystal structures when they exhibit flow birefringence (9). The growth of large-faceted cubic phase structures may be characterized under reflected light conditions (10). The use of Jamin-Lebedev optics offers the ability to map cubic (11) and sponge phase (12) boundaries. It is also possible to image these phases with electron microscopy. Freeze-fracture transmission electron microscopy (FF/TEM) is one of the premier electron microscopy techniques to measure concentrated surfactant phases, allowing visualization of complex organization of individual domains (13–15) including the "wormy" patterns associated with sponge phase structures (16). Cryo-transmission electron microscopy (cryo-TEM) provides very high resolution images of dispersed liquid crystalline particles, such as square-shaped cubosomes (17–19). Further, Talmon (20) suggests that the sponge phase has also been imaged by this technique. Newer techniques like atomic force microscopy have also provided insight into the behavior of dispersed cubic and other liquid crystalline phases (21). Finally, x-ray diffraction is a well-established tool for characterizing liquid crystalline structures (22) with recent work from Caffrey (23) and Landh (24) providing interesting insight into the temperature- and composition-dependent changes in the cubic liquid crystal lattices.

It is also important to note established work in the characterization of mechanical properties of bicontinuous phases. Bicontinuous cubic phases behave as solid-like viscoelastic fluids (25, 26) with relatively constant modulus prior to rupture of the lattice (27). The regular lattice structure of cubic phases provides one of the only models of yield stress in liquid crystalline systems based on the disruption of unit cell branches (27). Interestingly, disordered sponge phases exhibit significantly lower viscosity than cubic phases despite structural similarities, and are characterized by pronounced flow birefringence above a critical shear rate (9).

Although recent reviews exist (28), it is worth noting some of the seminal works of discovery and understanding in the area of dispersed bicontinuous phases given their role in stimulating much of the current research. Larsson is credited with discovery and certainly naming of cubosomes (29), as well as linkage of disparate work on microbiology (30), differential geometry (31), and fat digestion (32) into a common area. Later work is crucial to a comprehensive understanding of the interplay between equilibrium phase behavior of cubic phases (33) and their kinetic behavior when dispersed into cubosomes (19, 20). Some empirical work has also been conducted on high-energy dispersion of bulk cubic phases into cubosomes and the effective yield obtained (34).

REFERENCES

1. Von Schnering, H. G. and Nesper, R. Nodal Surfaces of Fourier Series: Fundamental Invariants of Structured Matter. *Z. Phys. B* 1991, *83* 407–412.
2. Wolfram, S. *The Mathematica Handbook*. 5th ed. Wolfram Media: Champaign, 2003.
3. Hyde, S.; Andersson, A.; Larsson, K.; Blum, Z.; Landh, T.; Lidin, S. and Ninham, B. W. *The Language of Shape*. Elsevier: New York, 1997.
4. Larson, R. G.; Monte Carlo Simulation of Microstructural Transitions in Surfactant Systems. *J. Chem. Phys.* 1992, *96*, 7904–7918.
5. Wènnerstrom, H; Daicic, J.; Olsson, U.; Jerke, G.; Schurtenberger, P. Sponge phases and balanced microemulsions: What determines their stability? *J. Mol. Liq.* 1997, *72* 15–30.
6. Le, T. D.; Olsson, U.; Wènnerstrom, H.; Schurtenberger, P. Thermodynamics of a nonionic sponge phase. *Phys. Rev. E* 1999, *60* 4300–4309.
7. Daicic, J.; Olsson, U.; Wènnerstrom, H.; Jerke, G.; Schurtenberger, P. Thermodynamics of the L3 (sponge) phase in the flexible surface model. *J. Phys.* II 1995, 5, 199–21.
8. Rosevear, F. B. The Microscopy of the Liquid Crystalline Neat and Middle Phases of Soaps and Synthetic Detergents. *J. Amer. Chem. Soc.* 1954, *31* 628–639.
9. Diat, O. and Roux, D. Effect of Shear on Dilute Sponge Phase. *Langmuir* 1995, *11* (4), 1392–1395.
10. Pieranski, P.; Sittler, L.; Sottaa, P. and Imperor-Clerc, M. Growth and Shapes of a Cubic Lyotropic Liquid Crystal. *Eur. Phys. J. E* 2001, *5* 317–328.
11. Lynch, M. L.; Kochvar, K. A.; Burns, J. L. and Laughlin, R. G. Aqueous-Phase Behavior and Cubic Phase-Containing Emulsions in the $C_{12}E_2$-Water System. *Langmuir* 2000, *16* 3537–3542.

12. Laughlin, R. G. "Recent Advances in Aqueous Surfactant Phase Science: Coexistence Relationships of the 'Sponge' Phase," in *Micelles, Microemulsions, and Monolayers*, D. O. Shah, Ed., Marcel Dekker, New York, 1998.

13. Gulik-Krzywicki, T.; Aggerbeck, L. P. and Larsson, K. The Use of Freeze-Fracture and Freeze-Etching Electon Microscopy for Phase Analysis and Structure Determination of Lipid Systems. In *Surfactants in Solution* K. Mittal and B. Lindman, Eds. Plenum Press, New York, 1984; 237–257.

14. Gulik-Krzywicki, T. and Delacroix, H. Combined Use of Freeze-Fracture Electron Microscopy and X-Ray Diffraction for the Structure Determination of Three-Dimensionally Ordered Specimens. *Biol. Cell* 1994, *80* 193–201.

15. Delacroix, H.; Gulik-Krzywicki, T.; Mariani, P.; Risler, *J. Liquid Crystal* 1993, 15, 605–625.

16. Strey, R.; Jahn, W.; Porte, G.; Bassereau, P.; Freeze Fracture Electron Microscopy of Dilute Lamellar and Anomalous Isotropic (L$_3$) Phases. 1990, *6*, 1635–1639.

17. Almgren, M.; Edwards, K. and Gustafsson, J. Cryotransmission Electron Microscopy of Thin Vitrified Samples. *Curr. Opin. Colloid Int. Sci.* 1996, *1* (2), 270–278.

18. Gustafsson, J.; Ljusberg-Wahren, H.; Almgren, M. and Larsson, K. Cubic Lipid-Water Phase Dispersed into Submicron Particles. *Langmuir* 1996, *12* (20), 4611–4613.

19. Gustafsson, J.; Ljusberg-Wahren, H.; Almgren, M. and Larsson, K. Submicron Particles of Reversed Lipid Phases in Water Stabilized by a Nonionic Amphiphilic Polymer. *Langmuir* 1997, *13* 6964–6971.

20. Talmon, Y. 2004 personal communication

21. Neto, C.; Aloisi, G.; Baglioni, P. and Larsson, K. Imaging Soft Matter with the Atomic Force Microscope: Cubosomes and Hexosomes. *J. Phys. Chem.* B 1999, *103* (19), 3896–3899.

22. Longley, W. and Mcintosh, T. J. A Bicontinuous Tetrahedral Structure in a Liquid-Crystalline Lipid. *Nature* 1983, *303* 612–614.

23. Qiu, H. and Caffrey, M. The Phase Diagram of the Monoolein/Water System: Metastability and Equilibrium Aspects. *Biomaterials* 2000, *21* 223–234.

24. Barauskas, J.; Landh, T. Phase Behavior of the Phytantriol/Water System. *Langmuir* 2003, 19, 9562–9565.

25. Jones, J. L. and Mcleish, T. Concentration Fluctuations in Surfactant Cubic Phases: Theory, Rheology, and Light Scattering. *Langmuir* 1999, *15* (22), 7495–7503.

26. Radiman, S.; Toprakcioglu, C. and Mcleish, T. Rheological Study of Ternary Cubic Phases. *Langmuir* 1994, *10* 61–67.

27. Warr, G. G. and Chen, C.-M. Steady Shear Behavior of Ternary Bicontinuous Cubic Phases. In *Structure and Flow in Surfactant*

Solutions C. A. Herb and R. K. Prud'homme, Eds. American Chemical Society, Washington, D.C., 1994;578, 306–319.

28. Spicer, P. T. Cubosomes: Bicontinuous Cubic Liquid Crystalline Nanostructured Particles. Cubosomes: Bicontinuous Cubic Liquid Crystalline Nanostructured Particles. In *Marcel Dekker Encyclopedia of Nanoscience and Nanotechnology* J. A. Schwarz, C. Contescu and K. Putyera, Eds. Marcel Dekker, New York, 2003.

29. Larsson, K. Cubic Lipid-Water Phases: Structures and Biomembrane Aspects. *J. Phys. Chem.* 1989, *93* 7304–7314.

30. Gunning, B. E. S. The Greening Process in Plastids 1. The Structure of the Prolamellar Body. The Greening Process in Plastids 1. The Structure of the Prolamellar Body. In *Protoplasma* K. Höfler and K. R. Porter, Eds. Springer-Verlag, New York, 1965;111–139.

31. Scriven, L. E. Equilibrium Bicontinuous Structure. *Nature* 1976, *263* 123–125.

32. Patton, J. S. and Carey, M. C. Watching Fat Digestion. *Science* 1979, *204* 145–148.

33. Landh, T. Phase Behavior in the System Pine Oil Monoglycerides-Poloxamer 407-Water at 20° C. *J. Phys. Chem.* 1994, *98* 8453–8467.

34. Siekmann, B.; Bunjes, H.; Koch, M. H. J. and Westesen, K. Preparation and Structural Investigations of Colloidal Dispersions Prepared from Cubic Monoglyceride-Water Phases. *Int. J. Pharm.* 2002, *244* (1-2), 33–43.

The Editors

Matthew Lynch has a Ph.D. and B.S. in chemistry from the University of Wisconsin and the Virginia Tech. He is currently a senior research scientist in the Colloid and Surfactant Group, Corporate Research Division, Procter & Gamble Company. He has written numerous publications and filed patents in the area of colloids, nanoparticles, liquid crystalline systems, solid-state behavior of soaps, and non-linear optics of surfaces. He is an adjunct assistant professor of chemistry at the University of Cincinnati in the College of Applied Science where he teaches a course in Surfactant and Colloid Science. Dr. Lynch is a member of the American Chemical Society (ACS), American Institute of Chemical Engineers (AIChE), and American Society for the Advancement of Science, and has worked in numerous outreach programs including the Minorities in Math, Science and Engineering (M2SE), the PACT Ambassador Program and the Institute of Chemical Education (ICE).

Patrick Spicer has a Ph.D. and B.S. in chemical engineering from the University of Cincinnati and the University of Delaware and is a technology leader in the Complex Fluids Group of the Corporate Engineering Division in the Procter & Gamble Company. His research interests include population balance modeling, crystallization, soft condensed matter, and complex fluids engineering. He is an adjunct assistant professor of chemical engineering at the University of Cincinnati. In 1998 he received the AIChE Best Ph.D. in Particle Technology Award and in 2001 was a participant in the National Academy of Engineering Frontiers of Engineering Program. Dr. Spicer is a member of the AIChE, AAAS, and the ACS.

Contributors

R. Beck
Universität Bayreuth
Bayreuth, Germany

Roland Bodmeier
Freie Universität Berlin
Berlin, Germany

Ben J. Boyd
Monash University
Parkville, Australia

Martin Caffrey
University of Limerick
Limerick, Ireland

Chin-Ming Chang
Allergan Inc.
Irvine, California

Hesson Chung
Korea Institute of Science and
   Technology
Seoul, Korea

Robert W. Corkery
Australian National University
Canberra, Australia
and
YKI, Institute for Surface
   Chemistry (Current address)
Stockholm, Sweden

Nissim Garti
The Hebrew University of
   Jerusalem
Jerusalem, Israel

Steven Hoath
University of Cincinnati
Cincinnati, Ohio

H. Hoffmann
Universität Bayreuth
Bayreuth, Germany

Michael C. Holmes
University of Central
  Lancashire
Preston, United Kingdom

Seo Young Jeong
Korea Institute of Science and
  Technology
Seoul, Korea

Ian W. Kellaway
University of London
London, England

Ick Chan Kwon
Korea Institute of Science and
  Technology
Seoul, Korea

Ehud M. Landau
University of Texas Medical
  Branch
Galveston, Texas

Kåre Larsson
Lund University
Lund, Sweden

Marc S. Leaver
University of Central
  Lancashire
Preston, United Kingdom

Jaehwi Lee
Chung-Ang University
Seoul, Korea

Björn Lindman
Lund University
Lund, Sweden

Lars Norlén
University of Geneva
Geneva, Switzerland
and
Karolinska Institute
Stockholm, Sweden

Stephen E. Rankin
University of Kentucky
Lexington, Kentucky

Valdemaras Razumas
Institute of Biochemistry
Vilnius, Lithuania

David P. Siegel
Givaudan, Inc.
Cincinnati, Ohio

Olle Söderman
Lund University
Lund, Sweden

Acknowledgments

We would like to gratefully acknowledge Arthur Hubbard for his encouragement in the genesis and preparation of this volume; Joseph Stubenrauch and Anita Lekhwani from Taylor & Francis for their patience and guidance; all the contributors for their hard work in preparing truly exceptional chapters; and Jeff Seeley and Rich Owens from the Procter & Gamble Company for providing a nurturing environment in which our research on these liquid crystals flourished.

Contents

SECTION 1   *Theoretical Aspects and Modeling of Bicontinuous Phases*

SECTION 2   *Physical Chemistry and Characterization*

SECTION 3   *Applications*

Section 1

Theoretical Aspects and Modeling of Bicontinuous Phases

1

Bicontinuous Cubic Liquid Crystalline Materials: A Historical Perspective and Modern Assessment

KÅRE LARSSON

CONTENTS

1.1 INTRODUCTION

The general structural features of the different phases in surfactant–water systems were revealed by classical work at the beginning of the 1960s by Luzzati and coworkers, cf. (1). Subsequent to that work, the only major unresolved issue concerned the structure of cubic phases.

From differences in x-ray diffraction, data with tentative space group determinations, and differences in the location of the actual cubic phases in phase diagrams, it was obvious early that different types of cubic structures existed. A first detailed structure determination of a cubic phase, that formed by anhydrous strontium myristate at 232°C, was reported in 1967 (2). The proposed structure was quite complex, with two interpenetrating rod networks in which the rods were joined three and three at their ends. The rods were proposed to consist of strontium ions and carboxyl groups embedded in a hydrocarbon chain matrix.

Bicontinuous cubic structures have been derived successively via a series of studies and assumptions in the interpretations, and their final structures became generally accepted during the 1980s. Some of the history of this work will be summarized below, with emphasis on the early steps.

After thesis work on the crystal structures of fatty acid glycerides, the present author began to study the aqueous phases of monoglycerides. Luzzati kindly offered an opportunity to come to his laboratory and learn the experimental technique for structural studies of liquid-crystalline phases that he had developed. Some results were presented at the Scandinavian Chemical Society Meeting in 1965, but shortly afterwards an extensive and conclusive study of the aqueous systems of a whole series of fatty acid monoglycerides was reported by Lutton (3). Our work in Sweden was then reoriented to cover short-chain members (down to C6) and focused on x-ray diffraction characterization of the phases. It was shown that the phase described as "viscous isotropic" by Lutton was a true cubic phase (4). All monoglycerides with chain length C14 and higher form cubic phases. The most important part of this work, however, did not involve the cubic phases but concerned a general characterization of the colloidal aqueous dispersions of the lamellar liquid-crystalline phase and the demonstration of spontaneous formation of closed spherically concentric aggregates at short chain lengths.

Initially it was natural to assume that cubic phases consisted of spherical micelles or the corresponding inverse structure. Assuming that the micellar arrangement followed the well-known face-centered or body-centered packing arrangements, there were already four alternatives (including the inverse structures). This was also the first approach in the work on monoolein–water cubic phases, indicating that the structure was a body-centered arrangement of water aggregates in a lipid continuum (4).

1.2 MONOOLEIN FORMS A CONTINUOUS BILAYER, WITH A MINIMAL SURFACE CONFORMATION, SEPARATING TWO CONTINUOUS WATER COMPARTMENTS

Many studies of cubic lipid phases at present involve monoolein because of its exceptionally large cubic existence range with regard to both water content and temperature. In a combined NMR and x-ray diffraction study of monoolein–water phases (5), it was reported that the structure was water continuous as well as lipid continuous, and a bilayer model of the organization was proposed. It was also realized that this structure could as well be described as the periodic minimal surface termed the P-surface (6). The possibility that minimal surfaces might occur in microemulsion and liquid-crystalline phases of surfactants had earlier been pointed out by Scriven (7). By that time, the general significance of periodic minimal surfaces for structure description of different kinds of organized matter had become obvious, cf. (8).

A minimal surface (MS) conformation is an attractive model of a lipid bilayer. Besides the planar bilayer structure known as the lamellar liquid-crystalline phase, curved bilayer conformations are also possible. Such curved infinite bilayers can be free from self-intersections, and the bilayer is as concave as it is convex everywhere (the center of the bilayer follows the MS). Three fundamental types of cubic MS occur in lipid–water systems: the P-surface (termed P as it was named after its primitive space group), the D-surface (D for its relation to the diamond lattice), and the G-surface (G for gyroid). The MS—the center of the bilayer—partitions space into two equivalent continuous labyrinth systems.

A new MS-type of structure of a cubic monoolein phase, which had been obtained at maximal swelling in water, was reported in 1983 by Longley and McIntosh (9). These authors determined the space group to be Pn3m, and the corresponding MS structure is that of the D-surface. Our earlier indexing of the monoolein–water phases (5) was based on the assumption that only one cubic phase exists over the whole region. The remarkable fact was now revealed (10) that there were two phases with such a perfect epitaxial relation that the x-ray diffraction spacing was perfectly linear against composition (c/1 – c). There was only one way to index these diffraction lines as one phase, which gives space group Im3m as the least common denominator of the space groups Ia3d and Pn3m.

Based on the knowledge that the cubic region consists of two different phases, it was finally possible to give a full description of

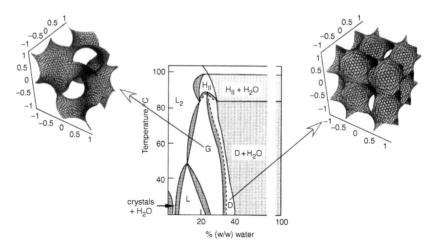

Figure 1.1   Phase diagram of the monoolein–water system as reported in ST Hyde, S Andersson, B Ericsson, K Larsson. *Z Kristallogr* 168:213–219, 1984, with the structure of the bilayer of the two cubic phases indicated by the corresponding nodal surface.

the monoolein–water cubic phases (11), as shown in Figure 1.1. Some features of this phase diagram, which have been modified according to work by Caffrey, are shown in his chapter of this book.

It is now generally accepted that the center of the bilayer of the cubic monoolein–water phases follows minimal surfaces with the D- and G-types of bilayer conformation. Also, the P-surface can be formed, for example at solubilization of proteins into monoolein–water phases (see next paragraph).

Some efforts have been made to determine these cubic structures from electron density calculations based on the reflection intensities. One problem is orientation effects on the intensities, as the cubic phases tend to form large single-crystal domains. An analysis of experimental diffraction data of monoolein and phospholipid cubic phases was recently reported by Garstecki and Holyst (12,13). Their model of these cubic phases consists of a bilayer of constant thickness (and water layers of constant thickness in the inverse structures). This feature might be questioned, however. The bilayer thickness should be expected to be considerably thicker at the regions with maximal absolute value of the Gaussian curvature compared to the regions with minimal value. Still, this would probably not change the structural conclusions that they reported.

A problem that remains to be solved is the exact epitaxial relation between the (211)-index of the G-surface monoolein structure and the (111)-index of the D-surface structure.

1.3 THE NODAL SURFACE DESCRIPTION OF CUBIC MINIMAL SURFACE STRUCTURES

In 1991, von Schnering and Nesper (14) reported that they had discovered that the roots of the Fourier series of the first structure factors give periodic nodal surfaces (PNS) that are closely related to the periodic MS. As the analytical calculation of minimal surfaces is extremely complicated, the nodal surface description is now routinely used; it provides simple equations for calculation of bilayer structure. The physical significance of PNS will be addressed below based on the electron density in a periodic structure.

The electron density ρ at a point within the unit cell defined by the vector $\mathbf{r}$ is obtained from the sum over all structure factors $F(\mathbf{H})$:

$$\rho(\mathbf{r}) = \Sigma F(\mathbf{H}) \exp(-2\pi i \mathbf{H}\mathbf{r})/V$$

where V is the volume of the unit cell and $\mathbf{H}$ is a vector in reciprocal space, defined by the indices (hkl). $F(\mathbf{H})$ is obtained from the intensities of the x-ray reflections. Every structure factor generates a wave function contributing to the total electron density distribution. These waves represent various levels in the electron density space. If we take the P-surface with space group Im3m, the F(100) structure factor will contribute to the electron density in a point (xyz) within the unit cell with:

$$F(100) \ (\cos 2\pi x + \cos 2\pi y + \cos 2\pi z)$$

when the symmetry is taken into account according to the *International Tables of Crystallography* (ed. 1976).

The nodal surface of this wave function is:

$$F(100) \ (\cos 2\pi x + \cos 2\pi y + \cos 2\pi z) = 0$$

The volume on each side of this PNS within the unit cell will be identical. It is also evident that this wave function will exhibit the smallest possible surface area as the index (100) represents the longest wavelength of the elements contributing to the electron density. It is therefore not surprising that this PNS will be very close to the minimal surface with the same symmetry, which is the P-surface.

The D-surface is obtained from the (111)-structure factor and the G-surface from the structure factor with index (110), when the actual symmetry is taken into account (directly available from *The International Tables of Crystallography*).

Another approach in our understanding of PNS as models of cubic bilayer structures is to consider the dynamic properties of the bilayer. In lamellar liquid-crystalline phases it is generally accepted that undulations of the bilayer occur. If similar bilayer motions should exist in a cubic bilayer phase, they must take place in phase with the corresponding motions in all the other unit cells (a direct consequence of the cubic periodicity evident by the x-ray diffraction). Thus these motions will form standing waves. If the P-surface equation above is replaced by

$$F(100) \ (\cos 2\pi x + \cos 2\pi y + \cos 2\pi z) = A \ \sin \omega t$$

the surface will move through the nodal surface with an amplitude equal to A and an angular frequency of ω (t is time). Such motions will reflect the elastic properties of the bilayer and density fluctuations in the aqueous medium.

Due to the thermal motions of the bilayer in cubic phases, the PNS description therefore seems to be an adequate description of the conformation; perhaps it can be regarded as better motivated then the minimal surface description.

1.4 SOLUBILIZATION OF AMPHIPHILIC MOLECULES INTO THE MONOOLEIN–WATER PHASE

The shape of the molecules that can be solubilized into a cubic monoolein phase will determine the ultimate bilayer conformation. An important application is fat digestion and absorption in the gastrointestinal tract. When lipase attacks a triglyceride oil such as triolein in the intestine, monoolein and free fatty acids are formed. Only very low concentrations of triolein can be solubilized into the bilayer of the monoolein cubic phase, and above this limit an inverse hexagonal phase is first formed and then an L2 phase (15). Recent studies of model systems corresponding to intestinal fat digestion by lipase have shown spontaneous formation of cubic particles (16), and formation of the cubic structure seems to be one significant feature of the metabolism of fats in foods.

In a study of the ternary phase diagram monoolein–lysozyme–water (17), it was seen that cubic phases consisting of monoolein bilayers can incorporate up to 34% (w/w) of lysozyme. Thermal analysis indicated that the lysozyme molecules kept their native structure, as the denaturation temperature (64°C) and enthalpy were the same as those in water solution. The structure of this cubic phase follows the P-surface (18). Various other proteins were also examined, and proteins with a molecular weight up to 150 kDa were seen to form cubic phases with monoolein.

Another interesting ternary phase diagram is that of pine needle oil monoglycerides–water–poloxamer 407 determined by Landh (19). All three cubic phases, corresponding to the minimal surface structures D, G, and P, were seen. This phase diagram has been particularly important due to its ability to give well-defined dispersions of cubic phases (see below).

Numerous recent studies of solubilization of amphiphilic molecules into the cubic monoolein–water phase, showing that induction of phase transitions can be predicted quite well from the molecular geometry (towards either the lamellar or the inverse hexagonal phase) have been published.

1.5 COLLOIDAL DISPERSIONS OF CUBIC PHASES

A study of mechanical fragmentation of the cubic monoolein phase in water in the presence of a lamellar phase of phosphatidylcholine showed that colloidal dispersions with good kinetic stability could be obtained (18). It was natural to term these particles of the cubic phase Cubosomes®. Dispersions of the cubic monoolein phase with much higher stability were obtained by Landh using Poloxamer 407. Such Cubosomes have been well characterized by cryo–transmission electron microscopy and x-ray diffraction (20,21) and also by NMR (22).

Cell membranes have been shown to sometimes form Cubosomes (18), and Landh has reported a wide range of examples (23). It seems that such cubic membrane structures often represent a vegetative state, without (or with reduced) metabolic activity (24).

In order to describe particles of finite periodicity, such as Cubosomes, the so-called exponential scale mathematics discovered by Sten Andersson, cf. (25), provides a useful tool, which will be demonstrated below. Functions defining hyperbolic geometry can be combined with functions giving elliptic geometry. In that way inner periodicity can be combined with outer shape by adding the corresponding functions on an exponential level into one single function. An example of a Cubosome with a P-surface core and outer cube shape is shown in Figure 1.2. As can be seen from the function given in the figure caption, the first expression gives three repetition units

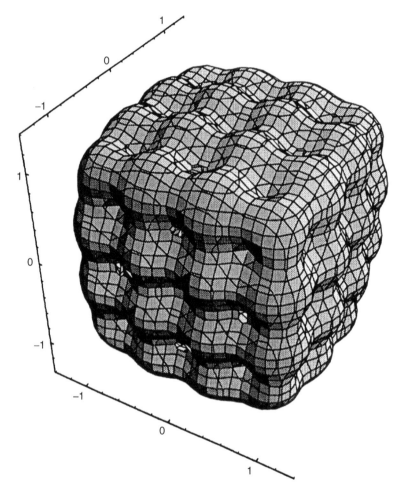

Figure 1.2 A P-surface type of Cubosome calculated by the function exp(cos 3πx + cos 3πy + cos 3πz) + exp(–/x/) + exp(–/y/) + exp(–/z/) = 8.7 using the exponential description, cf. S Andersson, K Larsson, M Larsson, M Jacob. *Biomathematics: Mathematics of Biostructures and Biodynamics.* Amsterdam, Elsevier, 1999, pp. 375–429, and the program Mathematica version 2.2 on Macintosh.

of the P-surface (as a PNS) and the other exponential functions give the outer cube. This Cubosome, like other cubic particles calculated in the same way, forms a closed bilayer structure. This is in agreement with the surface of Cubosomes observed experimentally, cf. (20,21).

The outer shape of Cubosomes reflects the inner structure, as in true crystals. Thus the P-surface type of structure of a cubic phase that is fragmented results in cubic shape, whereas the D-type tends to give particles with the shape of a dodecahedron, and the G-type of Cubosome usually exhibits a spherical outer surface (21).

1.6 FUTURE ASPECTS

Two features of bicontinuous cubic lipid–water phases, summarized below, are expected to be particularly important in the future in basic research as well as in technical applications.

Incorporation of proteins and other large biomolecules

The cubic lattice of monoolein–water D phase can incorporate protein molecules several hundred kDa in size. By solubilization of small amounts of charged lipids such as phosphatidylcholine into the monoolein bilayers (26), much larger aqueous compartments can be obtained, with a corresponding growth of the maximal size of incorporated macromolecules. Chemical reactions in thin films of such cubic phases are an interesting possibility that can be taken advantage of in biosensors (27). Incorporation of glucose oxidase in a surface region of a food package material, for example, will provide protection against oxidation by eliminating (or reducing) the oxygen content in a food product containing glucose.

Oxidative and hydrolytic degradation of monoolein might be a problem in technical applications. As an alternative to monoolein, phytoantriol could sometimes be considered, with an aqueous phase diagram very similar to that of monoolein (29).

The crystallization of membrane proteins in these cubic phases, treated elsewhere in this book, is expected to become a significant crystallographic tool with possibilities also in biotechnology.

Nanostructures based on dispersed cubic phases

An interesting question is how small particles of a cubic phase can be made under controlled conditions. Can particles equal to one unit cell be produced? When the lipid bilayer is closed during the formation of a cubic particle, there should ideally be no contacts between the hydrocarbon chain interior of the bilayer and water. This means that the cubic particle contains two kinds of water compartments. One is closed, like the core of a vesicle, and another is in open contact with the outside water medium. How is the balance between these compartments influenced by changes in the outside water phase, such as pH and ionic strength? These are examples of interesting basic problems that should be addressed by future research.

On the technology side the most interesting development can be expected within medical applications. The recently reported peroral administration of insulin, an important achievment in drug delivery (28), is treated elsewhere in this book. The mechanism termed enhanced permeability and retention (EPR) of colloidal particles within cancer tumors means that Cubosomes carrying anticancer drugs will be enriched in tumor tissue. The possibility of binding antibodies to a Cubosome surface represents another alternative method of drug targeting.

Cubosomes can also be prepared from food-grade components of low cost, for example monoolein and caseinates, and therefore broader controlled release applications, including applications in food technology and nutrition, can be expected in the future.

REFERENCES

1. V Luzzati, T Gulik-Krzywicki, A Tardieu. *Nature* 218:1031–1034, 1968.

2. V Luzzati, PA Spegt. *Nature* 215:710–715, 1967.

3. E Lutton. *J Am Oil Chem Soc* 42:1068–1070, 1965.

4. K Larsson. *Z Phys Chem* 56:173–198, 1967.

5. G Lindblom, K Larsson, L Johansson, K Fontell, S Forsén. *J Am Chem Soc* 101:5465–5470, 1979.

6. K Larsson, K Fontell, N Krog. *Chem Phys Lipids* 27:321–328, 1980.

7. LE Scriven. *Nature* 263:173–175, 1976.

8. S Andersson, ST Hyde, K Larsson, S Lidin. *Chem Rev* 88:221–242, 1988.

9. W Longley, TJ McIntosh. *Nature* 303:612–613, 1983.

10. K Larsson. *Nature* 304:664, 1983.

11. ST Hyde, S Andersson, B Ericsson, K Larsson. *Z Kristallogr* 168:213–219, 1984.

12. P Garstecki, R Holyst. *Langmuir* 18:2519–2528, 2002.

13. P Garstecki, R Holyst. *Langmuir* 18:2529–2537, 2002.

14. HG von Schnering, R Nesper. *Z Phys B Condensed Matter* 83:407–412, 1991.

15. M Lindström, H Ljusberg-Wahren, K Larsson, B Borgström. *Lipids* 16:749–754, 1981.

16. J Borné. Lipid self-assembly and lipase action. Thesis, Lund Univesity, Lund, Sweden, 2002.

17. B Ericsson, K Larsson, K Fontell. *Biochim Biophys Acta* 729:23–27, 1983.

18. K Larsson. *J Phys Chem* 93:7304–7314, 1989.

19. T Landh. *J Phys Chem* 98:8453–8467, 1998.

20. G Gustavsson, H Ljusberg-Wahren, K Larsson, M Almgren. *Langmuir* 12:4611–4613, 1996.

21. G Gustavsson, H Ljusberg-Wahren, K Larsson, M Almgren. *Langmuir* 13:6964–6971, 1997.

22. M Monduzzi, H. Ljusberg-Wahren, K Larsson. *Langmuir* 16:7355–7358, 2000.

23. T Landh. Cubic cell membrane architectures. Thesis, Lund Univesity, Lund, Sweden, 1996.

24. K Larsson. *Curr Opin Colloid Interface Sci* 5:64–69, 2000.

25. S Andersson, K Larsson, M Larsson, M Jacob. *Biomathematics: Mathematics of Biostructures and Biodynamics.* Amsterdam, Elsevier, 1999, pp. 375–429.

26. J Engblom, Y Miezis, T Nylander, V Razumas, K Larsson. *Progr Colloid Polym Sci* 116:9–15, 2000.

27. V Razumas, J Kanapieniene, K Larsson, T Nylander, S Engström. *Anal Chem Acta* 289:155–160, 1994.

28. H Chung, J Kim, JY Um, IC Kwon, SY Jeong. *Diabetologia* 45:448–451, 2002.

29. J Barauskas, T Landh. *Langmuir* 19:9562–9565, 2003.

2

Intermediate Phases

MICHAEL C. HOLMES AND MARC S. LEAVER

CONTENTS

2.1   INTRODUCTION

Lyotropic liquid crystalline phases can be formed in concentrated mixtures of amphiphilic molecules and water [1–6]. The normal hexagonal (H_1) and lamellar (L_α) are two such liquid crystalline phases that are well characterized [7]. The H_1 phase has been shown to consist of infinite cylindrical micellar rods, which possess a right circular cross section, hexagonally close packed onto a lattice that has p6m symmetry [7–9]. The rod radius in the phase is usually constant and always less than the fully extended hydrophobic chain of the amphiphilic entity [2]. The structural building block of the L_α phase is a bilayer, the thickness of which is never greater than 1.5 times the length of the fully extended hydrophobic moiety of the surfactant [2,7,10–12]. In this phase the bilayers are stacked one on top of another, with water in the inter bilayer region, to form a phase that possesses one-dimensional periodicity. These two phases are ubiquitous to many systems that form lyotropic liquid crystalline phases, with the H_1 phase always occurring at lower amphiphilic volume fractions than the L_α phase. The transition between these phases upon increasing surfactant concentration is marked both by a topological transition, with the mesogenic units going from discrete micellar aggregates having positive interfacial curvature to continuous bilayers with zero curvature, and a commensurate change in the symmetry of the resultant phase. The topological transformation at the H_1 to L_α transition requires an alteration in the packing of the amphiphilic molecules at the mesogen–water interface. If an interfacial organization of the amphiphilic monomers other than that required to form a continuous bilayer minimizes the interactions in the system and possesses a lower interfacial curvature than the H_1, that organization will be stabilized prior to the formation of the L_α.

Two types of phases can be formed in this H_1–L_α region. The first is a phase with a bicontinuous structure that possesses long-range three-dimensional cubic symmetry, the V_1 phase [2,13–15]. Such phases are viscous, isotropic, and therefore optically inactive. The properties of these phases are detailed extensively in other chapters of this book. The mesogenic unit of this phase is a surfactant bilayer that separates two distinct continuous solvent networks. This bilayer decorates a triply periodic surface, the midplane of which possesses cubic symmetry. This surface is locally highly curved, but the overall curvature of the phase is zero. The interfacial curvature of the V_1 phases is therefore uniform and smoothly varying over its unit cell. Three cubic symmetries for the V_1 phase have been observed from extensive x-ray scattering studies: Ia3d, Pn3m, and Im3m [16].

However, a number of phases that are of lower viscosity than the V_1 phase and that are optically birefringent have been observed in the H_1 to L_α region. Such phases have attracted the generic nomenclature "intermediate phases" [17]. Phases with noncubic structures were first identified in aqueous soap mixtures and anhydrous soap melts by Luzzati et al. [18,19] using x-ray scattering. The term intermediate was applied to a rectangular structure found in aqueous mixtures of potassium and sodium oleates and potassium laurate and palmitate. The structures found in the anhydrous soaps were reinterpreted as intermediate, tetragonal, rhombohedral, and ribbon structures [19] in 1968.

Intermediate is a generic term that covers three possible structures [20], see Figure 2.1(a) to Figure 2.1(d). Each can be considered

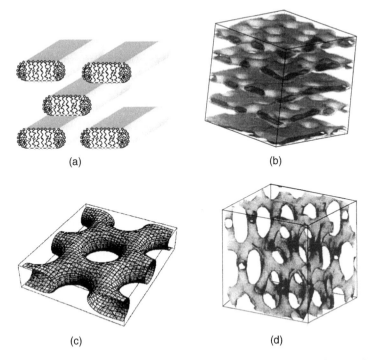

(a)   (b)

(c)   (d)

Figure 2.1 Some representations of possible intermediate phase structures: (a) centered rectangular ribbon phase, Rb_1(c2mm), (b) random mesh phase, $Mh_1(0)$, (c) rhombohedral mesh $Mh_1(R\bar{3}m)$ layer, which is stacked in an ABCA arrangement, and (d) bicontinuous rhombohedral phase structure, $Bc_1(R\bar{3}c)$. Figures 2.1(b) and 2.1(d) are reproduced from M Imai, A Saeki, T Teramoto, A Kawaguchi, K Nakaya, T Kato, K Ito. *J Chem Phys* 115: 10, 525–10, 531, 2001, copyright 2001 from *Journal of Chemical Physics* by M. Imai et al., with permission from Prof. Masyuki Imai and American Institute of Physics, http://www.aip.org/. Interested readers are directed both there and to ST Hyde. *Colloq de Phys* 51: C7-209–C7-227, 1990 for details of the predictive model used to generate the structures.

as a perturbation of the H_1, V_1, and L_α phase structures. The three groups of intermediate phases are ribbon, bicontinuous structures (which do not have cubic symmetry), and mesh respectively. The formation of the intermediate phases has been postulated to result from a perturbation of the balance of head group interactions and hydrophobic chain flexibility [21], a hypothesis supported by their formation in fluorocarbon systems and surfactants with long alkyl chains.

While there are three classes of intermediate phase with their own unique experimental properties, they possess a common property that is characteristic of this class of lyotropic phase: nonuniform interfacial curvature. This is a property of the constituent mesogenic units of the intermediate phases. In the ribbon phase the mesogenic unit is micellar, as in the H_1 phase, but it is elliptical in cross section rather than circular [22], see Figure 2.1(a). Mesh phases consist of surfactant bilayers like those of the L_α phase; however, they are pierced by water-filled pores [23–25]. The position of the pores may (or may not) be correlated between the layers, see Figure 2.1(b) and Figure 2.1(c). Finally, bicontinuous intermediate phases have an interconnected three-dimensional bicontinuous structure similar to that of the V_1 phase, see Figure 2.1(d). However, unlike the V_1 phase, which is isotropic in all directions, the bicontinuous intermediate phase is stretched in one of the three characteristic dimensions of the phase, so as to make it anisotropic in structure, resulting in a nonuniformity of the interfacial curvature [26].

Nonuniform interfacial curvature is not regularly reported in lyotropic phases; however, it is has been shown to be essential for the stabilizing of lyotropic nematic phases [27]. Such phases consist of anisotropic micelles, short rods, small biaxial ribbons, or discs that possess orientational order only [28]. While rare, considerable work has been carried out to prove that the micelles are anisotropic in structure, resulting in a body of work that has elucidated the characteristic properties of the mesogenic units and identified the experimental signatures of nonuniform interfacial curvature [29–31]. The addition of a third component to binary surfactant water mixtures was important in the stabilization of the anisotropic mesogenic units, with a number of authors suggesting an inhomogeneous distribution of surfactant and third component over the surface of the aggregates [32–38]. The geometrical requirements for the formation of the nonuniform interfacial curvature are restrictive, and therefore any phase that possesses this topological feature tends to have only limited temperature and compositional extension in phase diagrams.

In a review, one of us proposed a system of nomenclature for the numerous intermediate phases [20]. Holmes proposed introducing Rb, Mh, and Bc as symbols for ribbon, mesh, and noncubic bicontinuous phases, respectively, i.e., specifying the aggregate type, with the choice of two letters avoiding confusion with previous notations. The subscript 1 or 2 refers to normal or reversed versions of the phase. Since several possible structures exist for each type, the space group [39] could be appended in brackets, as often happens now with

V_1(Ia3d) for a bicontinuous cubic phase. Thus Mh_1 $(R\bar{3}m)$ is a normal rhombohedral mesh phase, and Rb_1(c2mm) a centered rectangular ribbon phase (note the use of a two-dimensional space group in the latter example). Where the structure is unknown or in doubt, one might write Mh_1 or Mh_1(rhombohedral). The random mesh phase, a lamellar phase pierced by water-filled holes, in other words a one-dimensional lattice of liquidlike layers, Figure 2.1(b), is represented by $Mh_1(0)$. A summary of the author's notation is provided in Table 2.1.

Table 2.1 Summary of Nomenclature for Lyotropic Phases and for Their Intermediate Phases

Phase	Symbols	Possible Space Groups	Notes
Isotropic	L_1, L_2, L_3, L_4		L_3, L_4 refer to bicontinuous isotropic and vesicle phases
Micellar cubic	I_1, I_2	Pm3n, Fd3m	
Hexagonal	H_1, H_2	p6mm	Two-dimensional space group
Bicontinuous cubic	V_1, V_2	Ia3d, Pn3m, Im3m	
Lamellar	L_α	pm	One-dimensional space group
Ribbon	Rb_1, Rb_2	c2mm, p2gg	Two-dimensional centered rectangular space groups
Mesh	Mh_1, Mh_2	0, I4mm, $(R\bar{3}m)$	0 represents the random mesh, while three-dimensional space groups I4mm and $(R\bar{3}m)$ represent tetragonal and rhombohedral mesh phases, respectively
Bicontinuous	Bc_1, Bc_2	$R\bar{3}c$	

Note: The subscripts 1 and 2 refer to normal (water-continuous) and reversed (surfactant-continuous) phases.

In this review, phases are referred to by this nomenclature and the authors' notation added in brackets { }.

2.2   RIBBON OR RECTANGULAR PHASES

Ribbon phases have been identified in systems where surfactant molecules aggregate to form long flat ribbons with an aspect ratio of circa 0.5 that are subsequently arranged on two-dimensional lattices of oblique, rectangular (primitive or centered), or hexagonal symmetry. Figure 2.1(a) illustrates the structure of one such ribbon phase, namely a centered rectangular ribbon phase, $Rb_1(c2mm)$. The key to establishing the structure of the phase was the demonstration of an aggregate shape that was noncircular in cross section.

In many systems the stabilization of ribbon-shaped aggregates requires the addition of a third component to facilitate a modification of the intermolecular interactions in order to achieve the preferred surface curvature. Hendrikx and Charvolin [35–38] showed that the addition of decanol to sodium decyl sulfate (SdS)/water decreases the curvature of the amphiphile– water interface, promoting the formation of ribbons from rods. These authors used optical microscopy and x-ray diffraction to reveal the presence of two centered rectangular phases in this system, one $Rb_1(c2mm)$, the other with "herringbone" packing, $Rb_1(p2gg)$. The aspect ratio of the ribbons was found to be circa 0.4. The tilt in the latter structure was about 9° from the *b* axis, which was shown to be energetically more favorable. Neutron scattering showed a micro segregation of the decanol to the flatter parts of the ribbon aggregate [36]. This concentration distribution creates an uneven charge density on the interface, which contributes to an anisotropic electrical potential of the aggregate and which could account for the structural changes as decanol is increased. In a similar system, and in the presence of decanol, NMR studies [40] have suggested that the flattening of rod shaped aggregates into ribbons occurs in the H_1 phases close to calimatic nematic phases composed of short ribbonlike micelles.

The first direct confirmation of ribbon phases came from Chidichimo et al. [41]. In a series of papers, Doane et al. [41–45] and Chidichimo et al. [41,43–48] investigated the ribbon phase and distorted hexagonal phase in ternary mixtures of potassium palmate (KP)/potassium laurate (KL)/water. X-ray diffraction [43] showed the ribbon phase to have a simple rectangular structure where, as the temperature was decreased toward the hexagonal phase, the long dimension of the rectangular cell decreased while the short dimension remained constant, and the ribbon aggregates became more circular. In reference [49], NMR experiments on specifically deuterated surfactants (KL and KP) were used to show that the KL is mainly located in the curved edges of the ribbons, while the KP is located in the flatter surfaces. Here the NMR signal was shown to be sensitive to the aggregate curvature in which the surfactant molecule was residing. Water diffusion measurements [48] confirmed previous structural identifications in this system. In addition the axial ratios, which can be calculated from [48], are consistent with x-ray studies of the phase and results presented by others [50,51]. This work [48] and that of Charvolin et al. [34–38] clearly indicate that the inhomogeneity of the aggregate surface curvature that facilitates the formation of the ribbon phases can be induced by an inhomogeneous distribution of amphiphilic molecules. In a binary system that forms mesogens with nonuniform interfacial curvature, the interfacial inhomogeneity must result from unbalanced interactions or an ability of the amphiphile to adopt discrete conformations that facilitate the stabilization of aggregate geometries with two possible curvature environments. Advances in NMR relaxation have facilitated more sensitive probes of nonuniform interfacial curvature in ribbonlike aggregates [52].

In the sodium dodecyl sulfate (SDS)/water system a two-dimensional lattice of ribbons $Rb_1(c2mm)$ $\{M_\alpha\}$ that had a centered rectangular structure was identified [26]. Interestingly, in this system the ribbon intermediate phase acted as an intermediate between the H_1 phase and a three-dimensional rhombohedral phase, $Bc_1(rhombohedral)\{R_\alpha\}$. This surfactant, in aqueous and nonaqueous solvents, was considered in more detail by Auvray et al. [53,54]. The centered rectangular phase in water was confirmed, although it was not observed in the non-aqueous solvents.

Hagslätt et al. [22,33,55,56] have investigated the formation of ribbon phases in a number of ternary systems. In reference [22], they reviewed the results from a wide variety of studies on ribbon phases and showed that none of these studies is inconsistent with the conclusion that all ribbon phases index to a centered rectangular cell, $Rb_1(c2mm)$. They go on to present results for the dodecyl-1,3-propylene bisamine (DoPDA)/HCl/water system, in which it is possible to vary the surfactant surface charge. Again this is fitted to a centered rectangular structure with aggregate aspect ratios in the range 0.83 to 0.5. A "hexagon-rod" model of the ribbon cross section is suggested, in which the ribbon structure is controlled by the competition between the requirement for a constant water layer thickness around each ribbon, the surface area per molecule, and the minimization of total surface area. Comparisons between experimental results and a Poisson–Boltzman cell model were made [56] using previously published phase diagrams. Both the model and experimental results showed that the axial ratio of the ribbon aggregates increased as temperature, surfactant concentration, or average surfactant charge was decreased. It was proposed that cylindrical aggregates of hexagonal phases generally have noncircular normal sections (with a low axial ratio) at the lowest water content that can be solubilized in the phase, and as the surfactant content was increased further the asymmetry of the aggregates increased until they could no longer pack onto a hexagonal lattice. However, in most of these systems a classical lamellar phase is formed at lower water content, even though the aggregate asymmetry required for the formation of a ribbon phase is present. This indicates that if the system cannot support the nonuniform interfacial curvature, via one of the mechanisms discussed earlier, the increase in axial ratio necessary for the formation of the ribbon phase cannot be supported in most systems, and a major structural reorganization to a phase with a uniform interfacial curvature is favored.

2.3 MESH AND BICONTINUOUS PHASES

Intermediate mesh phase structures were first identified by Luzzati et al. [18,19] from the measurements of Spegt and Skoulios on anhydrous soap melts. It was not until the work of Kékicheff and other workers [17,26,57–60] on ultrapure sodium dodecyl sulfate (SDS)/water and lithium perfluoro-octanoate (LiPFO)/water [23] that intermediate mesh phases were recognized in aqueous surfactant–water systems. The SDS/water system [17,26,58,59] shows a very rich intermediate phase behavior. Kékicheff and Cabane [26] used neutron scattering to study samples of SDS/water aligned by a stack of quartz plates. The change from hexagonal to lamellar phase takes place in a logical and sequential fashion; H_1 {H_α}, hexagonal phase; Rb_1(c2mm) {M_α}, two-dimensional monoclinic phase; Bc_1(rhombohedral) {R_α}, rhombohedral phase; V_1(Ia3d) {Q_α}, cubic phase; Mh_1(I4mm) {T_α}, tetragonal phase; and L_α phase, Figure 2.2. Modulations or defects in the lamellar structure are the final step in the process [61]. The only physical parameter to change in a consistent way throughout the transition from hexagonal to lamellar phase was the mean curvature of the interface, as each following structure had a lower mean surface curvature than its predecessor [17].

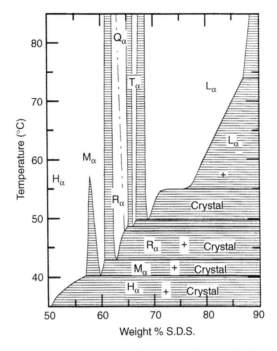

Figure 2.2 The binary phase diagram of the SDS/water system. The symbols associated with the single-phase regions are as follows, noting that the bracketed symbols are used in the original reference while those unbracketed are consistent with MC Holmes. *Curr Opin Colloid Interface Sci* 3: 485–492, 1998: H_1 {H_α}, hexagonal phase; Rb_1(c2mm) {M_α}, two-dimensional monoclinic phase; Bc_1(rhombohedral) {R_α}, rhombohedral phase; V_1 {Q_α}, cubic phase; Mh_1(I4mm) {T_α}, tetragonal phase. Reproduced from P Kékicheff. *Mol Cryst Liq Cryst* 198: 131–144, 1991, copyright 1991 from *Molecular Crystals and Liquid Crystals* by P. Kékicheff. Reproduced by permission of Taylor & Francis, Inc., http://www.routledge-ny.com.

The fluorination of the alkyl chains introduces additional rigidity and so restricts its conformations — an observation that has been postulated to favor intermediate phase formation over cubic phases [21]. Kékicheff and Tiddy [23] showed that a tetragonal mesh intermediate phase, Mh_1(I4mm) {T_α} exists in the lithium perfluorooctanoate (LiPFO)/water system. The lamellar phase above the Mh_1(I4mm) phase also appeared to contain random water-filled defects (an unpublished observation by the authors while studying the synchrotron scattering recorded in the Mh_1(I4mm) phase).

The authors noted the higher viscosity of the correlated mesh over the lamellar phase, explaining this in terms of the node associated with one junction zone fitting over the hole in the layer below, restricting the sliding of one mesh layer past another. They also postulated the possibility of the formation of inverse mesh phases, although to date none have been observed.

A good point to start in the consideration of mesh phases is with the random mesh phase, $Mh_1(0)$, Figure 2.1(b). Here the lamellae making up a lamellar phase are perforated by water-filled defects that are not correlated between the layers. The formation of the phase is marked by the appearance of a diffuse scattering feature in the SAXS pattern of the phase, which arises from the interference of X-rays scattered by the water-filled defects. This reflection is characteristic of the phase formation. The separation of the defects within the lamellae is similar to that between bilayers. In many cases, it has been observed that the reflection occurs to the low-scattering vector (Q) side of the scattering pattern, Figure 2.3. Since the

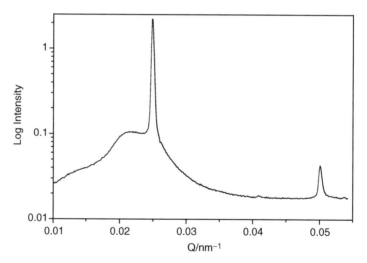

Figure 2.3   The x-ray scattering recorded from a powder sample of the random mesh phase [$Mh_1(0)$] in the tetramethylammonium (TMA) perfluorodecanoate/water system (S Puntambekar, MC Holmes, MS Leaver. *Liq Cryst* 27: 743–747, 2000). The scattering was recorded from a sample containing 70% by weight surfactant and at a temperature of 65°C. Under these conditions the sample is in a single $Mh_1(0)$ phase just above the Mh_1 ($R\bar{3}m$) phase.

reflection is isotropic (because there is a liquidlike distribution of water-filled defects in the lamellae) and diffuse, it can be easily missed and requires careful examination of the scattering behavior for confirmation. However, the swelling behavior of the phase will be different from that of the classical lamellar phase, and other experimental techniques such as optical microscopy or NMR can be shown to be sensitive to its formation.

The random mesh phase has been observed in a wide variety of systems, such as SDS/water [62], LiPFO/water, decyl-ammonium chloride/NH_4Cl/water [63] and cesium perfluoro-octanoate/water [64–66], both as a pretransitional fluctuations and as an independent phase. The fact that the random mesh phase is an independent phase and can be separated from a classical lamellar phase by a first-order phase transition was demonstrated in the nonionic surfactant system $C_{22}EO_6$/water [24]. Optically, the phase has the characteristic "oily streak" texture of a lamellar phase with a parabolic focal conic (PFC) texture, indicating the structure has been subject to dilative stress [67] as it cools from the classical lamellar phase via a pronounced two-phase region. As water-filled defects open up in the bilayer structure, the bilayer repeat distance, d_o, decreases. If the lamellae are constrained within a flat cell with the lamellae parallel to the cell walls, this decrease is equivalent to a dilative stress.

Random mesh structures analogous to those discussed above have been observed in several di-block copolymer systems [68–71]. Qi and Wang [72], working on lamellar phases of di-block copolymers and using a Leibler free-energy functional, showed that fluctuations in a lamellar phase develop as the spinodal is approached. These fluctuations occur at 90° to the layer normal and at a separation of approximately 1.15 times the interlayer separation, an observation consistent with the behavior of random mesh phases in the systems discussed above. In the wide variety of di-block polymer systems that have now been studied, this is found to be approximately general, although some variation occurs with system, temperature, and concentration.

As postulated earlier, competition between alkyl chain length and the surface packing of the surfactant molecules can result in the formation of intermediate phases [21]. Nonionic polyoxyethylene surfactants of the type C_mEO_n [$CH_3(CH_2)_{n-1} - (OCH_2CH_2)_mOH$] in water systems are ideal systems with which to study such systematic molecular perturbations and the resultant effects they may have on phase behavior since they are available pure and in a range of structures where the values of m and n can be varied. They have been extensively studied in the literature [73]. Finally, the surface area per molecule is fixed by the temperature since an increase in temperature has been postulated to drive a dehydration of the EO chains decreasing the surface curvature [74,75]. Hence, these molecules enable intermediate phases to be accessed by variation in molecular architecture and/or temperature.

The $C_{30}EO_9/^2H_2O$ system has a large stable intermediate phase region, as postulated if lengthening of the alkyl chain precludes the formation of bicontinuous cubic phases [21]. X-ray scattering shows up to seven reflections, allowing the indexation of the pattern to a rhombohedral mesh Mh_1 ($R\bar{3}m$) phase [76,77]. In other such systems, the $Mh_1(0)$ is found in association with other phases, L_α, H_1, $V_1(Ia3d)$ and more ordered intermediate phases [24], suggesting that the formation of the correlated mesh phases is associated with the formation of pores in a lamellar phase as a structural precursor to the mesh phases. Interestingly, in this system, the Mh_1 ($R\bar{3}m$) phase is separated by a first-order transition from an L_α phase, and there is no evidence for the formation of water-filled defects within the lamellar phase from which the Mh_1 ($R\bar{3}m$) phase is formed.

Where the alkyl and EO chains are of roughly comparable lengths, intermediate phases are found between L_α and H_1 phases, either as metastable precursors to a $V_1(Ia3d)$ phase or as independent stable phases [76,78–82]. The homologue $C_{16}EO_6$/water [80–82] system shows a number of interesting intermediate-phase features. A random mesh phase $Mh_1(0)$ is separated from the classical L_α phase by a second-order transition. There is a metastable (lifetime >3 months) rhombohedral mesh phase Mh_1 ($R\bar{3}m$) that has now been shown to be a three-connected structure, Figure 2.1(c) [82]. Both $Mh_1(0)$ and Mh_1 ($R\bar{3}m$) phases are distinguishable by their optical textures [81]. There is also a narrow $V_1(Ia3d)$ phase. On heating, the Mh_1 ($R\bar{3}m$) phase is totally replaced by the $V_1(Ia3d)$ phase, but the higher-temperature phase sequence is unaffected, showing no thermal hysteresis whatsoever. This indicates that the correlated mesh phase exists in a local free energy minimum corresponding to a metastable correlation of the water-filled defects between the layers before undergoing the structural reorganization necessary for the formation of the thermodynamic equilibrium bicontinuous cubic phase.

A novel phase behavior is observed in the tetramethylammonium (TMA) perfluorodecanoate/water system [83], which exhibits extensive $Mh_1(0)$ and Mh_1 ($R\bar{3}m$) phases. The Mh_1 ($R\bar{3}m$) phase is stable over a wide range of temperature (0 to 62°C) and concentration

(50 to 80 weight %). In a systematic study, Puntambekar [84] showed that as the counter ion was varied from ammonium to tetrabutylammonium, the phase behavior of the resultant surfactant changed dramatically. Only in the TMA system was a correlated mesh phase stabilized. For TMA the counter ion is closely associated with the interface, modifying the head group interaction and affecting the surface area per molecule. Increasing or decreasing the hydrophobicity of the counter ion causes the Mh_1 ($R\bar{3}m$) phase to be lost and replaced by either a $Mh_1(0)$ phase (less hydrophobic) or L_α phase (more hydrophobic).

Solubilization of an extra component (i.e., salt, alcohol, or oil) into a lyotropic liquid-crystalline–forming system introduces an added degree of freedom into the interfacial organization of the system. The possible variation in the interfacial packing of amphiphilic molecules could facilitate the stabilization (or destabilization) of intermediate phases. Recently, we have investigated the affect of adding oils to the $C_{16}EO_6$/ water system [85] and oil, cosurfactant, and counter ions to the tetramethylammonium perfluorodecanoate/water system [86]. Additives, which lead to a weakening of the interaggregate interactions, promote a loss of correlation between the layers and therefore the loss of the Mh_1 ($R\bar{3}m$) phase and its replacement by a $Mh_1(0)$ phase. The transition from the Mh_1 ($R\bar{3}m$) to the $Mh_1(0)$ phase may be regarded as an order–disorder transition. The fundamental aggregate structure changes little across the transition. Additives that are wholly or partially solubilized in the alkyl chain region (oil or cosurfactant) cause only small changes in the mesh dimensions. Once critical values of volume and/or aggregate thickness are reached, a transition from the Mh_1 ($R\bar{3}m$) phase to a $Mh_1(0)$ phase is triggered, often with a rapidly following transition to a L_α phase. This suggests that these additives are modifying the interfacial curvature of the bilayer aggregate. As interfacial curvature is reduced first, the Mh_1 ($R\bar{3}m$) phase and then the $Mh_1(0)$ phase structure are destabilized with respect to the L_α phase.

Imai et al. [78,79] have investigated the transition region between L_α and $V_1(Ia3d)$ in the $C_{16}EO_7$/water system. They found experimental evidence for the formation of a $Mh_1(0)$ phase within the lamellar phase at low temperature, followed by a metastable rhombohedral intermediate phase. In the first of these papers, this was interpreted as a rhombohedral mesh Mh_1 ($R\bar{3}m$) phase, Figure 2.1(c) [78]. However, in the latter the authors reinterpreted the structure as Bc_1 ($R\bar{3}c$) [79], in other words a bicontinuous structure with rhombohedral symmetry, Figure 2.1(d). This structure has the advantage of being structurally very similar to the V_1(Ia3d) cubic phase, for which it would act as precursor. This reassessment of their scattering data was driven by the results of a computer simulation technique that indicated that the such a bicontinuous phase could be nucleated directly out of a $Mh_1(0)$ phase.

Bicontinuous intermediate structures have been reported previously [26]. In many papers, the identification of tetragonal or rhombohedral intermediate phases, mesh or bicontinuous, is ambiguous because information is usually insufficient to make a definitive identification, and consequently only a few examples exist where authors have claimed bicontinuous phases. For example, the rhombohedral R_α phase in the SDS/ water system was identified as bicontinuous [26] because it was adjacent to a bicontinuous cubic phase V_1(Im3m) $\{Q_\alpha\}$ and separated from it by a second-order transition. Auvray et al. [54] have reinvestigated this sequence of intermediate phases in some detail and identify the rhombohedral phase as a distorted Im3m cubic phase, although they noted that the predicted radius of the hydrocarbon rods was larger than the length of the all-*trans* surfactant alkyl chain.

Since a number of intermediate rhombohedral phases with up to nine diffraction lines have now been reported, it is worth looking at their interpretation as either $Mh_1(R\bar{3}m)$ or $Bc_1(R\bar{3}c)$ [79,82,83]. The x-ray scattering patterns from all the rhombohedral intermediate phases reported are rather similar, Figure 2.4. Reference to the *International Tables of X-ray Diffraction Data* [39] shows that the conditions for diffraction maxima for these two space groups are almost identical. Experimentally, the patterns are always characterized by three strong lines at low scattering vector (Q), lines 1, 2, and 3, Table 2.2. Lines 2 and 7 develop directly from the first- and second-order reflections of the bilayer spacing in the L_α phase or $Mh_1(0)$. These two reflections can be identified as first- and second-order reflections from the $(0\,0\,l)$ planes of the rhombohedral structure. They therefore correspond to either the $(0\,0\,3)$ and $(0\,0\,6)$ planes in $Mh_1(R\bar{3}m)$ or to $(0\,0\,6)$ and $(0\,0\,12)$ planes in $Bc_1(R\bar{3}c)$. In the two papers of Imai et al. [78,79], these two reflections were correctly assigned as being from the $(0\,0\,3)$ and $(0\,0\,6)$ planes of an $Mh_1(R\bar{3}m)$ phase in the first paper [78]. In the second paper, they were incorrectly assigned as the $(0\,0\,3)$ and $(0\,0\,6)$ planes in a $Bc_1(R\bar{3}c)$ phase, making the assignment of other reflections and the dimensions suspect [79]. These two lines enable the lattice parameter c to be calculated directly. Reflection 4 is quite weak but is useful in that it arises from planes of a type (hk0) in both systems enabling a

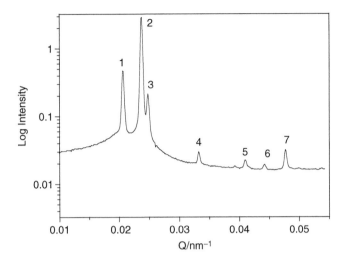

Figure 2.4 The x-ray scattering recorded from a powder sample of the rhombohedral intermediate phase in the tetramethylammonium (TMA) perfluorodecanoate/water system. The sample is 70% by weight surfactant at a temperature of 20°C. The labeled peaks are referred to in Table 2.2. (From S Puntambekar, MC Holmes, MS Leaver. *Liq Cryst* 27: 743–747, 2000).

to be calculated directly. The rest of the lines follow from the symmetries of the two structures. Lines around 5 and 6 are generally weak and can represent reflections from a number of possible planes depending upon the ratio c/a. In general, it is possible to obtain satisfactory fits to both Mh_1 ($R\bar{3}m$) and Bc_1 ($R\bar{3}c$) structures, although the fit to the latter is usually less good. In particular the

Table 2.2 The Principal Peaks in the X-Ray Scattering Pattern from a Rhombohedral Intermediate Phase and Their Assignment to Both Mh_1 ($R\bar{3}m$) and Bc_1 ($R\bar{3}c$) Structures

Peak No.	Mh_1 ($R\bar{3}m$) Indices	Bc_1 ($R\bar{3}c$) Indices	Relative Intensity
1	101	012, 1$\bar{1}$2, 102	s
2	003	006	vs
3	012, 1$\bar{1}$2, 10$\bar{2}$	1$\bar{1}$4, 104	vs
4	1$\bar{2}$0, 110	1$\bar{2}$0, 110	m
7	0 0 6	0 0 12	s

Note: This table should be read in conjunction with Figure 2.4.

$(1\,0\,1)$ reflection from a Bc_1 $(R\bar{3}c)$ structure is always missing, which would have to be explained by a form factor effect suppressing the intensity of this reflection.

Turning to other techniques helps to distinguish between these two possibilities. In the $C_{16}EO_6$/water system, the transition from rhombohedral intermediate phase to $Mh_1(0)$ is second order as indicated by the continuous evolution of the ^2H NMR splitting, while that from the intermediate phase to $V_1(Ia3d)$ phase is first order with a two-phase coexistence region detectable using the same technique [81]. Secondly, the optical texture of the intermediate phase is characteristic of a layered structure, with a viscosity commensurate with that of the $Mh_1(0)$ phase but much lower than that of the $V_1(Ia3d)$ phase [81]. Finally, calculations of the structure factor from the Mh_1 $(R\bar{3}m)$ [82] show a good qualitative match of line intensities to those that are observed. While the Bc_1 $(R\bar{3}c)$ structure for these intermediate phases cannot be dismissed, the experimental evidence is more consistent with the Mh_1 $(R\bar{3}m)$ structure.

Theoretically, Hyde [87] has cast doubt on noncubic bicontinuous phases because periodic minimal surfaces with tetragonal or rhombohedral symmetries are expected to have a higher associated bending energy cost than their cubic phase counterparts do. He suggests that the second-order phase transition from Mh_1 $(R\bar{3}m)$ to Bc_1 can be achieved by extra tunnels connecting the mesh layers. Others [89–91] have proposed that both rhombohedral and tetragonal bicontinuous structures are the most likely intermediate phase structures with these symmetries. Clearly, the existence of bicontinuous noncubic phases still remains an open question.

2.4 THEORETICAL IDEAS

Self-assembly of amphiphilic molecules in water is driven by the competition of two processes: the exclusion of water from the hydrophobic alkyl chain region and the packing of these chains into the volume dictated by closure and the total volume of the chains in the aggregate. The balance of these two processes dictates the aggregate size and shape in solution. At a molecular level, the closure of the aggregate imposes a curvature on the hydrophilic head-group–water interface. The interfacial curvature is controlled by the molecular architecture as well as by the presence of additional components. This concept can be conveniently expressed in term of the surfactant parameter, $N_s = v/al$. Simple closed aggregates, which possess uniform interfacial curvature (spheres, rods, and bilayers), possess values of N_s that are discrete (1/3, 1/2, and 1 respectively). Experimentally, the concentrated phase behavior of amphiphilic systems is dominated by phases that consist of mesogenic units that possess uniform curvature, namely hexagonal (rod), lamellar (bilayer), or bicontinuous cubic (bilayer that decorates a multiply connected surface that in addition possesses zero mean curvature) phases, implying a preference for structures that possess uniform curvature. Deviations away from these discrete values give rise to aggregates with nonuniform interfacial curvature, which is a prerequisite for the formation of intermediate phases as well as nematic and some micellar cubic phases.

The surfactant parameter, N_s, may be also written in terms of the mean (H) and Gaussian (K) curvature of the interface [16],

$$N_s = 1 + Hl + \frac{Kl^2}{3}$$

It is apparent that while N_s may be fixed, H and K can be varied. Thus, for a given N_s, the equation can be satisfied by a number of surface shapes. The surfaces can be distinguished by their topology, indeed the higher the topology the larger the area that the surface possesses. Since the surface area per molecule is determined for an amphiphile under a given set of experimental parameters (temperature, water content, salt concentration, etc), then the surface-to-volume ratio of the mixture, and hence the interfacial topology, is set by the concentration of the surfactant. It is possible to relate the surface-to-volume ratio to the geometry of the interface even for structures more complex than spheres, rods, and bilayers. This particular approach has had considerable success in the modeling of bicontinuous cubic phases, where these phases are considered to consist of bilayers ($N_s \sim 1$), which decorate one of three periodic minimal surfaces (D, G, and P) corresponding to cubic symmetries Pn3m, Ia3d, and Im3m [13,92]. The scattering data recorded from the amphiphilic systems in the bicontinuous cubic phases always index to one of these symmetries. In addition, it is possible to model the D, G, and P surfaces as a function of H and K, and hence if the concentration of amphiphile is known then it is possible to predict the cubic phase's range of existence and calculate its structural parameter with considerable success [93]. Hyde [94] and Fogden [95] have extended this approach by allowing uniform nonzero mean curvature mesh structures with tetragonal, rhombohedral, and

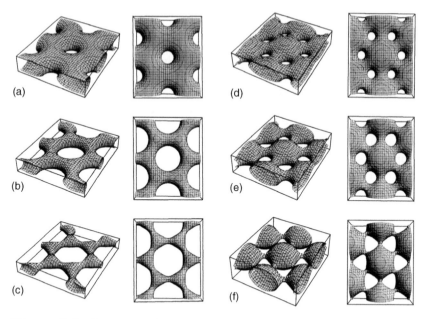

Figure 2.5 Representations of individual rhombohedral mesh layers that possess uniform mean curvature but that differ in connectivity. The symmetry of the two structures is identical, but the topology differs. The surfaces are allowed to evolve by varying the uniform mean curvature (H^*). Figure 2.5(a) to Figure 2.5(c) illustrate the three-connected mesh, while Figure 2.5(d) to Figure 2.5(f) represent the six-connected variants while retaining the same viewpoints. H^* in Figure 2.5(a) to Figure 2.5(c) is increasing, taking values of 0.695, 2.11, and 3.71 respectively. The corresponding H^* values in Figure 2.5(d) to Figure 2.5(f) are 0.907, 1.49, and 2.10. The H^* values chosen for the two conectivities span the stability region of the surface. The main difference between the connectivity at a comparable H^* is that the Gaussian curvature distribution of the six-connected mesh is always significantly less homogeneous than that of the three-connected analogue. Fits of available experimental data for a number of systems to the predicted surfaces precluded the formation of a six-connected mesh (MS Leaver, A Fogden, MC Holmes, CE Fairhurst. *Langmuir* 17: 35–46, 2001; S Puntambekar. Molecular Self Assembly in Fluorocarbon Surfactant/ Water Systems, Ph.D. Thesis, University of Central Lancashire, Preston, U.K., 2000). The figures are reproduced from AS Fogden, M Stenluka, CE Fairhurst, MC Holmes, MS Leaver. *Prog Colloid Polym Sci* 108: 129–138, 1998, copyright 1998 from Progress in Colloid and Polymer Science by A. Fogden et al. Reproduced with permission from Springer-Verlag, http://www. springer.de/. Readers interested in the development of the surfaces are directed to AS Fogden, M Stenluka, CE Fairhurst, MC Holmes, MS Leaver. *Prog Colloid Polym Sci* 108: 129–138, 1998 and the literature cited therein.

monoclinic symmetry, although this is neither as simple nor as accurate as for the bicontinuous cubic phases.

Bicontinuous nonsymmetric intermediate phases have been characterized by a parameterization process to reveal their structural properties [96]. Surfaces that possess uniform mean curvature but are not bicontinuous do exist in the form of mesh phases, but their explicit construction is nontrivial. Fogden [96] has extended the approach used for the anisometric bicontinuous intermediate phase to obtain idealized mesh phase models that possess the highest possible symmetry and interfacial smoothness. The problem is solved either by numerical or Fourier decomposition techniques; the latter was favored by Fogden in the construction of uniformly curved three- and six-connected mesh phases [96]. The structural parameters of two layers of a mesh intermediate phase can be explored for a range of mean curvatures. Pictorial representations of the three- and six-connected mesh phases possessing the correct layer topology to form a rhombohedral mesh are shown in Figure 2.5. Further, if the unit cell dimensions and the hydrophobic content (or water content) of the phase are known, the layer thickness and/or surface area per amphiphile can be calculated and compared to adjacent simple phases.

Following careful reexamination of experimental data, Leaver et al. [82] were able to explore the structural alternatives, including bicontinuous phases, for the structure of two rhombohedral mesh intermediate phases. They showed that the structure was commensurate with a three-connected mesh layer in both systems. Comparison between the phase in these systems, as well as a similar phase found in a fluorocarbon system [83], indicated that within the stability regions of the phase many structural parameters are similar; for example, the ratio of the unit cell dimensions, c/a, is about two. Although relatively rare, the phase structure is apparently independent of the type of amphiphilic system within which it is formed.

It is possible to predict the stability of these phase structures and concomitantly the stability of the liquid crystalline phases. By assigning an elastic cost to a deformation of the structure away from its preferred N_s, it is possible to estimate the energy for a local perturbation. Such a bending energy approach has been pioneered by Helfrich [97] and has allowed Hyde [87] to calculate the stability of the monolayer geometry as a function of surfactant parameter and the hydrophobic volume fraction, see Figure 2.6. One can infer from this the stability of a given lyotropic liquid crystalline phase given the stability of the mesogenic unit predicted. It is possible

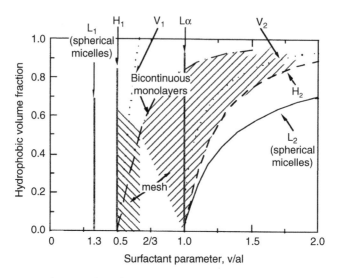

Figure 2.6 A pseudo phase diagram representing the stability regions for monolayer formation in surfactant–water mixtures, where the hydrophilic moiety is 1.4 nm long. Here V_1 and V_2 are bicontinuous cubic phases, H_1 and H_2 hexagonal phases, L_α a lamellar phase, and L_1 and L_2 micellar liquid phases. Note the hashed regions indicating the areas of the phase diagram where mesh phases could be stable. The figure is taken from S Hyde, S Andersson, K Larsson, Z Blum, T Landh, S Lidin, BW Ninham. *The Language of Shape*, Amsterdam, Elsevier, 1997, reprinted from Chapter 4, pp. 141–199, Beyond flatland: the geometric forms due to self-assembly, Copyright 1997, with permission from Elsevier.

then to compare real amphiphilic phase behavior against the "pseudo" phase diagram described above and shown in Figure 2.6.

2.5 CONCLUSION

Intermediate phases provide a wealth of possible complex two- and three-dimensional phase structures for surfactant water mixtures. Although they require rather specific experimental conditions to be achieved, they probably occur more commonly than is supposed since their experimental signatures are often very subtle. The range of structures that is physically possible as against theoretically possible has not been fully elucidated, and the existence of a bicontinuous rhombohedral structure is still ambiguous. Intermediate phases can form where a third component allows microsegregation within the aggregate and thereby nonuniform interfacial curvature or where the alkyl chain is longer or more rigid than normal, disfavoring bicontinuous cubic phases. The structures that these phases present are often close to structures that have been found in biological membrane systems. For example, mesh phases are examples of "holey" bilayers, which may play a role in some biological processes. The study of these phases and their subtle changes in surface curvature and structure may provide an insight into important biological processes.

REFERENCES

1. RG Laughlin. *The Aqueous Behaviour of Surfactants,* London, Academic Press, 1994.

2. CE Fairhurst, S Fuller, J Gray, MC Holmes, GJT Tiddy. In D Demus, JW Goodby, GW Gray, HW Spiess, V Vill, Eds. *The Handbook of Liquid Crystals,* Weinheim, Germany, Wiley-VCH, 1998, Vol. 3, Chapter VII, pp. 341–392.

3. K Fontell. *Mol Cryst Liq Cryst* 63: 59–82, 1982.

4. C Tanford. *The Hydrophobic Effect. Formation of Micelles and Biological Membranes.* 2nd ed., New York, John Wiley & Sons, 1980.

5. B Jönsson, B Lindman, K Holmberg, B Kronberg. *Surfactants and Polymers in Aqueous Solution.* New York, John Wiley & Sons, 1998.

6. GJT Tiddy. *Phys Rep* 57: 1–46, 1980.

7. V Luzzati. In D Chapman, Ed. *Biological Membranes,* London, Academic Press, 1968. pp. 71–123.

8. LQ Amaral, A Gulik, R Itri, P Mariani. *Phys Rev A* 46(6), 3548–3550, 1992.

9. R Itri, LQ Amaral, P Mariani. *Phys Rev E* 54(5), 5211–5216, 1996.

10. D Fennel Evans, H Wennerström. *The Colloidal Domain,* New York, VCH Publishers, 1994, Chapter 6, pp. 239–283.

11. K Fontell, A Khan, D Maciejewska, S Puang. *Colloid Polym Sci* 269: 727, 1991.

12. MC Weiner, SH White. *Biophys J* 61: 434, 1991.

13. JM Seddon. *Biochem Biophys Acta* 1031: 1–69, 1990.

14. K Fontell. *Advan Colloid Interface Sci* 41: 127–147, 1992.

15. JM Seddon, RH Templer. *Phil Trans R Soc Lond A* 344: 377–401, 1993.

16. S Hyde, S Andersson, K Larsson, Z Blum, T Landh, S Lidin, BW Ninham. *The Language of Shape,* Amsterdam, Elsevier, 1997.

17. ID Leigh, MP McDonald, RM Wood, GJT Tiddy, MA Trevethan. *J Chem Soc Farad Trans I* 77: 2867–2876, 1981.

18. V Luzzati, H Mustacchi, A Skoulios, F Husson. *Acta Crystallogr* 13: 660–667, 1960.

19. V Luzzati, A Tardieu, T Gulik-Krzywicki. *Nature* 217: 1028–1030, 1968.

20. MC Holmes. *Curr Opin Colloid Interface Sci* 3: 485–492, 1998.

21. C Hall, GJT Tiddy. In KL Mittal, Ed. *Surfactants in Solution,* New York, Plenum Press, 1989, Vol. 8, pp. 9–23.

22. H Hagslätt, O Söderman, B Jönsson. *Liq Cryst* 12: 667–688, 1992.

23. P Kékicheff, GJT Tiddy. *J Phys Chem* 93: 2520–2526, 1989.

24. SS Funari, MC Holmes, GJT Tiddy. *J Phys Chem* 96: 11,029–11,038, 1992.

25. E. Blackmore, GJT Tiddy. *J Chem Soc Farad Trans II* 84: 1115–1127, 1988.

26. P Kékicheff, B Cabane. *Acta Crystallogr* B44: 395–406, 1988.

27. N Boden, PH Jackson, K McMullen, MC Holmes. *Chem Phys Lett* 65: 479, 1979.

28. B Luhmann, H Finkelmann, G Rehage. *Macromol Chem* 186: 1059, 1985.

29. MC Holmes, N Boden, K Radley. *Mol Cryst Liq Cryst* 100: 93–102, 1983.

30. P-O Quist, B Halle, I Furó. *J Chem Phys* 96: 3875–3891, 1992.

31. I Furó, B Halle. *Phys Rev E* 51: 466–477, 1995.

32. P Ekwall. In GH Brown, Ed. *Advances in Liquid Crystals,* New York, Academic Press, 1975.

33. H Hagslätt, K Fontell. *J Colloid Interface Sci* 165: 431–444, 1994.

34. Y Hendrikx, J Charvolin. *J de Phys* 42: 1427–1440, 1981.

35. S Alpérine, Y Hendrikx, J Charvolin. *J de Phys Lett* 46: L27–L31, 1985.

36. Y Hendrikx, J Charvolin. *Liq Cryst* 3: 265–273, 1988.

37. Y Hendrikx, J Charvolin. *Liq Cryst* 11: 677–698, 1992.

38. J Charvolin. *J de Phys II* 13: 829–842, 1993.

39. T Hahn. *International Tables For Crystallography,* 4th ed., Dordrecht, Kluwer Academic, 1995.

40. P-O Quist, B Halle, I Furó. *J Chem Phys* 95: 6945–6961, 1991.

41. G Chidichimo, NAP Vaz, Z Yaniv, JW Doane. *Phys Rev Lett* 49: 1950–1954, 1982.

42. JW Doane. *Isr J Chem* 23: 323–328, 1983.

43. JW Doane, G Chidichimo, A Golemme. *Mol Cryst Liq Cryst* 113: 25–36, 1984.

44. G Chidichimo, A Golemme, JW Doane. *J Chem Phys* 82: 4369–4375, 1985.

45. J W Doane. In JW Emsley, Ed. *Nuclear Magnetic Resonance of Liquid Crystals*. D. Reidel, Dordrecht, Holland, 1985, pp. 413–419.

46. G Chidichimo, D De Fazio, GA Ranieri, M Terenzi. *Chem Phys Lett* 117: 514–517, 1985.

47. G Chidichimo, D De Fazio, GA Ranieri, M Terenzi. *Mol Cryst Liq Cryst* 135: 223–236, 1986.

48. G Chidichimo, A Golemme, GA Ranieri, M Terenzi. *Mol Cryst Liq Cryst* 132: 275–288, 1986.

49. G Chidichimo, A Golemme, JW Doane, PW Westerman. *J Chem Phys* 82: 536–540, 1985.

50. C Kang, O Söderman, PO Eriksson, J Stael Von Holstein. *Liq Cryst* 12: 71, 1992.

51. P-O Quist, B Halle. *Molec Phys* 65: 547, 1988.

52. S Gustafsson, B Halle. *J Chem Phys* 107: 1460–1469, 1997.

53. X Auvray, C Petipas, I Rico, A Lattes. *Liq Cryst* 17: 109–126, 1994.

54. X Auvray, T Perche, R Anthore, C Petipas, I Rico, A Lattes. *Langmuir* 7: 2385–2393, 1991.

55. H Hagslätt, O Söderman, B Jonsson. *Langmuir* 10: 2177–2187, 1994.

56. H Hagslätt, O Söderman, B Jonsson. *Liq Cryst* 17: 157–177, 1994.

57. O Söderman, G Lindblom, LB-A Johansson, K Fontell. *Mol Cryst Liq Cryst* 59: 121–136, 1980.

58. P Kékicheff, B Cabane. *J de Phys* 48: 1571–1583, 1987.

59. P Kékicheff. *J Colloid Interface Sci* 131: 133–152, 1989.

60. P Quist, K Fontell, B Halle. *Liq Cryst* 16: 235–256, 1994.

61. X Auvray, C Petipas, A Lattes, I Ricolattes. *Colloid Surface A* 123: 247–251, 1997.

62. P Kékicheff. *Mol Cryst Liq Cryst* 198: 131–144, 1991.

63. MC Holmes, J Charvolin. *J Phys Chem* 88: 810–818, 1984.

64. MS Leaver, MC Holmes. *J de Phys II* 3: 105–120, 1993.

65. MC Holmes, AM Smith, MS Leaver. *J de Phys IV* 3, C8: 177–180, 1993.

66. MC Holmes, MS Leaver, AM Smith. *Langmuir* 11: 356–365, 1995.

67. P Boltenhagen, O Lavrentovich, M Kléman. *J de Phys II* 1: 1233–1252, 1991.

68. T Hashimoto, S Koizumi, H Hasegawa, T Izumitani, ST Hyde. *Macromolecules* 25: 1433–1439, 1992.

69. IW Hamley, KA Koppi, JH Rosedale, FS Bates, K Almdal, K Mortensen. *Macromolecules* 26: 5959–5970, 1993.

70. M Laradji, AC Shi, RC Desai, J Noolandi. *Phys Rev Lett* 78: 2577–2580, 1997.

71. M Laradji, AC Shi, J Noolandi, RC Desai. *Macromolecules* 30: 3242–3255, 1997.

72. S Qi, Z-G Wang. *Macromolecules* 30: 4491–4497, 1997.

73. DJ Mitchell, GJT Tiddy, L Waring, T Bostock, MP McDonald. *J Chem Soc Farad Trans I* 79: 975–1000, 1983.

74. M Andersson, G Karlström, *J Phys Chem* 89: 4957–4962, 1985.

75. G Karlström. *J Phys Chem* 89: 4962–4964, 1985.

76. J Burgoyne, MC Holmes, GJT Tiddy. *J Phys Chem* 99: 6054–6063, 1995.

77. CE Fairhurst, MC Holmes, MS Leaver. *Langmuir* 12: 6336–6340, 1996.

78. M Imai, A Kawaguchi, A Saeki, K Nakaya, T Kato, K Ito, Y Amemiya. *Phys Rev E* 62: 6865–6874, 2000.

79. M Imai, A Saeki, T Teramoto, A Kawaguchi, K Nakaya, T Kato, K Ito. *J Chem Phys* 115: 10,525–10,531, 2001.

80. SS Funari, MC Holmes, GJT Tiddy. *J Phys Chem* 98: 3015–3023, 1994.

81. CE Fairhurst, MC Holmes, MS Leaver. *Langmuir* 13: 4964–4975, 1997.

82. MS Leaver, A Fogden, MC Holmes, CE Fairhurst. *Langmuir* 17: 35–46, 2001.

83. S Puntambekar, MC Holmes, MS Leaver. *Liq Cryst* 27: 743–747, 2000.

84. S Puntambekar. Molecular Self Assembly in Fluorocarbon Surfactant/Water Systems, Ph.D. Thesis, University of Central Lancashire, Preston, U.K., 2000.

85. Y Wang, MS Leaver, MC Holmes, A Fogden. In preparation.

86. R Zhou. Counter Ion Identity Effects on the Self Assembly Processes in a Series of Perfluorinated Surfactant–Water Mixtures, Ph.D Thesis, University of Central Lancashire, Preston, U.K., 2003.

87. ST Hyde. *Colloq de Phys* 51: C7-209–C7-227, 1990.

88. DM Anderson, P Ström. In MA El-Nokaly, Ed. *Polymer Association Structures: Microemulsions and Liquid Crystals,* ACS Symposium Series 384. Washington, American Chemical Society, 1989, pp. 204–224.

89. DM Anderson. *Colloq de Phys* 51: C7-1–C7-18, 1990.

90. DM Anderson, HT Davis, LE Scriven, JCC Nitsche. In I Prigogine, SA Rice, Eds. *Advances in Chemical Physics,* Vol. LXXVII. New York, John Wiley & Sons, 1990, pp. 337–396.

91. AH Schoen. NASA Technical Note D-5541, 1970.

92. ST Hyde. *J Phys Chem* 93: 1458–1464, 1989.

93. A Fogden, ST Hyde, G Lundberg. *J Chem Soc Farad Trans* 87: 949–955, 1991.

94. ST Hyde. *Pure Appl Chem* 64: 1617–1622, 1992.

95. AS Fogden. *Phil Trans R Soc London A* 354: 2159–2172, 1996.

96. AS Fogden, M Stenluka, CE Fairhurst, MC Holmes, MS Leaver. *Prog Colloid Polym Sci* 108: 129–138, 1998.

97. W Helfrich. *Z Naturforsc* 28C: 693–703, 1973.

3

Cubic Phases and Human Skin: Theory and Practice

STEVEN HOATH AND LARS NORLÉN

CONTENTS

3.1 INTRODUCTION

The skin is an information-rich interface that forms the foundation for multiple practical applications including drug delivery, adhesion, bathing, massage, and biomedical monitoring (1). In the future, it is foreseeable that such applications will be enhanced or enabled by the creative juxtaposition of selected cubosomal gel/polymer systems with the skin and/or mucosal surfaces (Table 3.1). The choice and design of such gel/polymer systems requires detailed understanding of the biological surfaces they encounter, specifically the electrical, chemical, mechanical, and thermal properties of the outermost portion

Table 3.1 Potential Clinical Applications of Skin-Based Cubosomal Gels

Polymer interface regulating skin/mucosal hydration
Vehicle for transdermal drug delivery
Contact gel for noninvasive bioelectronic monitoring/sensing
Wound dressing containing bioactive molecules (e.g., lysozyme)
Cosmetic formulation (e.g., topical gel containing chromophores)
Smart adhesive for skin or mucosal surfaces
Regulator of topical pH (skin or mucosa)
Barrier against skin irritants or environmental toxins
Absorbent material for cosmetic pollutants or noxious agents
Protective agent against skin friction and mechanical trauma
Gel patch for use in transcutaneous vaccination

of the epidermis; i.e., the stratum corneum. In humans, the stratum corneum is constituted by a thin (~20 microns thick), conformal, highly organized biopolymeric film that undergoes continual replenishment. In material engineering terms, polymer films that interact with the environment in seemingly intelligent ways are typically called "smart materials" (2,3). This designation seems particularly appropriate for the stratum corneum insofar as the epidermis is derived from the same embryonic germ tissue as the brain (both are ectodermal derivatives). The formation of the stratum corneum is the "raison d'être" of the epidermis (4). The stratum corneum, in this view, is the surface of the brain and manifests a close and obvious link to perception throughout life (5,6).

In this chapter, we build upon the above viewpoint and explore the concept that cubosomal gel preparations may be usefully applied to human skin. The feasibility of this concept is strengthened by the recognition that the *in vivo* organization of the epidermal barrier itself involves cubic-to-lamellar phase transitions (7–11). These results open the door to the development of bioengineered product designs and clinical approaches wherein the cubic organization of the developing epidermal barrier can be "matched" to exogenous cubic nanoarchitectures applied on the skin surface (Figure 3.1). In order to realize this concept in practical form, we propose the specific example of developing cubosomal contact gels for the purpose of noninvasive electrical monitoring of the skin surface. It is well known, for example, that the stratum corneum has a high endogenous electrical resistance (12). Current electrical contact gels often contain abrasives such as small embedded glass particles that damage the stratum corneum and thereby lower the effective skin resistance.

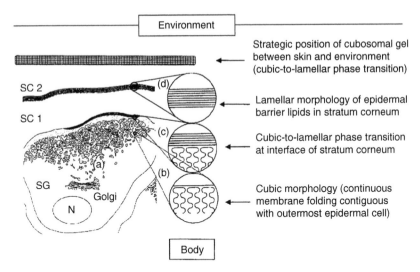

Figure 3.1 Schematic diagram of recently proposed theoretical model of a cubic-to-lamellar phase transition in the formation of the epidermal barrier (7,8,11). This model is extended to include the application of a cubosomal gel overlying the stratum corneum. According to this concept, the lamellar bilayers of the stratum corneum are "sandwiched" between the external cubosomal gel preparation and the internal cubic nanoarchitecture of the nascent barrier lipids. A noninvasive sensor system applied to the skin would occupy the position of the environment in this schema, whereas a cellular target (e.g., Langerhans cell) or a bioelectric signal (e.g., electrical potential from the brain) would occupy the position of the body. This conceptual schema is further extended and strengthened by the recent demonstration of a putative cubic architecture for the intracorneocyte keratin domain (28) coupled with the known ability of keratins to function as piezoelectric/pyroelectric elements (30) that can be oriented by physiologically relevant tensions (5). By hypothesis, seamless integration of exogenous cubosomal materials with endogenous biological cubic architectures is the *sine qua non* for a variety of practical applications ranging from drug delivery to skin protection to noninvasive monitoring (Table 3.1).

Similar destructive approaches are often used for drug and gene delivery systems wherein the stratum corneum barrier is physically ablated by ultrasound, electroporation, mechanical trauma, or chemical exposure. In contrast, we propose that seamless interfacing of an electrode interface with human skin will be enabled by contact gels with cubic nanostructures and/or contained functional groups that are *designed* to mesh with this biological structure. The stratum corneum/sensor interface, in this view, is not so much a barrier as a "window."

The organization of this chapter is as follows: An overview of skin surface electrical measurement using cubosomal contact gels juxtaposed to adult human volar forearm skin is presented. These results correspond to the top half of Figure 3.1 and are contrasted with new data on the cubic architecture of the developing epidermal barrier (lipid + keratin domains) as revealed by cryo-transmission electron microscopy (cTEM) performed on vitrified *native* (i.e., fresh, fully hydrated, nonstained) human epidermal biopsies. The cTEM results correspond to the lower half of the diagram in Figure 3.1. A major challenge for the future is to develop applications linking all these structural levels. We believe such applications are dependent upon better understanding of cubic architectures in biological systems.

3.2 ELECTRICAL CHARACTERIZATION OF TOPICAL CUBOSOMAL GELS

In a series of experiments, the electrical properties of bulk cubosomal gel preparations (75% monoolein and 25% water) were examined *in vitro* (freestanding state) and *in vivo* (applied to human skin). These preliminary experiments established a quantitative framework for investigating applications wherein bulk cubosomal gels are juxtaposed to human skin for purposes of monitoring, drug delivery, adhesion, or protection (Table 3.1). All experiments were performed at room temperature using a prototype skin sensor designed specifically for electrical measurement of the stratum corneum (13). This device utilizes a flat probe head containing 18 separate anodic contact sites and 18 separate cathodic contact sites (14) arranged in facing parallel rows and confined within a circular solid-state array. A linearly increasing direct current (DC) voltage is applied to the multielectrode sensing pad, and the instrument records the nonlinear dependence between current and voltage until a threshold amperage has been reached. Upon reaching the threshold current, the instrument cycles back to zero. Hysteresis curves can thereby be generated and integration of voltage × current × time calculations performed to obtain energy plots in microJoules. The prototype instrument allows repetitive cycling with the production of serial electrical energy plots. Figure 3.2 illustrates the results of an experiment in which bulk cubosomal gel composed of approximately 75% monoolein and 25%

water was applied to the probe head of a prototype Cortex SkinSensor (13,14). Specifically, the gel was applied as a bridging sheet connecting the electrode contact sites of the SkinSensor and repetitive five-cycle voltage ramps were performed increasing 1.5 V/sec to a current threshold of 2.0 μA. As the cubosomal gel preparation lost water to evaporation (thereby undergoing a spontaneous phase transition from cubic to lamellar phase), the electrical properties were monitored as a function of time, as shown in Figure 3.2.

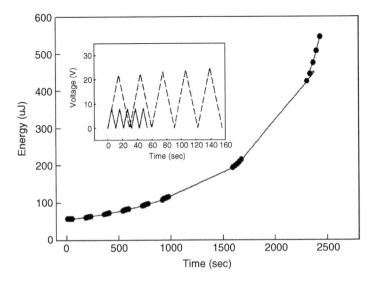

Figure 3.2   Energy profile (voltage × current × time) of free-standing monoolein–water cubosomal gel over time. The cubosomal gel was spread evenly over the probe head of a prototype Cortex SkinSensor (A Pinyayev. 2003. U.S. Patent No. #20030004431 http://www.cortex.dk/; SkinSensor. Cortex Technology, 2003) and conductivity (inverse resistance) was measured at eight separate time intervals over approximately 42 min. The voltage ramp rate was set at 1.5 V/sec with a current threshold of 2.0 μA. Five voltage cycles were completed at each time interval. The insert depicts the voltage data for the first five-cycle run (solid line) and the last five-cycle run (dashed line). As shown, both voltage and integrated energy profiles show progressive increases with time. The time frame of this experiment is sufficient to allow evaporation of water from the cubosomal gel and phase transition from cubic to lamellar phases. The electrical resistance of the gel increases as a function of time, water evaporation, and phase transition.

Both voltages and integrated energy profiles showed exponential increases over the first 2500 sec.

Clearly, the electrical properties of the polymer gel are changing as a function of time, phase, and water content (Figure 3.2). This technique provides a quick method for standardizing and comparing contact gels and topical ointments used in clinical care. Figure 3.3 shows the electrical properties of selected conducting gels measured before application to human skin; i.e., applied to the sensor probe head only. These measures are contrasted with measurements on native skin in contact with the specific conducting interface materials (Figure 3.4). The skin contact materials included: a) Aquasonic® conducting gel (Dynatronics Corporation), b) monoolein cubosomal gel, and c) petrolatum. Thus, the current profiles for a ramp rate of 1.5 V/sec and a current threshold of 1.0 μA are shown in Figure 3.3

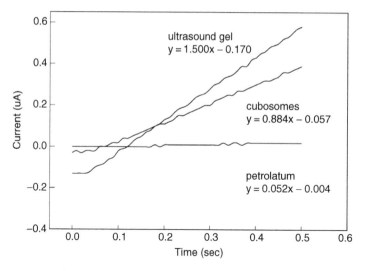

Figure 3.3   Current profiles of interfacial gels without skin contact (nascent measurements). The rate of change of current over time was measured as a function of contact gel application to the probe head at the beginning of a steady DC voltage ramp of 1.5 V/sec. The SkinSensor probe head was covered evenly with petrolatum (flat line), monoolein–water bulk cubosomal gel (middle line), and Aquasonic ultrasonic gel (steepest line). In this experiment there was no contact with skin. The resulting current profiles are shown for the initial 500 msec of data acquisition.

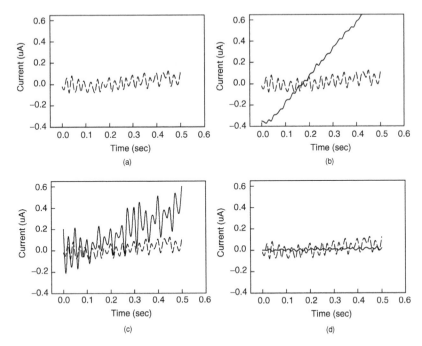

Figure 3.4   Current profiles of interfacial gels with skin contact (nascent measurements). As in the preceding figure, the rate of change of current over time was measured as a function of contact gel application at the beginning of a steady DC voltage ramp of 1.5 V/sec. Panel a shows the results obtained with skin only (no contact gel). Panels b to d illustrate the current profiles when contact gels were used to bridge the sensor probe head with the skin (adult volar forearm). The contact gels included petrolatum (Panel b), cubosomal gel (Panel c), and ultrasonic gel (Panel d). The resulting current profiles are shown for the initial 500 msec of data acquisition. Native skin and skin/cubosomal gel exhibit a highly variable baseline. By hypothesis, such variability is not mere "noise" but contains information on the fine structure of the epidermal barrier in contact with the sensor/environment.

for the gel preparations alone and in Figure 3.4 following contact with human skin (adult volar forearm).

Compared to petrolatum, the Aquasonic gel and the cubosomal gel had similar, relatively high conductivity profiles prior to application to skin (slope of current/time regression line = 1.500 and

0.884 µA/sec, respectively). Following application to skin, both the Aquasonic gel and the cubosomal gel exhibited increases in conductivity (Figure 3.4, panels b and c). The current profile in the latter case now exhibited a high degree of background "noise". The similar appearance of a noisy baseline in the current plots is seen with skin only (slope = 0.1867 µA/sec) and, to a much lesser extent, with petrolatum (slope = 0.0356 µA/sec) (Figure 3.4, panels a and d).

This experiment raises the question of whether cubosomal gel preparations are suitable for function as contact gels in skin electrical measurements. This application is highly dependent upon the assignment of biological meaning to the fluctuating current profiles for the skin only and the skin + cubosomal gel measurements (Figure 3.4, panels a and c). The question thus arises whether the "fine structure" information in the measurement profiles of current versus time is noise versus signal. Stated in other words, can cubosomal contact gels be designed to specifically interrogate the local barrier structure of human skin? Achievement of seamless interfacing with the stratum corneum/epidermal barrier of human skin is the *sine qua non* for development of high-fidelity sensors designed to measure the electrical activity of underlying tissues such as the brain. The hypothesis advanced in this chapter is that achievement of the latter goal is dependent upon the former; i.e., one must first seamlessly interface with the local tissue architecture. To this end, the remainder of the chapter focuses to the exposition of the developing epidermal barrier in terms of cubic "phases"; i.e., cellular lipid or protein organizations with a small lattice parameter (<30 nm) and cubic-like symmetry.

3.3 CUBIC "PHASES" IN THE SKIN

A better understanding of the formation and molecular organization of the stratum corneum intercellular lipid matrix may provide new insights into cell membrane trafficking events, cell dynamics, and cell homeostasis. The epidermal lipids are generally considered to constitute that portion of the epidermal barrier that is largely responsible for water impermeability and high electrical resistance (15). In the standard "bricks-and-mortar" model of the epidermal barrier, the lipid matrix forms the mortar surrounding the embedded corneocyte "bricks" (16). Delineation of the structure of the lipid matrix may provide for a better understanding of the origin of skin diseases displaying impaired barrier function (e.g., atopic dermatitis) and suggest new approaches to optimize transdermal drug delivery, barrier-restorative manipulations (e.g., skin protection, cosmetic formulations, etc.), and skin biosensing techniques.

The conventional view of the formation of the stratum corneum intercellular lipid matrix is essentially that "lamellar bodies" (i.e., discrete spherical lipid bilayer vesicles), containing in their turn "lamellar disks" (i.e., discrete flattened lipid bilayer vesicles), bud off from the trans-Golgi network and diffuse towards the plasma membrane of the differentiating stratum granulosum cells (i.e., topmost viable epidermal cells facing the stratum corneum). After fusion of the limiting membrane of the "lamellar bodies" with the plasma membrane of the stratum granulosum transition cell, the lamellar body lipid content is thought to be discharged into the intercellular space, where the "lamellar discs" merge into intercellular lamellar sheets via a second massive fusion process (17).

A schematic illustration of the end-stage differentiation of the epidermal intercellular space, i.e., skin barrier morphogenesis, based on new high-resolution cryo–transmission electron microscopic findings of native human forearm epidermis, is given in Figure 3.5. It was recently proposed (7–11) that the skin barrier formation process may be viewed as a lamellar "unfolding" of a small lattice parameter (<30 nm) lipid "phase" with cubiclike symmetry (A-C) with subsequent "zipping" (i.e., "crystallization" or "condensation," including lamellar reorganization) of the intercellular lipid matrix (D-F).

In accord with the standard "bricks and mortar" model of epidermal barrier development, the "bricks" are constituted by the anucleated cellular corneocyte component containing a rich complement of keratin intermediate filaments (16). The structural organization of the keratin intermediate filament-dominated stratum corneum corneocyte matrix is of major importance for the barrier properties of skin, for the water-holding capacity of skin, for the appearance (i.e., optical properties) of skin, for the mechanical strength and elastic resilience of skin, and for skin pathologies characterized by alterations of one or more of these properties (e.g., dry skin, atopic dermatitis, psoriasis, ichthyosis). Consequently, the three-dimensional higher-order organization of keratin intermediate filaments in the stratum corneum has been the subject of much debate over the last 50 years. Initially it was suggested that groups of seven 10-nm keratin intermediate filaments aggregated into 25-nm fibrils with an 8-nm lipid layer covering the surface of each fibril, thus forming lipoprotein fibrils with a total diameter of ~40 nm. The fibrils were proposed to be oriented in a plane parallel to the plane

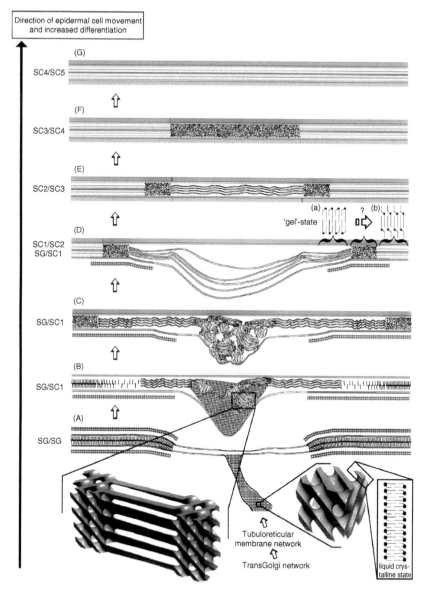

Direction of epidermal cell movement and increased differentiation

(G)
SC4/SC5

(F)
SC3/SC4

(E)
SC2/SC3

(a) 'gel'-state ? (b)

(D)
SC1/SC2
SG/SC1

(C)
SG/SC1

(B)
SG/SC1

(A)
SG/SG

Tubuloreticular membrane network

TransGolgi network

liquid crys-talline state

Figure 3.5 Schematic overview of the end-stage differentiation process of the epidermal intercellular space, i.e., skin barrier formation (11). SG/SG: Interface between cells of stratum granulosum; SG/SC: stratum granulosum/stratum corneum interface; SC/SC: Interface between cells of stratum corneum. (A): Extrusion of

of the flattened stratum corneum cells (18–21). The presence of intracellular lipids in the corneocyte matrix was later contested (22,23), and then reaffirmed (24). Today, the leading opinion seems to favor the absence of substantial amounts of intracellular membrane lipids. Further, the cell matrix is most often regarded as a network of randomly oriented keratin intermediate filaments embedded in a filaggrin-rich protein/water ground substance (25–27). However, a comprehensive model capable of explaining keratin intermediate filament structure, function, and formation when newer findings regarding the native structural organization of fully hydrated epidermis have been taken into account, has until recently (28) been lacking. Such a theoretical model may provide for a rational design of experimental studies on skin diseases, skin permeability, topical drug administration, skin protection, cosmetic formulations, skin biosensing techniques, etc.

Figure 3.5 (Continued) a cubic lipid "phase" (i.e., membrane network with cubiclike symmetry with a lattice parameter of <30 nm) derived from the trans-Golgi network of the uppermost stratum granulosum cells. (B-C): Cubic-to-lamellar "phase" transition of the newly extruded cubic lipid "phase" in the intercellular space between the stratum granulosum and the stratum corneum. Concomitant progressive reorganization of desmosome intercellular adhesion proteins and intracellular desmosome plaques together with a widening of the desmosome extracellular core domain from ~30 to ~43 nm and the formation of the protein "cell envelope" beneath the corneocyte plasma membrane. (D): The newly "unfolded" (via the now completed cubic-to-lamellar "phase transition") loosely packed membrane bilayers of the intercellular space have entered the transition desmosome extracellular core domain and become close packed, and an internal lipid reorganization/condensation/crystallization has begun. Tentatively, this internal lipid reorganization may partly involve a flip-flop from hairpin to splayed chain conformation of a fraction of the stratum corneum ceramides (a,b). (E): Close-packing of newly "unfolded" loosely packed membrane bilayers of the intercellular space of lowermost stratum corneum. (F): Final internal lamellar lipid reorganization/ condensation/ crystallization of the intercellular lipid matrix (cf. Da-Db). (G): Mature "zipped" skin barrier lipid organization of the major part (i.e., ~43 nm wide regions) of the intercellular space of stratum corneum.

A schematic overview of the recently proposed cubic rod-packing and membrane templating model for epidermal keratin structure, function, and formation (28) is given in Figure 3.6. Note that this detailed description is but one out of many possible, all sharing the basic concepts of membrane templating and cubiclike rod-packing of keratin intermediate filaments:

Step I: The keratinocyte cytoplasm is proposed to contain an extended "endoplasmatic reticulum" with a small lattice parameter (~20 nm) and cubiclike symmetry. Note that balanced primitive (lower left) and balanced gyroid (lower right) cubic surfaces (i.e., "the membrane mid-surface") can be isometrically transformed into each other via the Bonnet transformation (29). The transition *in vivo* between small-lattice-parameter (<30 nm) cubic membrane systems with primitive and gyroid symmetry could therefore occur at a low bending energy cost, a fact that is further underlined by the low enthalpy difference between (and thereby coexistence, in e.g., cubosome suspensions, of) primitive (P), gyroid (G), and diamond (D) cubic lipid/water "equilibrium" phases. Note further that a balanced cubic surface (lower right "membrane mid-surface") is transformed into an imbalanced cubic surface (upper "membrane mid-surface") simply by parallel displacement of the membrane mid-surface (i.e., selective shrinkage/swelling of one of the two subvolumes separated by the hyperbolic membrane surface).

Step II: Particulate keratin and metabolic machinery association to a small-lattice-parameter cubic membrane "template" surface.

Step III: Fast, highly dynamic, reversible transitions between hexagonal ("filamentous") and cubic ("particulate") forms of the membrane/keratin/water(/filaggrin) complex.

Step IV: Final irreversible "template" transition of the membrane/keratin/water(/filaggrin) complex into a nanocomopsite pseudo-gyroid chiral (here right-handed) membrane/filament structure at the interface between stratum granulosum and stratum corneum (T).

Step V: Possible keratin filament close-packing and disparition/degradation of the "templating" cubic membrane surface, yielding the mature "isotropic" cubic-like keratin organization of the corneocyte matrix.

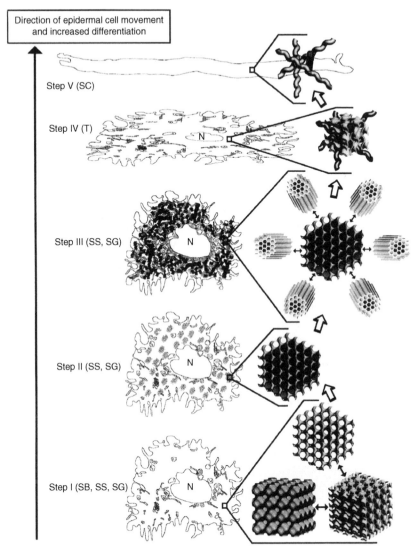

Figure 3.6 Schematic overview of the cubic rod-packing and membrane templating model for epidermal keratin structure, function, and formation (L Norlén, A Al-Amoudi. *J Invest Dermatol* 123:715–732, 2004). Double-headed black arrows: reversible transition; N: nucleus; SB: stratum basale; SS: stratum spinosum; SG: stratum granulosum; T: transition cell layer (between SG and SC); SC: stratum corneum. Note that the scales are not normalized throughout the schematic drawing, neither within nor between steps I to V.

3.4 SUMMARY

This chapter aims ultimately at the development of intelligent noninvasive skin-based sensor systems incorporating functionalized cubic gels for purposes of sensory evoked potential testing of human brain function (Figure 3.1). Hydration state and physicochemical phase emerge as critical time-dependent variables affecting the behavior of simple (nonfunctionalized) monoolein–water cubic gels in contact with sensor interfaces and the skin (Figure 3.2). Meaningful application of cubic-phase contact gels with the skin requires the ability to distinguish biologically meaningful signals from background noise (Figure 3.3 and Figure 3.4). Current techniques requiring skin surface detection of weak (microvolt) signals generated by distant organs, such as the brain, typically employ abrasive contact gels designed to lower the high electrical resistance by wounding the skin surface (stratum corneum). Hypothetically, a better understanding of local tissue architecture and the development of the epidermal barrier will lead to seamless sensor interfacing with the skin and meaningful signal acquisition. Critical to this approach is the notion that local, largely structural factors determine the quality/fidelity of global functional signals.

To this end, an overview of new morphological data on epidermal barrier structure is presented that focuses on the two major compartments of the stratum corneum; i.e., the "external" barrier lipids (Figure 3.5) and the "internal" intermediate filament (keratin) complex (Figure 3.6) (7–11,28). Use of cTEM on vitrified human epidermis allows for nanoscale resolution of epidermal architecture without artifacts associated with chemical fixation, dehydration, and staining of biological specimens. The cTEM results provide for the first time qualitative physical evidence for the possible involvement of cubic morphologies in the development of both the lipid and protein domains of the nascent stratum corneum (11,28). Molecular strategies and maturational schemes wherein such cubic architectures explain and enable epidermal barrier development are presented (Figure 3.5 and Figure 3.6). New concepts relevant to epidermal biological organization include:

1. The presence of a continuous extracellular cubic lipid "phase" intercalating into the cytoplasm of the uppermost nucleated epidermal cell
2. Cubic-to-lamellar phase transitions accompanying maturation of the extracorneocyte lipid barrier

3. Formation of a cubic paracrystalline three-dimensional network for keratin filament organization in the nucleated (living) keratinocyte
4. The concept of "membrane templating" to account for the cubiclike organization of the elongated, polar keratin dimer molecules

What emerges from this approach is a major putative role for cubic nanostructures in the organization of the epidermal barrier. Specifically, "artifact-free" techniques such as cTEM provide morphological evidence for the possible involvement of cubic "phase" systems in the biogenesis of the lipid and protein domains of the stratum corneum. These domains regulate the electrical resistance, water handling, mechanical stress distribution, drug permeability, and other important functions of the epidermis. A better understanding of the spatiotemporal (specifically cubic) organization of the stratum corneum will enhance the design of sensor systems, transdermal drug delivery devices, topical adhesives, and other practical interfaces with the skin surface. In this chapter, we present a practical example of a bicontinuous (monoolein–water) cubosomal contact gel utilized for purposes of skin surface electrical measurement. The conceptual design of the sensor/contact gel/stratum corneum system is dependent upon successful interfacing between the external sensor elements and the presumptive cubic and lamellar nanoarchitecture of the epidermis. According to this approach, the stratum corneum is only a barrier insofar as its structure and design are unknown. The notion of seamless interfacing of external device, coupling medium, and internal stratum corneum cubic/lamellar nanostructures overrides the notion of a barrier and recasts the usual distinction of body and environment. The requirement for seamless interfacing implies identification of all components and the development of an information-rich window, rather than a barrier, for sensing purposes and other noninvasive applications.

ACKNOWLEDGMENTS

We thank Alex Pinyayev of the Procter & Gamble Company, Cincinnati, Ohio, for useful discussions and a review of the data generated with the SkinSensor. We thank William Pickens of the Skin Sciences Institute for technical assistance in data generation with the SkinSensor.

REFERENCES

1. SB Hoath. In RA Polin, WW Fox, SH Abman, Eds. *Physiologic Development of the Skin.* Philadelphia: Saunders, 2004, pp. 597–612.

2. O Ikkala, GT Brinke. *Science* 295:2407–2409, 2002.

3. W Cao, HH Cudney, R Waser. *PNAS* 96:8330–8331, 1999.

4. KC Madison. *J Invest Dermatol* 121:231–241, 2003.

5. S Hoath, M Donnelly, R Boissy. *Biosens Bioelectron* 5:351–366, 1990.

6. S Hoath. *NeoReviews* 2:e292–e301, 2001.

7. L Norlén. *J Invest Dermatol* 117:823–829, 2001.

8. L Norlén. *J Invest Dermatol* 117:830–836, 2001.

9. L Norlén, A Al-Amoudi, J Dubochet. *J Invest Dermatol* 120:555–560, 2003.

10. L Norlén. *Skin Pharmacol Appl Skin Physiol* 16:203–211, 2003.

11. A Al-Amoudi, L Norlén. *J Invest Dermatol* (in press).

12. TJ Faes, HA van der Meij, JC de Munck, RM Heethaar. *Physiol Meas* 20:R1–10, 1999.

13. A Pinyayev. 2003. U.S. Patent No. #20030004431

14. http://www.cortex.dk/. SkinSensor. Cortex Technology, 2003.

15. ML Williams, K Hanley, PM Elias, KR Feingold. *J Investigative Dermatol Symp Proc* 3:75–79, 1998.

16. Z Nemes, PM Steinert. *Exp Mol Med* 31:5–19., 1999.

17. L Landmann. *J Invest Dermatol* 87:202–209, 1986.

18. G Swanbeck. *Acta Derm Venereol (Stockh)* 39:1–37, 1959.

19. G Swanbeck. *J Ultrastructure Res* 3:51–57, 1959.

20. GL Wilkes, AL Nguyen, R Wildnauer. *Biochim Biophys Acta* 304:265–275, 1973.

21. HP Baden, LA Goldsmith. *J Invest Dermatol* 59:66–76, 1972.

22. PM Elias, DS Friend. *J Cell Biol* 65:180–191, 1975.

23. AS Breathnach, T Goodman, C Stolinski, M Gross. *J Anat* 114:65–81, 1973.

24. JC Garson, J Doucet, JL Leveque, G Tsoucaris. *J Invest Dermatol* 96:43–49, 1991.

25. I Brody. *J Ultrastructure Res* 2:482–511, 1959.

26. AG Matoltsy. *J Invest Dermatol* 67:20–25, 1976.

27. F Odland. In LA Goldsmith, Ed. *Structure of the Skin.* New York: Oxford University Press, 1991, pp. 3–62.

28. L Norlén, A Al-Amoudi. *J Invest Dermatol* 123:715–732, 2004.

29. S Hyde, S Andersson, K Larsson, Z Blum, T Landh, S Likin, B Ninham. *The Language of Shape: The Role of Curvature in Condensed Matter: Physics, Chemistry and Biology.* Amsterdam: Elsevier, 1997, pp. 27–31.

30. H Athenstaedt, H Claussen, D Schaper. *Science* 216:1018–1020, 1982.

4

The Relationship between Bicontinuous Inverted Cubic Phases and Membrane Fusion

D.P. SIEGEL

CONTENTS

4.1 INTRODUCTION

Membrane fusion is the process by which two closed lipid membrane vesicles form a single, larger vesicle. The aqueous contents and the membranes of the two original vesicles mix, and no significant leakage of the contents to the suspending medium occurs. Fusion is a ubiquitous process in biology. It is a necessary step in many critical functions in single cells (e.g., exocytosis, vesicular transport), in multicellular organisms (e.g., fertilization, immuno- and neuro-secretion), and viral infection. The process is of intense biomedical interest and is the basis of some drug delivery systems. The purpose of this chapter is to review the ways in which studies of inverted cubic (Q_{II}) phase behavior help us understand the mechanism of this biological process.

Specialized proteins have evolved to control and induce fusion *in vivo* [1–5]; the mechanism by which these proteins act is still unclear. However, membrane fusion occurs in many model membrane systems composed only of lipids (for reviews, see [6–11]). There is good evidence that fusion proteins act by inducing formation of intermembrane intermediates composed largely of the membrane lipids, as opposed to primarily the proteins (see [8,9,12] for reviews). The susceptibility of biomembranes to fusion is strongly influenced by the liquid-crystalline phase behavior of the lipids in the membranes: increasing the fraction of inverted phase-forming lipids increases the susceptibility to fusion, and adding lipids that tend to inhibit inverted phase formation reduces it [8,9,12]. Moreover, at least one mechanism of membrane fusion is closely related to the mechanism of lamellar/inverted phase transitions in phospholipids. Hence, a detailed understanding of lipid phase behavior and lipid phase transitions is necessary for an understanding of membrane fusion mechanisms. The bicontinuous inverted cubic phases are especially relevant.

Studies of lipid phase behavior contribute in three ways to the study of membrane fusion mechanisms:

First, the intermediates that form in membrane fusion seem to be the same intermediates that form in the phase transitions between lamellar (L_α), bicontinuous inverted cubic (Q_{II}), and inverted hexagonal (H_{II}) lipid liquid-crystalline phases. This has been shown by a combination of electron microscopic, NMR, x-ray diffraction, and fluorescence techniques (see [8,9,11,12] for reviews). In particular, lipid membrane fusion is an early step in the L_α/Q_{II} phase transition. It is not easy to study the structure of fusion intermediates in biomembrane systems because they are small (<10 nm in size) and transient (lifetimes <0.1 msec in some cases) and represent only a very small fraction of the total membrane area in the system at any given instant. It is easier to study the structure of fusion intermediates in lipid model membrane systems, especially pure phospholipids. Intermediates in these phase transitions can be created in relatively large numbers and under controlled conditions by manipulation of variables such as the lipid composition, temperature, water activity, and salt content of the suspending medium. This will be discussed in Section 4.2

Second, the effect of exogenous agents (e.g., fusogenic peptides and other parts of fusion-catalyzing proteins) on fusion intermediates can be inferred from their effects on the position of the phase boundaries, and on the kinetics of lamellar/inverted phase transitions. The fusion rate of membranes increases rapidly as systems approach the L_α/Q_{II} or L_α/H_{II} phase boundary (see [8,9,11,12] for reviews), so a good correlation exists between phase behavior and fusion kinetics. It is easier to study these effects in static systems, such as liquid-crystalline phases, rather than in fusing membrane systems. These effects will be discussed in Section 4.3.

Finally, studying the relative stability of the L_α and inverted phases under different conditions teaches us the rules determining the energies of lipid structures of different geometries. These rules can be used to estimate the energies of different hypothetical fusion intermediates so we can arrive at models that best explain the experimental data. The relative stability of lamellar and inverted phases can be understood in terms of the Helfrich curvature elastic energy of the lipid monolayers [13]. Most attention has been paid to the influence of the bending elastic energy. The bending elastic modulus for the Helfrich model is determined by measuring the unit cell dimensions of H_{II} phases as a function of osmotically applied stress (e.g., [14]) and also provides a value for the tilt modulus under some circumstances [15]. Studies of Q_{II} and rhombohedral phase stability have recently yielded values for the Gaussian curvature elastic modulus (or saddle splay modulus) [16–18], which was neglected in previous fusion intermediate models. Accurate descriptions of intermediate energies allow us to test models of fusion intermediate structures and look for those that require the least activation energy to form (see [12,19] for recent reviews). This is especially important in determining the mechanisms of protein-induced fusion. These proteins do not have well-defined "active sites" as do enzymes, and they act by modifying the physical environment of small patches of lipids. Even with high-resolution protein structures, it is often unclear how the different conformations of fusion-inducing proteins act to form fusion intermediates. Data obtained from studies of Q_{II} phase formation have provided valuable insights into fusion intermediate energetics. These will be outlined in Section 4.4.

In the last few years, studies of lipid phase behavior have had especially profound effects on our models of membrane fusion and, by extension, how proteins induce fusion. It can be said that the rhombohedral phase [20–22] consists of structures identical to the first intermembrane intermediate to form in membrane fusion, and that the Q_{II} phase assembles from intermediates identical to the completed intermembrane connection in the fusion process. Recent work also indicates that the transmembrane domains of proteins have a substantial effect on intermediate formation energies, focusing new attention on domains of the proteins previously thought to be comparatively unimportant in fusion (see Section 4.3).

4.2 RELATIONSHIP OF MEMBRANE FUSION TO LAMELLAR/INVERTED PHASE TRANSITION MECHANISMS

Early in the study of inverted phases in biologically relevant lipids, isotropic ^{31}P-NMR resonances and peculiar freeze-fracture morphology were observed when lipid compositions were close to the lamellar/inverted phase boundary and when liposomal dispersions of the same lipids underwent membrane fusion (growth of lipid vesicles) under the same conditions (e.g., [23–29]). The structures corresponding to the freeze-fracture morphology were variously interpreted as intramembrane inverted micelles or as connections between opposed lamellae (later known as ILAs, or interlamellar attachments [30]; Figure 4.1D). Subsequent freeze-fracture and time-resolved cryoelectron microscopy studies [31–35] demonstrated that the isotropic

^{31}P-NMR resonance and the morphology arose from ILAs. (The isotropic resonances are observed because the lipid molecules diffuse around the surfaces of the ILAs on the NMR time scale, experiencing all orientations with respect to a given director. While the correlation between isotropic resonance detection and ILAs or Q_{II} phases in multilamellar lipid preparations is very good, it must be remembered that in theory other structures, such as small unilamellar vesicles and rhombohedral phases, can give rise to such resonances.) Time-resolved cryoelectron micrographs of ILAs are shown in Figure 4.2. The ILA is clearly a structure corresponding to the end point of membrane fusion: if an ILA forms between two liposomes, both the membranes and the aqueous compartments of the two liposomes have become continuous. ILAs are thought to arise from earlier intermembrane intermediates (stalks and hemifusion intermediates; Figure 4.1B and Figure 4.1C). The structures of these earlier intermediates are inferred from the details of lipid mixing versus membrane fusion behavior of lipid suspensions and biomembranes, time-resolved cryoelectron microscopy data, theoretical models of lipid intermediate energies, and the existence of similar structures in a lipid liquid-crystalline phase at low water activity (see [12] for a review).

ILAs were also demonstrated to be Q_{II} phase precursors [31–33]. Arrays of ILAs were observed forming in systems incubated under conditions where Q_{II} phases formed. As proposed in [36], as the number of ILAs increases in a stack of L_α phase bilayers, ILAs should be able to form arrays that spontaneously generate Q_{II} phase unit cells (Figure 4.3). One bicontinuous Q_{II} phase structure (Q_{II} – Im3m) arises from a square planar array of ILAS (Figure 4.3C and 4.3D) and Q_{II} – Pn3m arises from a hexagonal array [36a]. Experimentally it has been shown that Q_{II} – Im3m, Q_{II} – Pn3m, and Q_{II} – Ia3d all readily interconvert [36a, 36b]. It has been shown theoretically that these three structures can interconvert by a process that involves only bending transformations with zero mean curvature conserved throughout [36c], and that all three are very close in energy [37]. These results show that membrane fusion is a step in the L_α/Q_{II} phase transition.

Membrane fusion rates were also measured at different temperatures using fluorescent assays, and the results were correlated with lipid phase behavior as determined by ^{31}P NMR, differential scanning calorimetry, x-ray diffraction, and combinations thereof. These studies verified that liposomes fuse much more readily under conditions where the equilibrium phase of the lipids is either the H_{II} or Q_{II} phase [31,32, 38–42]. In particular, there was a strong linkage of the onset

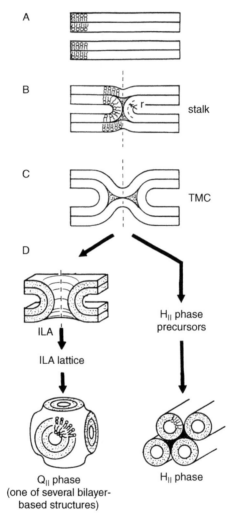

Figure 4.1 The modified stalk mechanism of membrane fusion and inverted phase formation. (A) Planar lamellar (L$_\alpha$) phase bilayers. (B) The stalk intermediate. The stalk is cylindrically symmetrical about the dashed vertical axis. The stalk is also the structural unit of the rhombohedral phase (L Yang, HW Huang. *Science* 297:1877–1879, 2002). (C) The TMC (*trans* monolayer contact) or hemifusion structure. The TMC can rupture to form a fusion pore, also referred to as an ILA, or interlamellar attachment, shown in cross section to the left in (D). If ILAs accumulate in large numbers they can rearrange to

temperature and intensity of the isotropic ^{31}P NMR resonance with the onset and increase in rate with temperature of membrane fusion [31,31a]. An example of the correlation of the fusion rate with the detection of isotropic NMR resonances in multilamellar preparations is reproduced in Figure 4.4. The very slow kinetics and hysteresis in the L_α/Q_{II} phase transition in some phospholipids complicates demonstration of the association of fusion with Q_{II} phase formation *per se* [43–48]. This has been done for DOPE -Me [31,31a,42,48], but more often one must rely on detection of ILAs as Q_{II} precursors, via ^{31}P NMR detection of isotropic resonances.

Mechanistically, the association of fusion and lamellar/inverted phase transitions is not surprising. The L_α/Q_{II} and L_α/H_{II} phase transitions require drastic changes in lipid monolayer topology, and extensive interactions between opposed lamellar phase bilayers are required to make these changes. Like fusion, the transitions start with formation of intermembrane structures. It is clear that interaction of opposed membranes is critical: when liposomes made of H_{II} phase-preferring lipids are incubated at temperatures near the L_α/H_{II} phase transition temperature, T_H, liposomes do not leak until they come into contact [36]. Nor are nonlamellar structures observed via cryoelectron microscopy in the membranes of isolated unilamellar liposomes when they are incubated under conditions where nonlamellar phases are at equilibrium [32–35,49,50]. In particular, the association of fusion with appearance of ILAs (Figure 4.3) makes mechanistic sense, since the ILAs are fusion pores, the end point of membrane fusion.

In lipid compositions that form only H_{II} phases, some liposome fusion is observed at temperatures near the lamellar/H_{II} transition temperature, T_H [40], but predominantly lipid mixing and leakage occur [38–40]. The rates of these processes increase rapidly as a function of temperature as T_H is approached [41]. The ILA precursors in such systems are thought to form H_{II} nuclei by an aggregation process more rapidly than they form ILAs [11,12,48,49]. However, these

Figure 4.1 (Continued) form Q_{II} phases (see Figure 4.2). For systems close to the L_α/H_{II} phase boundary, TMCs can also aggregate to form H_{II} precursors and assemble H_{II} phase domains (lower right in D). The balance between Q_{II} and H_{II} phase formation is dictated by the value of the Gaussian curvature elastic modulus of the bilayer (see Section 4.4). (Reprinted from DP Siegel. *Biophysical J* 76:291–313, 1999, with permission.)

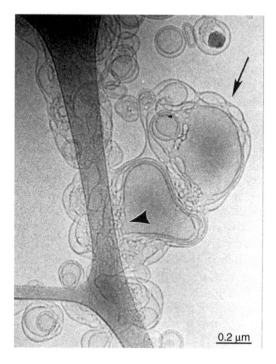

0.2 µm

Figure 4.2 Cryoelectron micrograph showing ILAs forming in a suspension of DOPE-Me unilamellar vesicles. The liposomes were originally at pH 7.4, and formation of fusion intermediates was triggered by a flashtube-induced temperature jump to approximately 90°C about 10 msec before vitrification. By that time, ILAs (fusion pores) had formed. One of these, seen from the side, is indicated by an arrow, while another, viewed down the pore axis, is indicated by an arrowhead. What appear to be holes at the edges of folds of membrane elsewhere in this image are projections of ILAs (see Figure 1 of DP Siegel, JL Burns, MH Chestnut, Y Talmon, *Biophys J* 56:161–169, 1989). The scale bar is 200 nm. Adapted from DP Siegel, RM Epand. *Biophys J* 73:3098–3111, 1997.

systems still produce small numbers of ILAs near T_H, as observed via cryoelectron microscopy [34,49,50], consistent with the observation of some fusion. Moreover phosphatidylethanolamines (the principal components of the H_{II}-phase forming systems studied most often) form Q_{II} phases if they are temperature-cycled through T_H many times [44–47], which is also consistent with formation of small numbers of ILAs and fusion around T_H. The propensity of lipids

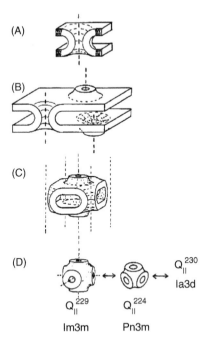

Figure 4.3   Schematic diagram of Q_{II} phase assembly from fusion pores (ILAs). (A) Isolated ILA, shown in cross section in the plane of the vertical axis. (B) Within a stack of planar bilayers, as many ILAs are likely to project "up" as "down" from a given bilayer. Here the bases of two ILAs, one each on the top and bottom of a pair of planar bilayers, are depicted with the ILAs bisected in planes perpendicular to the vertical axes. (C) As the number of ILAs increases, the ILAs will eventually form a primitive tetragonal array, in which the bases of the ILAs pointing "up" or "down" in a given bilayer form a square array. The primitive tetragonal array is shown here, with the vertical axes of the ILAs that compose it indicated by dashed lines. This structure has the same connectivity as the Q_{II} phase of Im3m symmetry (D, left), and the bilayers have to undergo only minor bending to form this Q_{II} phase. Q_{II}-Im3m can convert into other bicontinuous Q_{II} phases by bending processes (D, right). Reprinted from DP Siegel, JL Banschbach. *Biochemistry* 29:5975–5981, 1990, with permission.

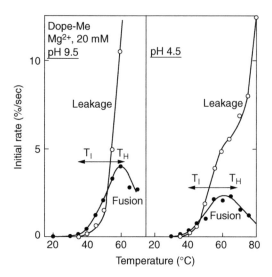

Figure 4.4 The initial rate of fusion and leakage of large unilamellar DOPE-Me liposomes induced to fuse by addition of 20 m*M*
Mg^{2+} at pH 9.5 (left); and by a drop of the pH from 9.5 to 4.5 in the
absence of divalent cations (right). The temperature range in which
isotropic ^{31}P NMR resonances were detected in multilamellar dispersions under the same final conditions is indicated by the bar
marked T$_I$ in both figures. Isotropic resonances may have persisted
at higher temperatures. Fusion is promoted starting at temperatures where isotropic ^{31}P NMR resonances appear. The leakage
process may be related to the formation of H$_{II}$ phase aggregates
from the liposomal aggregates, which is the dominant process at
higher temperatures, and which ruptures the liposomes (see DP
Siegel. In PL Yeagle, Ed. *The Structure of Biological Membranes*,
2nd ed. CRC Press, 2005 Chapter 8 for a review). Reprinted from
H Ellens, DP Siegel, D Alford, PL Yeagle, L Boni, LJ Lis, PJ Quinn,
J Bentz. *Biochemistry* 28:3692–3703, 1989., with permission.

to form H$_{II}$ phases is also associated with increased propensity for
membrane fusion because such lipids (e.g., phosphatidylethanolamine and monogalactosyldiglyceride) have poorly hydrated head
groups. Membranes rich in such lipids have comparatively weak
short-range repulsive forces and are more susceptible to close
approach and formation of intermembrane intermediates. Moreover,
as will be discussed in Section 4.3.1, addition of H$_{II}$-preferring lipids
to membranes that are otherwise stable in the lamellar phase reduces
the energies of the intermembrane intermediates themselves.

The association of membrane fusion with the L_α/Q_{II} phase transition may seem peculiar when one compares the end states of fusion and the phase transition. The end state of fusion is a membrane-enclosed volume of water, while the unit cell structures of bicontinuous Q_{II} phases contain continuous water channels. If unilamellar liposomes are incubated under conditions where Q_{II} phase is the equilibrium phase, the liposomes will eventually rearrange into bulk Q_{II} phase, and the aqueous contents of the original liposomes will leak out. Fusion without leakage is only observed in the first few rounds of liposomal interactions. However, fusion can occur when the system is close to the L_α/Q_{II} phase boundary, but when L_α is still the equilibrium phase. For example, intermediates in lamellar/inverted phase transitions are observed by cryoelectron microscopy at temperatures as much as 20°C below T_H in DiPoPE and T_Q DOPE -Me [34,49,50], and many authors have shown that ILAs begin to form slowly, as detected via ^{31}P-NMR, tens of degrees below T_H and T_Q in DOPE-Me and other phospholipid systems [31,42,51–56]. Moreover, when Q_{II} is the equilibrium phase, fusion occurs as the first steps in the phase transition process, which can be extremely slow and hysteretic in phospholipids, taking hours [43,45,48,51]. The initial interactions between unilamellar liposomes result in fusion, but Q_{II} phase can form only after extensive aggregation and rearrangement into multilamellar liposomes [32–34,49,50,57]. This transition between the two kinetic regimes (fusion in early liposomal encounters and concerted leakage after bigger aggregates form at longer times) has been observed ("collapse"; [31]). Fusion events *in vivo* usually involve only two membrane-enclosed structures, so bulk inverted phase formation is unlikely even if Q_{II} is the equilibrium phase of the membrane lipid composition.

Biomembrane lipid compositions may be close to lamellar/nonlamellar phase boundaries under physiological circumstances. Several classes of lipids that occur frequently in biomembranes adopt inverted phases under physiological conditions when the acyl chains are unsaturated [58–62]. The principal ones are phosphatidylethanolamine (PE) and monogalactosyldiglyceride, which is prominent in chloroplast thylakoid membranes. Cardiolipin (CL) and phosphatidic acid (PA) can also form inverted phases in the presence of divalent cations, and phosphatidylserine (PS) and PA both form inverted phases at low pH. Multilamellar dispersions of PE and PC with cholesterol [62a], PC and cholesterol [62b], and polyunsaturated PC with or without cholesterol [62c], exhibit prominent isotropic [31]P

NMR resonances near physiological temperatures. The results for PC and cholesterol are a surprising indication that these components of cell plasma membranes also form ILAs and possibly Q_{II} phases under physiological conditions, even though neither component forms nonlamellar phases under physiological conditions in isolation (except possibly for the polyunsaturated PC). Many biomembranes or biomembrane lipid extracts form nonlamellar phases if incubated above the physiological temperature [63–68], dehydrated [69–71], or treated with divalent cations [72–74]. Q_{II}-like structures have been described in some organelles of mammalian cells [75]. Finally, prolamellar bodies in chloroplasts have a Q_{II} phase-like structure [77,78]. Some prokaryotic organisms have been shown to regulate their lipid compositions so as to maintain the membrane lipids at a temperature within 20°C or less of the lamellar/inverted phase boundary ([76]; for a review see [79]). This is interesting because this temperature interval corresponds roughly with the onset of ILA formation in pure phospholipids (DOPE-Me and DOPE-DOPC mixtures; DOPE-Me with physiological levels of diglycerides [31,42]). This adjustment in membrane lipid "phase behavior" seems to be necessary to regulate the function of membrane-bound enzymes and perhaps to regulate membrane fusion/membrane fission [79].

4.3 USING LIPID PHASE BEHAVIOR TO STUDY MEMBRANE FUSION

The linkage between membrane fusion and the lamellar/inverted phase boundary permits us to infer the effects of exogenous agents on membrane fusion rates from their effects on the positions of these phase boundaries. In order to interpret results of these studies, it is necessary to consider how exogenous agents can affect the position of lamellar/inverted phase boundaries in membrane lipids. Unless specifically noted, we consider effects on L_α/Q_{II} and L_α/H_{II} phase boundaries in the presence of excess water because most biomembranes function in the presence of excess water.

4.3.1 Ways in Which Exogenous Additives Move Lamellar/Inverted Phase Boundaries: Effects of Lipid Additives

The principal factors determining the stability of inverted phases versus the lamellar phase are being determined. The principles are discernible from the response of inverted phase stability and unit cell dimensions to the presence of lipid additives. The principal factors are the curvature elastic energy [13] and the chain-packing energy. Here we define these factors and show how lipid additives can act to stabilize or destabilize inverted phases. In Section 4.3.2, we will discuss the effects of peptides and proteins on the phase boundaries and fusion. Peptides and proteins can induce fusion through some of the same effects as those of lipid additives and through some qualitatively different mechanisms.

4.3.1.1 Spontaneous Curvature

The L_α/H_{II} and L_α/Q_{II} transitions in the most common inverted phase–preferring lipids in biomembranes, PE and glycolipids, are thermotropic: the transitions occur as the lipid/water mixture is heated. Lamellar/inverted transitions in some acidic lipids (e.g., CL and PA) also occur upon divalent cation binding. Both these effects can be thought of crudely as a change in lipid molecular shape induced by increased acyl chain conformational freedom with increasing temperature, changes in the degree of head group hydration, divalent cation binding, or a combination. These effects can be thought of a little more precisely as changes in the spontaneous curvature, C_0, of the lipid monolayers [13]. The spontaneous curvature is the curvature that the monolayers would curl up into in the absence of other constraints. If the head group area is less than the acyl chain area, then C_0 is negative, and the monolayers will tend to bend to form water-filled cavities, like the water channels in the H_{II} and Q_{II} phases. In the converse situation ($C_0 > 0$), the lipid monolayers will be concave, like the surface of a detergent micelle in water. The C_0 of lipid mixtures with $C_0 < 0$ can be measured by x-ray diffraction experiments in the presence of excess water and long-chain alkanes [80,81]. The C_0 of a lipid mixture appears to be a mole fraction–weighted average of the spontaneous curvatures of the components [82–85]. (Actually, it is more properly an area-weighted average [85a].) The C_0 decreases with increasing temperature.

Adding agents that are "cone"-shaped molecules (that have negative values of C_0 when pure) to lipid mixtures tends to lower the T_H of the mixture, and adding "rod"-shaped molecules (that have $C_0 > 0$) tends to increase it. Examples of "cone"-shaped molecules are diglycerides (DAGs), which lower the spontaneous curvature [42,84] as well as T_Q and T_H [42,53,86–88]. In general, decreases in C_0 of the lipid composition accelerate fusion in model membrane systems and lower the energy of intermediates in membrane fusion (see [8,9,12] for reviews). Accordingly, DAGs stimulate membrane fusion and contact-mediated liposome leakage [42,53]. Ceramides [89,90] and fatty acids [91] are other biologically important lipids that lower T_Q and T_H. In contrast, lysophosphatidylcholine (LPC) forms normal micelles in dilute solution ($C_0 > 0$). LPC increases (makes more positive) the spontaneous curvature of phospholipids [85], and raises T_Q and T_H [86,92]. Accordingly, Yeagle et al. [93] showed that LPC inhibited fusion and leakage of N-methylated-DOPE LUVs as well as ILA formation (detected via ^{31}P NMR). There are many specific examples of lipids that lower C_0 that have also been shown to accelerate the rates of lipid mixing and fusion between liposomes (see [12,94] for the effects of lipid additives on C_0 and on fusion [12]).

4.3.1.2 Chain-Packing Energy

The chain-packing energy is a factor that opposes formation of inverted phases. The chain-packing energy arises from an entropically driven need to pack the hydrophobic moieties of the lipids as uniformly as possible within the structure of the phases (for reviews see [62,95]). In the H_{II} phase, there are linear hydrophobic interstices where the rod micelle–like tubes pack together. These are stabilized in three ways. If the tubes are cylindrical, acyl chains of the molecules in the curved monolayers lining the interstices must stretch to fill them, which decreases the entropy and raises the free energy. Alternatively, the circular cylinders of monolayers may distort to a more hexagonal cross section to decrease the size of the interstices, which increases the curvature free energy. (According to x-ray diffraction studies, the tube cross section is hexagonally distorted; [96,97].) More recent theories indicate that gradients in the tilt of lipid molecules in the monolayers stabilize a hexagonal cross section for the H_{II} phase (Section 4.3). Adding long-chain alkanes [53,86,87,98,99], squalane [87], triglycerides [100], or other nonpolar oils [101–103] fills the interstices and greatly reduces T_H without significantly affecting C_0 [104]. When present at low mole fractions, long-chain alkanes reside primarily in the interstices, as determined by neutron diffraction [105] and NMR techniques [87].

The chain-packing energy in bicontinuous Q_{II} phases has a more subtle origin. In these phases the bilayer midplanes conform to one of a family of infinite periodic minimal surfaces, which have zero mean curvature. However, the monolayers lie on surfaces that are displaced from the bilayer midplanes, and these surfaces are not surfaces of constant curvature. The tendency of the monolayers to reduce the curvature energy by complying with the constant curvature surfaces would require variation in bilayer thickness across the unit cell, which would result in an unfavorable chain-packing energy [106,107]. The two requirements cannot be fulfilled simultaneously, which is referred to as "frustration." For Q_{II} phases forming in phospholipids, the free energy penalty for bilayer thickness variation is estimated to be comparatively large, and bilayer thickness variation in Q_{II} phases is estimated to be very small [106,107]. The chain-packing energy of the Q_{II} phase might be reduced by additives that stabilize gradients in bilayer thickness within the phase. To the knowledge of the present author, this effect has not been demonstrated, although this may be a way in which hydrophobic transmembrane peptides stabilize Q_{II} phases (see Section 4.3.2). The effects of nonpolar oils on Q_{II} phase stability have not been well studied. Addition of 1 to 2 mole% of hexadecane to DOPE-Me destabilizes Q_{II} phases in favor of H_{II} at equilibrium (DP Siegel and JL Banschbach, unpublished observations). However, eicosane at 2 wt% appears to reduce both the temperature at which isotropic ^{31}P NMR resonances are observed and T_H [53].

Intermediates in fusion and lamellar/inverted phase transitions also have chain-packing energies (i.e., the simple representations of the intermediates in Figure 4.1B and Figure 4.1C show hydrophobic interstices similar to those in the H_{II} phase). Accordingly, it has been found that nonpolar oils accelerate lipid mixing and fusion rates in model membrane systems [103,108,109], as expected.

4.3.2 Effects of Peptides and Enzymes on Inverted Phase Behavior and Fusion

Peptides and proteins can have three different effects on the susceptibility of lipid membranes to fusion that can be studied in the context of lipid phase behavior. Proteins can change the lipid composition through enzymatic action, which will affect the curvature and chain-packing energies. Peptides and proteins can also change the curvature and chain-packing energies of the lipids directly by the way they bind to the lipid–water interface and insert into the lipid bilayers. Finally, proteins can impose a curvature on the membrane as a whole through conformational changes or protein–protein interactions within a protein layer that is affixed to the lipid bilayer [19,110,111]. This may be one role of the fusion-inducing proteins in virus/membrane fusion, and is probably also a major role of proteins that induce membrane fission in endocytosis [112]. Many researchers have tried to infer the roles of different parts of proteins known to induce membrane fusion by adding peptides derived from their sequences to lipids and measuring their effects on lipid phase behavior.

4.3.2.1 Enzymatic Production of Lipids that Change Spontaneous Curvature and Chain-Packing Energy of Membrane Lipids

Proteins can change the lipid composition in several ways to change their susceptibility to membrane fusion (e.g., [113,114]). The role of diacylglycerol (DAG) has been studied especially closely in connection with its tendency to induce formation of Q_{II} phases. Goñi, Alonso, and their coworkers studied the effects of DAG produced *in situ* by enzyme action on liposomal fusion and leakage rates. Diacylglycerol stabilizes inverted phases and reduces C_0. It is produced as a second messenger *in vivo* by phospholipase C action on membrane phospholipids. Such stimulated DAG production may be one way that fusion of secretory vesicles with cellular plasma membranes (exocytosis) is coupled to cell stimulus. Studies on model systems showed that phospholipase C action induces membrane fusion in a fashion that correlates with the amount of DAG produced with increasing time (for reviews, see [115,116]). The effects of other lipid additives on the lamellar/inverted phase transitions and liposome leakage and fusion were also studied.

In enzyme-free samples corresponding to the lipid composition after different extents of phospholipase C action, it was shown that maximal fusion activity occurred when the liposomal system was in a part of the pseudo-phase diagram where Q_{II} precursors (indicated by isotropic ^{31}P-NMR resonances) coexisted with the lamellar phase, and where H_{II} phase was not the predominant phase [117]. Figure 4.5 is a reproduction of this pseudo-phase diagram. Fusion was most associated with the appearance of bicontinuous Q_{II} phases, while Q_{II} phases of micellar structure were associated with leakage. Under the latter circumstances intermediate formation may be so fast that the formation of coexisting H_{II} phase and liposomal disruption is simply too fast for liposomes to endure long enough to register fusion signals. Interestingly, although heat- or inhibitor-inactivated phospholipase C did not produce DAG, fusion, or leakage, the presence of the enzyme resulted in a faster fusion rate than observed in its absence at the same DAG concentration [118,119].

The effects of *in situ* production of ceramide by sphingomyelinase were also studied. Ceramide also stabilizes inverted phases. *In situ* production of ceramide induced leakage rather than fusion [120,121]. However, joint action of phospholipase C and

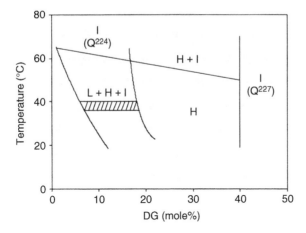

Figure 4.5 Pseudo phase diagram of egg PC/egg PE/cholesterol/ diacylglycerol in excess water, constructed from ^{31}P NMR data. L, H, and I indicate lamellar, inverted hexagonal, and isotropic phases; respectively. "Isotropic phase" refers to ILAs or Q_{II} phases. The ratios of the components egg PC/egg PE/cholesterol were fixed at 2/1/1 mol/mol/mol. Q_{II} phases were detected by x-ray diffraction (50% w/w lipid/water) at 10 mole% DAG after temperature cycling, and in samples with 50 mole% DAG. The space groups of the Q_{II} phases identified by x-ray diffraction are indicated in parentheses. Only the Q_{II} phase detected at lower DAG concentrations is a bicontinuous Q_{II} phase. The shaded area corresponds to the region of temperature and composition at which optimal liposome fusion was induced by phospholipase C action (the concentration of DAG was determined as a function of time after addition of the enzyme). These data are consistent with fusion pores being the precursors to bicontinuous Q_{II} phase formation. Predominance of the H_{II} phase or the inverted micellar Q_{II} phase at higher DAG concentrations and temperatures is associated with rapid rupture (leakage) of the vesicles. (Reprinted from JL Nieva, A Alonso, G Basáñez, FM Goñi, A Gulik, R Vargas, V Luzzati. *FEBS Lett* 368:143–147, 1995., with permission.)

sphingomyelinase was synergistic, producing more fusion than in the presence of either enzyme alone [120,122].

4.3.2.2 Effects of Viral Fusion Peptides

Viral fusion peptides are amphipathic fragments of fusion-inducing proteins from viruses that bind to lipid–water interfaces and are critical to the membrane fusion activity of the proteins (for reviews, see [123–125]). These peptides act early in the fusion process and may be involved in formation of the first intermembrane intermediates. Many researchers have studied the effects of these peptides on lamellar/inverted phase transitions. The rationale in most cases was that if these peptides stabilized inverted phases by reductions in bending energy, they would be likely to reduce the energy of fusion intermediates in parallel.

Fusion peptides from four different viruses have been found to reduce the temperature (T_I) at which isotropic ^{31}P NMR resonances form in lipids such as DOPE-Me [52,54–56]. The reductions are as much as 20K per mole% of peptide [52], and the effect is bigger at the pH at which the fusion peptide is most active *in vivo* [53]. Moreover, fusion peptides can induce isotropic resonances even when such resonances are not observed in the pure lipid (DiPoPE, which forms H_{II} and not Q_{II} phases; [56]). X-ray diffraction experiments have shown that fusion peptides from three different viruses stabilize Q_{II} phases in preference to H_{II} phases [50,56,126–128] — again, even in DiPoPE [50,56,127a,128]). The formation of Q_{II} phase precursors instead of H_{II} phase with increasing influenza fusion peptide concentration was also confirmed by cryo-TEM [50]. The studies using DOPE-Me report slightly different onset temperatures for isotropic resonance and Q_{II} and H_{II} phase formation in pure DOPE-Me: this may be due to the slow and hysteretic formation of these phases in DOPE-Me and the effects of low levels of impurities [45,48].

However, the correlation of fusogenicity of different peptides with effects on T_Q is not straightforward. In a detailed study of the effects of a series of mutants of the influenza hemagglutinin fusion peptide, Epand et al. [129] reported that the relative membrane fusion activity of the mutant peptides was not well correlated with the ability of the peptides to induce isotropic ^{31}P NMR resonances in DiPoPE, although the authors acknowledged that the signal/noise ratio of the ^{31}P NMR spectra was low. In addition, there is at least one case where a nonfusogenic mutant of the SIV fusion peptide was more effective in stabilizing Q_{II} phases than was the fusogenic wild-type peptide, in one of the two lipid systems investigated [128]. However, in studies with influenza and SIV fusion peptides, nonfusogenic mutants of fusion peptides either did not reduce the onset temperature for isotropic resonance detection as much, or induced lower-intensity isotropic resonances, than the fusogenic peptides did [54,55].

Epand et al. concluded that the effects of influenza fusion peptide on T_H were better-correlated with the fusogenicity of the different mutants [129]. Several groups report that fusion peptides from five viruses and a mammalian fertilin lower T_H of a pure PE [54–56,130–133]. In some cases fusogenic peptides are reported to reduce T_H by more than nonfusogenic mutants of the same peptide do at low peptide concentration [54,55,129,131,133]. However, there are some reasons for caution in assuming that a peptide's effect on T_H is proportional to its fusion activity.

First, the effect of a fusion peptide from feline leukemia virus on T_H is biphasic [56]: with increasing peptide concentration, T_H decreases only up until a peptide/lipid mole ratio of ca. 0.0015; T_H then increases with further increases in concentration. Since the local concentration of fusion peptides is anticipated to be high in viral protein-induced fusion, this would argue against a correlation of T_H with fusion activity *in vivo*. Second, with HIV fusion peptides, the effect of the peptide on the lipid phase behavior is a function of the length of the peptide used (i.e., the length of the corresponding native protein sequence); the sign of the effect on T_H can be reversed by selecting a longer sequence in at least one case (e.g., [131]). Measuring the influence of a part of the native protein sequence on lipid phase behavior is at best a crude way of inferring the role of a portion of that sequence. It may be that the correlation of fusogenicity with reductions in T_H and T_Q in a homologous series of peptides and peptide mutants reflects small differences in the way that the different peptides associate with the lipid–water interface, and that the critical structure–function relationships are only indirectly related to effects on lipid phase behavior. Similar suggestions have been made by many authors (e.g., [123,124,125a,129,131,133]). Qualitatively similar correlations have been made between the effectiveness of different peptide inhibitors of fusion and the extent to which the peptides raise T_H [134,135] or change either the onset temperature or width of the isotropic resonances [135,136]. The effects of the inhibitor peptides on Q_{II} phase formation via x-ray diffraction are more complicated: under some circumstances, the absence of coherent scattering obscured the relevant phase transitions [92, 127]. This may be due to formation of isotropic phases (dense accretions of ILAs without sufficient long-range order to form a Q_{II} phase lattice). The assignment of some of the diffraction data as representing H_I and Q_I phases in [92] is almost certainly incorrect. In H_I and Q_I phases, lipid rods would be surrounded by water. These would only form if C_0 had a large positive value, as opposed to the large negative values found in the pure lipid. The unit cell constants of the putative H_I and Q_I phases in [92] are approximately the same as for the pure lipid H_{II} and Q_{II} phases. It would be extremely surprising if a few percent of the peptides could produce a change in sign of C_0 and fortuitously form phases with nearly the same unit cell constants as in the H_{II} and Q_{II} phases.

It is not clear how fusion peptides act on the lipids to decrease T_H, T_I, and T_Q. This is an important question because the answer provides insight into how the fusion peptides may act in (presumably) lowering the energy of fusion intermediates. The effect of fusion peptides on T_H and T_I has often been rationalized as a decrease in C_0 of the lipids in the presence of the peptides (e.g., [54,55,56,129,133]). Reducing C_0 generally reduces the free energy of lipid fusion intermediates, so the inference was that this might be one role of fusion peptides in the fusion process. However, the effect of the peptides on C_0 can be inferred more directly from x-ray diffraction measurements of the H_{II} phase lattice constant in the presence of peptides and excess water: one would expect the lattice constant to decrease with increasing peptide concentration. Several authors have noted that peptides from influenza and SIV viruses do not significantly decrease the H_{II} phase lattice constant relative to the pure lipid [50,127–128]. In their work, in which they used higher peptide concentrations of several mole%, Siegel and Epand noted that the influenza peptide might actually increase C_0 [50]. Moreover, it is not clear why the peptides induce Q_{II} phases in a lipid that does not form them in the absence of peptide [50,56,127a,128]). It seems that the peptides are affecting a property or properties of the lipid other than C_0 that determines the relative stability of H_{II} and Q_{II} phases with respect to the lamellar phase, and perhaps of fusion intermediates as well. We will return to this question in Section 4.4 in the context of findings about the Gaussian curvature energy of fusion intermediates and Q_{II} phases.

4.3.2.3 Membrane-Spanning Peptides

There is evidence that membrane-spanning domains of viral fusion proteins are required for full activity of the proteins. Compared to the wild-type proteins, mutants with lipid (instead of protein) anchors either induce only lipid mixing and not fusion, or form fusion pores that are unstable and tend to close quickly [137–141]. Conflicting evidence exists about the sensitivity of fusion activity to mutations in the membrane-spanning regions for different proteins [140–143]. Armstrong et al. performed an especially complete study of the influenza virus hemagglutinin (HA) [141]. They made mutant

HAs with serial deletions from the C-terminal end of the membrane-spanning domain. A membrane-spanning domain of 17 or more residues of the wild-type domain was required for fusion activity of HA. HA mutants with shorter domains had very little fusion activity, probably because such domains are too short to span the bilayer. No specific sequences appeared to be necessary in the membrane-spanning domain for fusion activity [141]. This indicates that the transmembrane domains play an important role in the HA-mediated fusion process. There is also evidence that membrane-spanning domains from viral fusion proteins facilitate fusion in protein-free lipid membranes by stabilizing transient pores that otherwise close soon after formation [144,145]. This suggests that the membrane-spanning domains may stabilize the nascent fusion pores by changing some property of the lipids.

Since fusion pores (ILAs) are the first step in assembly of Q_{II} phases, this activity can be studied by determining the effects of membrane-spanning peptides on the stability of Q_{II} phases. Some natural, very hydrophobic membrane-spanning peptides (e.g., gramicidin [146]) can induce phospholipids that are normally in the lamellar phase to form inverted phases. Alamethicin [147] and gramicidin S [148–150] both induce Q_{II} phase formation. More detailed studies have been done using well-characterized synthetic peptides that are models of such domains (for recent reviews, see [151,152]). Many such studies show that membrane-spanning peptides induce Q_{II} and H_{II} phase formation or stabilize Q_{II} phase precursors (as detected via ^{31}P NMR). This even occurs in PCs (e.g., [153–156]), which usually form only lamellar phases in excess water, although substantial concentrations of peptide are required (a few mole% on a lipid basis). In PEs or PE-PG mixtures that are near the lamellar/ inverted phase boundary, even 0.1 mole% of peptide substantially changes T_I [156–161]. The formation of Q_{II} phases was confirmed by x-ray diffraction in two cases [157,158] (see also the preliminary report below).

It is not clear how the peptides stabilize nonlamellar phases and especially the Q_{II} phase. The effect of peptides of different lengths on lipid phase behavior has been explained in terms of hydrophobic mismatch between the peptide and the host lipid (reviewed in [151]). The bilayer is postulated to locally change in thickness to keep hydrophobic moieties on the membrane-spanning peptide from being exposed to water. This would induce a local interfacial curvature, either positive or negative, depending on whether the hydrophobic region of the peptide is longer or shorter, respectively, than the average thickness of the hydrophobic region of the bilayer. A good working definition of the hydrophobic region of the lipid bilayer is the distance between the two layers of lipid carbonyl groups (2.7 nm for DOPC; [162]). The hydrophobic length of membrane-spanning peptides is often inferred from the peptide sequence and measurements of the effects of homologous series of peptides on lipid phase behavior (e.g., [159,163]). Relatively "short" peptides should favor formation of H_{II} and Q_{II} phases, since these phases have average interfacial curvatures that are negative, and this is generally observed [151,153–161]. However, it is not easy to explain a few data points where the hydrophobic regions of the peptides are either about the same length or longer than the hydrophobic thickness of the host lipid bilayer, yet the peptides stabilize Q_{II} phases (WALP 23 and WALP27 in some PE-based systems [157,158]; L24 in DEPE [160]; and KALP31 in a mixture of PE and PG [161]). One must take care in assessing the relative lipid/peptide length mismatch for peptides with different types of anchoring groups [162,163], and account for the possible influence of peptide tilt with respect to the bilayer normal [163,164]. However, even with these allowances, the behavior of these "long" peptides is hard to explain.

The curious peptide length–dependent effect on Q_{II} phase formation is shown in a study by DP Siegel, V Cherezov, DV Greathouse, RE Koeppe II, JA Killian, and M Caffrey (manuscript in preparation), preliminary results of which are reported here. WALP peptides of different lengths (see Table 4.1) were incorporated into DOPE-Me. The peptides were incorporated, and T_Q determined via time-resolved rotating anode x-ray diffraction, by the procedures described in [48]. Under conditions of this study, when heated at a rate of 1.5°C/h or less, DOPE-Me began to form Q_{II} phase at a well-defined

Table 4.1 Sequences and Lengths of the WALP Peptides Incorporated into DOPE-Me

Peptide	Sequence[a]	Length as an α-Helix (Å)[b]
WALP19	Ac-GWW-$(LA)_6$ L-WWA-e	28.5
WALP23	Ac-GWW-$(LA)_8$ L-WWA-e	34.5
WALP25	Ac-GWW-$(LA)_9$ L-WWA-e	37.5
WALP27	Ac-GWW-$(LA)_{10}$ L-WWA-e	40.5
WALP31	Ac-GWW-$(LA)_{12}$ L-WWA-e	46.5

[a] Ac = acetyl; e = ethanolamine; A = alanine; G = glycine; L = leucine; W = tryptophan.
[b] Calculated by multiplying the number of residues by 1.5 Å.

temperature, T_Q, of 59.5 ± 0.4°C [48]. Coexisting H_{II} phase began to form at 63.7 ± 0.2°C [48]. This well-defined T_Q permitted a systematic study of the effects of different-length peptides as a function of peptide concentration. There are four principal results.

First, the peptides listed in Table 4.1 each reduced T_Q and T_H linearly as a function of peptide concentration in the range 0 to 0.50 mole% peptide. This shows that the peptides are not phase-separating from the lipid in the lamellar phase. The data were obtained using peptides shown to be more than 80% pure via HPLC.

Second, the slope of the T_Q vs. concentration plot for the peptides changed in a biphasic fashion as a function of peptide length. Peptides of intermediate length (WALP25 and WALP27) had small effects on T_Q and T_H. Peptides that were shorter (WALP19 and WALP23) were more effective in reducing T_Q and T_H. Surprisingly, the longest peptide (WALP31) also substantially reduced the transition temperatures by almost as much as the shortest WALP19 and WALP23. This result is summarized in Figure 4.6, which shows the

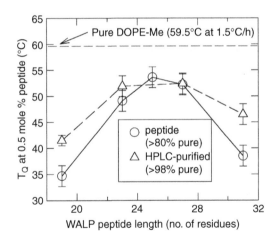

Figure 4.6 Plot of T_Q of DOPE-Me containing 0.50 mole% of WALP peptides of the indicated sequence lengths (horizontal axis). The WALP peptide sequences (WALP19, WALP23, WALP25, WALP27, and WALP31) are given in Table 4.1. Circles: peptides of >80% purity. Triangles: HPLC-purified peptides at 98% purity. The "impurities" in the less-pure peptides are chiefly peptides that are one or two residues shorter than the target sequence. Data from DP Siegel, V Cherezov, DV Greathouse, RE Koeppe II, JA Killian, and M Caffrey (manuscript in preparation).

T_Q determined for DOPE-Me samples containing 0.5 mole% of each peptide. As discussed above, it is not easy to explain the reduction in T_Q by WALP31 in terms of the bilayer mismatch hypothesis. The hydrophobic length of helix between the innermost two tryptophan residues of WALP31, which interacts most closely with the hydrophobic interior of the bilayer [159,163], is 3.8 nm. This is comparable to the total bilayer thickness (3.9 nm, [43]) and much longer than the thickness of the hydrophobic region of the bilayer, which is assumed to be the same as in DOPC (2.7 nm [162]). Tilt of the helical axis cannot significantly reduce the length mismatch. The maximum tilt of WALP31 in bilayers with similar degrees of length mismatch is 20° [163]. Such a tilt would still result in a projected hydrophobic length of 3.5 nm for the hydrophobic part of the helix. A complicating factor is that there is indirect evidence for formation of small aggregates of WALP31 in DOPE-Me. In the plot of T_Q vs. WALP31 concentration, a line extrapolating the T_Q vs. concentration data from higher to lower concentrations intersects the transition temperature axis 4.8°C below T_Q for pure DOPE-Me. At concentrations <0.05 mole%, WALP31 may be present in a form that is more effective in reducing T_Q on a mole-for-mole basis than the form of WALP31 present at higher concentrations. This suggests that small aggregates of WALP31 exist at concentrations >0.05 mole%. T_Q values were checked at some concentrations using more highly-purified peptides (98% pure via HPLC). These data are indicated in Figure 4.6 by triangles. The length-dependent effects are less prominent, but still present.

Third, the H_{II} phase is observed at higher temperatures than the onset of Q_{II} phase formation up until some peptide concentration for all the peptides. However, for all the peptides, the slope of T_H vs. peptide concentration is similar to the slope for T_Q. Moreover, H_{II} formation is observed in different concentration ranges for the different peptides: it is observed through 0.5 mole% for WALP27, but only through 0.2 mole% for WALP19 and WALP23, and only up until 0.05 mole% for WALP31. These two aspects of the data will be discussed in Section 4.4.

Fourth, both in the absence [48] and presence of these WALP peptides, DOPE-Me Q_{II} phases initially form with very large unit cell constants (>30 nm) that shrink rapidly with time and temperature after the onset of Q_{II} phase formation. This will be discussed below in connection with postulated effects of membrane-associated peptides on chain-packing energies in Q_{II} phases.

How then do membrane-spanning peptides, especially "long" ones, stabilize H_{II} and Q_{II} phases? This is especially perplexing when one considers that trivial amounts of peptide (0.1 to 1 mole%) can move lamellar/isotropic phase boundaries by as much as 10°C [158]. It has been shown that gramicidin, a very hydrophobic channel-forming peptide, can decrease C_0 of PE when present at concentrations less than a few mole% [165]. This appears to be due in part to the displacement of gramicidin molecules away from the lipid–water interface. Whether such displacement is possible for other sequences of membrane-spanning peptides has not been determined (WALP peptides are thought to form rather rigid helices with pairs of tryptophan residues at each end to act as particularly strong interfacial "anchors" [159,161,163]). Again, this is even harder to envision for the longer WALP peptides such as WALP31.

It has been suggested [151,158] that membrane-spanning peptides stabilize gradients in thickness in the bilayers of the Q_{II} phase, reducing chain-packing stress. However, the chain-packing stresses are thought to be negligible for Q_{II} phases with unit cells of ca. 30 nm or more [166,167]. The Q_{II} phases that first form in WALP peptide-containing samples (preliminary study of Siegel et al., summarized above) form with such very large unit cell constants. Thus this probably does not explain the WALP/DOPE-Me data.

Moreover, again in the study of WALP peptides in DOPE-Me, all the peptides reduce T_H at about the same rate as T_Q with increasing peptide concentration. It would be rather odd if phases of these very different geometries were affected in almost the same way by the addition the same substances. One possibility is that the peptides exert their principal effect on the free energy of the lamellar phase, rather than the H_{II} or Q_{II} phase. The peptides may increase the free energy of the lamellar phase through a hydrophobic mismatch effect more than they affect the free energy of either the H_{II} or Q_{II} phase. Thus the free energy difference between the phases would still decrease with increasing peptide concentration. This would be consistent with the near parallel decrease in T_H and T_Q with increasing peptide concentration observed in the WALP/DOPE-Me study. However, by itself, this principle cannot explain why H_{II} phases are destabilized across a wide concentration range by some peptides and not by others. It is possible that both membrane-spanning and peripheral peptides (e.g., fusion peptides) also act by affecting the Gaussian curvature (saddle splay) modulus of the lipid monolayers. This will be discussed in Section 4.4.

Clearly, more research into how membrane-spanning peptides stabilize Q_{II} phases and Q_{II} phase precursors is needed. However, it is already clear that at least some membrane-spanning peptides do stabilize fusion pores (lower T_I and T_Q) through an effect on lipids. This is consistent with the observed role of transmembrane domains of viral fusion proteins. More work with more faithful models of intrinsic membrane protein domains is warranted.

4.4 STUDYING LAMELLAR/Q_{II} PHASE TRANSITIONS TO REFINE MODELS OF FUSION INTERMEDIATE ENERGETICS

It appears that the curvature energy with respect to the initial bilayers is the most important contribution to the energy of fusion intermediates with respect to the lamellar phase [12]. The curvature energy for a unit area of lipid monolayer is [13]

$$f = (k/2) \cdot (C - C_0)^2 + \kappa \cdot K, \tag{4.1}$$

and the curvature energy for a lipid bilayer with the same lipid composition in each monolayer is [13]

$$f_{\text{bilayer}} = (k_b/2) \cdot (C_b)^2 + \kappa_b \cdot K_b. \tag{4.2}$$

In Equation 4.1, k is the bending elastic constant, C is the curvature, and K is the Gaussian curvature. The Gaussian curvature is

$$\begin{aligned} K &= 1/C^2, \\ K_b &= 1/C_b^2 \end{aligned} \tag{4.3}$$

and κ is the Gaussian curvature elastic modulus. In Equation 4.2 the corresponding properties of the bilayer bear the subscript b. C_0 is the spontaneous curvature of the monolayer lipids (Section 4.3.1.1). In evaluating f_{bilayer}, the curvature is evaluated at the bilayer midplane (i.e., the dividing surface between the two monolayers), and there is no spontaneous bilayer curvature because the lipids on both sides of the bilayer are the same.

The energy of fusion intermediates with respect to the initial bilayers has usually been evaluated using Equation 4.1 either with the assumption that κ is negligible (e.g., [168–171]) or with a crude estimate of κ [11]. This is because it is difficult to measure κ and because until very recently there was no measurement of this constant for phospholipids or lipid compositions resembling most biomembranes. Measurements of κ were made by careful measurements of the swelling behavior of Q_{II} phases, mostly in systems containing large mole fractions of monoglycerides [172–176]. These measurements are not very accurate for Q_{II} phases with small unit cell constants [175], and cannot be performed in systems where more than one Q_{II} phase forms [176].

Note that for the lamellar phase (flat bilayers) and the H_{II} phase, $K = 0$ because at least one of the two principal curvatures is zero in each of these phases. In contrast, the Gaussian curvature of ILAs and Q_{II} phases is less than zero since these surfaces have two nonzero principal curvatures, of opposite sign, at each point. This is also true of the first intermediates to form in the fusion process, stalks and TMCs (Figure 4.1B and Figure 4.1C). Therefore the Gaussian curvature energy (the second term on the right-hand side of Equation 4.1) changes when fusion intermediates, ILAs, or Q_{II} phases form from the lamellar phase, but not when the H_{II} phase forms. This shows that the magnitude of κ has an important influence on the energies of formation of fusion intermediates and that κ can in principle be evaluated by studying the stability of the Q_{II} phase with respect to the lamellar phases.

4.4.1 Evaluation of the Gaussian Curvature Elastic Coefficient by Measurement of T_Q

It was recently shown [17] that κ could be evaluated in any one-component lipid system with a well-defined T_Q in excess water, if the unit cell constant of the nascent Q_{II} phase is large (ca. 30 nm for common biomembrane lipids). Extension to multi-component systems may be possible. There are two basic elements of the treatment. The first is a recognition that both ILAs and Q_{II} phases are based on minimal surfaces. This means that the total bilayer curvature at each point on these structures is zero. (ILAs have the lowest total curvature energy if they adopt shapes similar to the catenoid minimal surface, with $C_b \approx 0$; [17,170].) The result is that the bilayer curvature energies (Equation 4.2) of ILAs and Q_{II} phases only contain the Gaussian curvature energy term. This means that the total curvature energies of ILAs and Q_{II} phase with respect to the lamellar phase are directly proportional to κ_b. For example, making use of the Gauss–Bonnet theorem [177], the free energy of ILAs with respect to the lamellar phase is

$$F_{ILA} = -4\pi \cdot \kappa_b. \qquad (4.4)$$

ILAs and Q_{II} phase can form from the lamellar phase whenever

$$\kappa_b \geq 0. \qquad (4.5)$$

The second principal element of the analysis is recognition that, from the mathematical properties of surfaces that are everywhere a fixed distance δ from a minimal surface [178], the bilayer Gaussian elastic modulus κ_b can be expressed in terms of monolayer properties [17];

$$\kappa_b = 2\kappa - 4 \cdot k \cdot C_0 \cdot \delta. \qquad (4.6)$$

In Equation 4.6, δ is the distance between the bilayer midplane and the neutral plane of the monolayers, and again (Equation 4.1) k is the monolayer bending elastic constant and C_0 is the monolayer spontaneous curvature. Applying Equation 4.5, we see that at $T = T_Q$,

$$\frac{\kappa}{k} = 2 \cdot C_0 \cdot \delta \ (\text{at } T = T_Q). \qquad (4.7)$$

All the terms on the right-hand side of Equation 4.7 can be measured by x-ray diffraction experiments. Thus κ (or more easily, the ratio κ/k) can be evaluated by measuring C_0 at the temperature for the onset of Q_{II} phase formation. We assume that the neutral plane is near the interface between the hydrophilic and hydrophobic regions of the monolayer. Hence the position of the Luzzati dividing surface is a good approximation. The value $\delta = 1.3$ nm was determined for monomethylated egg PE [176a]. The ratio κ/k is -0.83 ± 0.08 at 55°C for DOPE-Me [17]. A few ILAs should be able to form at temperatures a few degrees below T_Q, but the onset temperature for ILA formation can be taken as T_Q within the accuracy of the technique (a 10-degree error in T_Q changes the value of κ/k by less than 5% for DOPE-Me; [17]). To the knowledge of the authors of [17], this is the first value of this ratio measured for systems containing only phospholipids. Equation 4.6 shows how the stability of ILAs and Q_{II} phases changes with temperature in systems with thermotropic transitions. C_0 generally decreases with increasing temperature. ILAs and Q_{II} phases form when the C_0 becomes negative enough for the second term on the right-hand side of Equation 4.6 to overcome the negative contribution of 2κ.

In [17], it was shown that pure PEs should have values of κ/k of about −0.9 at T ≈ T_H. Moreover, many physiologically relevant phospholipid compositions develop isotropic ^{31}P NMR resonances and Q_{II} phases around 40°C; including DOPC/DOPE [31,179], DOPC/DOPE/cholesterol [180], DOPC/cholesterol [181], and polyunsaturated PC [182]. While the occurrence of isotropic ^{31}P NMR resonances in multilamellar preparations is not a firm phase assignment, this strongly suggests that these lipid compositions form ILAs and possibly Q_{II} phases in the physiological temperature range. ILAs and Q_{II} phases have been observed in the DOPC/cholesterol system by cryo-TEM (B van Duyl and JA Killian, personal communication; DP Siegel and B Tenchov, manuscript in preparation). Therefore, since k for DOPE and DOPC and their mixtures with low mole fractions of cholesterol are about the same, and the net C_0 values in these systems are about the same as for DOPE-Me at T_Q [104,183,184], these lipid compositions probably have comparable values of κ at their respective T_Q values (Equation 4.7). Thus, the value of κ/k measured for DOPE-Me [17] is probably typical of many phospholipid compositions near physiological temperature. However, we do not know how κ changes as a function of temperature far away from T_I or T_Q for a given lipid composition [17].

4.4.2 Implications for Membrane Fusion

κ is comparable to the value of k in DOPE-Me [17]. This means that κ has a large effect on the energy of stalks, TMCs, and ILAs. Using the κ value for DOPE-Me, if we calculated the energy of an ILA with Equation 4.1 and assumed that κ is essentially zero, as has been done previously [11,168–171], we would underestimate the energy of the ILA by almost 200 $k_B T$ [17] (where k_B is Boltzmann's constant) and of stalks and TMCs by almost half that amount. Clearly one cannot omit Gaussian curvature energy from calculations of fusion intermediate energies. This large Gaussian curvature elastic contribution removes some apparent paradoxes that had troubled fusion intermediate calculations [12]. First, ILAs were predicted (e.g., [11,170]) to be stable at temperatures far below temperatures at which they actually are observed, when their energies were calculated according to Equation 4.1 with $\kappa = 0$. The Gaussian curvature elastic energy accounts for this discrepancy. Second, the size of the Gaussian curvature elastic modulus also dictates whether Q_{II} or H_{II} phase predominates in lipids at temperatures near the lamellar/inverted phase boundary (Figure 4.1D); the factors determining this had not previously been clear [11]. Third, newer theories of fusion intermediate energies [168–171] predict that stalks are thermodynamically stable in the presence of excess water for lipid compositions in which no such "stalk phases" are observed (e.g., for any lipid with spontaneous splay $\approx C_0 < -0.26$ nm^{-1}; [168]). This discrepancy is also probably due to neglect of the (positive) Gaussian curvature energy contribution under those circumstances. Kozlovsky et al. [18] have recently shown that inclusion of Gaussian curvature elastic energy and hydration energy accounts for the observed range of stability of the rhombohedral ("stalk") phase observed at reduced water activity in DOPC/DOPE mixtures. Thus we are approaching a description of intermediate energetics that is consistent with the details of observed lipid phase behavior.

4.4.3 Implications for the Effects of Peptides on Membrane Fusion and Q$_{II}$ Phase Formation

In Section 4.3.2 we discussed the possibility that peptides may affect k and C_0 of the membrane lipids. Gramicidin has been observed to reduce C_0 and increase k at low concentration in some lipids [165]. It is possible that other peptides have similar effects in some concentration ranges. It has been suggested [17] that peptides may affect κ as well. This could explain some effects of membrane-associated peptides (such as fusion peptides and membrane-spanning peptides, Section 4.3.2.2 and Section 4.3.2.3) on membrane fusion and Q$_{II}$ phase formation. For instance, if fusion peptides and membrane-spanning peptides increase κ, this may explain why Q$_{II}$ phases form in the presence of peptides in preference to H$_{II}$ phases in systems that do not form Q$_{II}$ phases in the absence of peptides (Section 4.3.2). ILAs and Q$_{II}$ phases have negative Gaussian curvature, while the H$_{II}$ phase has zero Gaussian curvature. If addition of peptides makes $\kappa_b > 0$, the Gaussian curvature elastic energy favors formation of ILAs and Q$_{II}$ phase. Although peptides can lower the energy of H$_{II}$ phases through an effect on k and C_0, Equation 4.6 shows that peptides can lower the energy of Q$_{II}$ phases both through a reduction of the bending elastic constants k and C_0, and through a direct effect on κ. Hence, it is not surprising that the addition of peptides may stabilize Q$_{II}$ phases more than H$_{II}$ phases in some systems.

Moreover, one should expect the addition of membrane-associated peptides to change the Gaussian curvature elastic modulus of the monolayers, at least to some extent. k, C_0, and κ all arise from the same property of lipid monolayers — namely, the details of the lateral stress profile as a function of depth z in the monolayer, $\sigma(z)$. The monolayers have different lateral stresses as a function of depth to account for the different lateral interactions of head groups and acyl chains. The integral of the stress through the entire monolayer must be zero since the monolayer is at equilibrium with respect to area. However, the moments of $\sigma(z)$ determine the elastic constants (Seddon [62] gives a very accessible description of the origin and effects of the stress profile):

$$\int z \cdot \sigma(z) \cdot dz = -k \cdot C_0 \qquad (4.8)$$

$$\int z^2 \cdot \sigma(z) \cdot dz = \kappa. \qquad (4.9)$$

Thus, if peptides affect k and/or C_0 (the first moment of $\sigma(z)$; Equation 4.8) they are likely to affect κ (the second moment of $\sigma(z)$; Equation 4.9), at least to some degree. Since κ is the integral of $\sigma(z)$ weighted by z^2, it can be more sensitive than k and C_0 are to small differences in the distribution of mass within the peptide–lipid monolayer. Hence κ can be more sensitive to the effects of peptides than k and C_0 would be. The effects of a peptide on the elastic constants will be different depending on the size and structure of the peptide and where it resides in the monolayers. This may explain some of the differences between effects of different-length membrane-spanning peptides on T_Q and T_H (e.g., in systems such as WALP peptides in DOPE-Me; Section 4.3.2.3). For example, it may be part of the explanation of the length dependence of the peptide effects on T_Q as compared at constant peptide concentration (Figure 4.6). It may also explain why DOPE-Me H_{II} phases are destabilized across a wide concentration range by some WALP peptides and not by others (DP Siegel, V Cherezov, DV Greathouse, RE Koeppe II, JA Killian, and M Caffrey; manuscript in preparation).

4.5 CONCLUSION

The work summarized here shows that studies of lipid–peptide phase behavior and phase transition kinetics are central to an understanding of membrane fusion. Fundamentally, membrane fusion occurs through local changes in the physical interactions between lipids and between lipids and peptides. We are unlikely to explicate the roles of different parts of fusion-inducing proteins unless we understand those interactions in detail. It is the balance of these lipid–lipid and lipid–peptide interactions that determine the details of lipid–peptide phase behavior. We are just starting to understand the nature and consequences of the interactions between membrane-associated peptides and lipids, especially of their possible effects on the Gaussian curvature elastic energy of fusion intermediates. The nature and size of the Gaussian curvature elastic energies of Q_{II} is still only vaguely known even in pure lipid systems. Work on the factors that determine the rhombohedral phase is just starting [18], and this is a phase that is the closest link to the first, critical intermediate in membrane fusion. What is required is accurate and detailed work on the phase behavior and phase transition kinetics of intelligently chosen lipid–peptide systems.

REFERENCES

1. R Jahn, T Lang, TC Sudhof. *Cell* 112:519–533, 2003.

2. R Blumenthal, MJ Clague, SR Durrell, RM Epand. *Chem Rev* 103:53–69, 2003.

3. G Basáñez. *Cell Molecular Life Sci* 59:1478–1490, 2002.

4. BR Lentz, V. Malinin, ME Haque, K Evans. *Curr Opin Struct Biol* 10:607–615, 2000.

5. T Stegmann. *Traffic* 1:598–604, 2000.

6. J Bentz, H Ellens. *Colloids Surf* 30:65–112, 1988.

7. J Wilschut. In J Wilschut and D Hoekstra, Eds. *Membrane Fusion.* Marcel Dekker, New York, 1990, pp. 89–126.

8. LV Chernomordik, MM Kozlov, J Zimmerberg. *J Membr Biol* 146:1–4, 1995.

9. LV Chernomordik, J Zimmerberg. *Curr Opin Struct Biol* 5:541–547, 1995.

10. KNJ Burger. *Traffic* 1:605–613, 2000.

11. DP Siegel. *Biophysical J* 76:291–313, 1999.

12. DP Siegel. In PL Yeagle, Ed. *The Structure of Biological Membranes*, 2nd ed. CRC Press, Boca Raton, FL, 2005, Chapter 8.

13. W Helfrich. *Z Naturforsch C* 28:693–703, 1973.

14. RP Rand, NL Fuller, SM Gruner, VA Parsegian. *Biochemistry* 29:76–87, 1990.

15. M Hamm, MM Kozlov. *Eur J Biophys B* 6:519–528, 1998.

16. RH Templer, BJ Khoo, JM Seddon. *Langmuir* 14:7427–7434, 1998.

17. DP Siegel, MM Kozlov. *Biophys J* 87:366–374, 2004.

18. Y Kozlovsky, DP Siegel, MM Kozlov. *Biophys J* 87:2508-2521, 2004.

19. LV Chernomordik, MM Kozlov. *Annu Rev Biochem* 72:175–207, 2003.

20. L Yang, HW Huang. *Science* 297:1877–1879, 2002.

21. L Yang, HW Huang. *Biophys J* 84:1808–1817, 2003.

22. L Yang, L Ding, HW Huang. *Biochemistry* 42:6631–6635, 2003.

23. PR Cullis, MJ Hope. *Nature* 271:672–674, 1978.

24. AJ Verkleij, C Mombers, WJ Gerritsen, L. Leunissen-Bijvelt, PR Cullis. *Biochim Biophys Acta* 555:358–361, 1979.

25. AJ Verkeli, CJA Van Echteld, WJ Gerritsen, PR Cullis, B de Kruijff. *Biochim Biophys Acta* 600:620–624, 1980.

26. RG Miller. *Nature* 287:166–167, 1980.

27. SW Hui, TP Stewart. *Nature* 290:427–428, 1981.

28. SW Hui, TP Stewart, LT Boni, PL Yeagle. *Science* 212:921–923, 1981.

29. RP Rand, TS Reese, RG Miller. *Nature* 293:237–238, 1981.

30. DP Siegel. *Biophys J* 49:1155–1170, 1986.

31. H Ellens, DP Siegel, D Alford, PL Yeagle, L Boni, LJ Lis, PJ Quinn, J Bentz. *Biochemistry* 28:3692–3703, 1989.

31a. LC van Gorkom, SQ Nie, RM Epand. *Biochemistry* 31:671–677, 1992.

32. DP Siegel, JL Burns, MH Chestnut, Y Talmon, *Biophys J* 56:161–169, 1989.

33. PM Frederik, KNJ Burger, MCA Stuart, AJ Verkleij. *Biochim Biophys Acta* 1062:133–141, 1991.

34. DP Siegel, WJ Green, Y Talmon. *Biophys J.* 66:402–414, 1994.

35. M Johnsson, K Edwards. *Biophys J* 80:313–323, 2001.

36. DP Siegel. *Chem Phys Lipids* 42:279–301, 1986.

36a. AM Squires, RH Templer, JM Seddon, J Woenckhaus, R. Winter, S Finet, N Theyencheri. *Langmuir* 18:7384–7392, 2002.

36b. A Squires, RH Templer, O Ces, A Gabke, J Woenckhaus, JM Seddon, R Winter. *Langmuir* 16:3578–3582, 2000.

36c. A Fogden, ST Hyde. *Eur Phys J* B7:91–104, 1999.

37. US Schwarz, G Gompper. *Langmuir* 17:2084–2096, 2001.

38. H Ellens, J Bentz, FC Szoka. *Biochemistry* 23:1532–1538, 1984.

39. J Bentz, H Ellens, MZ Lai, FC Szoka. *Proc Natl Acad Sci USA* 82:5742–5745, 1985.

40. H Ellens, J Bentz, FC Szoka. *Biochemistry* 25:285–294, 1986.

41. J Bentz, H Ellens, FC Szoka. *Biochemistry* 26:2105–2116, 1987.

42. DP Siegel, J Banschbach, D Alford, H Ellens, LJ Lis, PJ Quinn, PL Yeagle, J Bentz. *Biochemistry* 28:3703–3709, 1989.

43. SM Gruner, MW Tate, GL Kirk, PTC So, DC Turner, DT Keane, CPS Tilcock, PR Cullis. *Biochemistry* 27:2853–2866, 1988.

44. E Shyamsunder, SM Gruner, MW Tate, DC Turner, PTC So, CPS Tilcock. *Biochemistry* 27:2332–2336, 1988.

45. DP Siegel, JL Banschbach. *Biochemistry* 29:5975–5981, 1990.

46. J Erbes, C Czeslik, W Hahn, M Rappolt, G Rapp. *Ber Bunsenges Phys Chem* 98:1287–1293, 1994.

47. B Tenchov, R Koynova, G. Rapp. *Biophys J* 75:853–866, 1998.

48. V Cherezov, DP Siegel, W Shaw, SW Burgess, M Caffrey. *J Membr Biol* 195:165–182, 2003.

49. DP Siegel, RM Epand. *Biophys J* 73:3098–3111, 1997.

50. DP Siegel, RM Epand. *Biochim Biophys Acta* 1468:87–98, 2000.

51. J Gagné, L Stamatatos, T Diacovo, SW Hui, PL Yeagle, JR Silvius. *Biochemistry* 24:4400–4408, 1985.

52. PL Yeagle, RM Epand, CD Richardson, TD Flanagan. *Biochim Biophys Acta* 1065:49–53, 1991.

53. LCM van Gorkum, SQ Nie, RM Epand. *Biochemistry* 31:671–677, 1992.

54. RM Epand, RF Epand. *Biochem Biophys Res Commun* 202:1420–1425, 1994.

55. RF Epand, I Martin, JM Ruysschaert, RM Epand. *Biochem Biophys Res Commun* 205:1938–1943, 1994.

56. SMA Davies, RF Epand, JP Bradshaw, RM Epand. *Biochemistry* 37:5720–5729, 1998.

57. AJ Verkleij. *Biochim Biophys Acta* 779:43–63, 1984.

58. PR Cullis, J Hope, CPS Tilcock. *Chem Phys Lipids* 40:127–144, 1986.

59. CPS Tilcock. *Chem Phys Lipids* 40:109–125,1986.

60. RNAH Lewis, DA Mannock, RN McElhaney. In RM Epand, Ed. *Lipid Polymorphism and Membrane Properties. Current Topics in Membranes,* Vol. 44, Academic Press, New York, 1997, chap. 2.

61. G Lindblom, L Rilfors. *Biochim Biophys Acta* 988:221–256, 1989.

62. JM Seddon. *Biochim Biophys Acta* 1031:1–69, 1990.

62a. CPS Tilcock, MB Bally, SB Farren, PR Cullis. *Biochemistry* 21:4596–4601, 1982.

62b. RM Epand, DW Highes, BG Sayer, N Borochov, D Bach, E Wachtel. *Biochim Biophys Acta* 1616:196–208, 2003.

62c. RM Epand, RF Epand, AD Bain, BG Sayer, DW Hughes. *Magnetic Resonance Chem* 42:139–147, 2003.

63. WJ de Grip, EHS Drenthe, CJA van Echteld, B de Kruijff, AJ Verkleij. *Biochim Biophys Acta* 558:330–337, 1979.

64. E Burnell, L van Alphen, A Verkleij, B de Kruijff. *Biochim Biophys Acta* 597:492–501, 1980.

65. K Gounaris, DA Mannock, A Sen, APR Brain, WP Williams, PJ Quinn. *Biochim Biophys Acta* 732:229–242, 1983.

66. JL Ranck, L Letellier, E Shechter, B Krop, P Pernot, A Tardieu. *Biochemistry* 23:4955–4961, 1984.

67. G Lindblom, I Brentel, M Sjolund, G Wikander, A Wieslander. *Biochemistry* 25:7502–7510, 1986.

68. S Morein, AS Andersson, L Rilfors, G Lindblom. *J Biol Chem* 271:6801–6809, 1996.

69. S Gruner, KJ Rothschild, NA Clark. *Biophys J* 39:241–251, 1982.

70. LM Crowe, JH Crowe. *Arch Biochem Biophys* 217:582–587, 1982.

71. WJ Gordon-Kamm, PL Steponkus. 1984. *Proc Natl Acad Sci USA* 81:6373–6377, 1984.

72. PR Cullis, B de Kruijff, MJ Hope, R Nayar, A Rietveld, AJ Verkleij. *Biochim Biophys Acta* 600:625–635, 1980.

73. AD Albert, A. Sen, PL Yeagle. *Biochim Biophys Acta* 771:28–34, 1984.

74. K Nicolay, R van der Neut, JJ Fok, B de Kruijff. *Biochim Biophys Acta* 819:55–65, 1985.

75. T Landh. *FEBS Lett* 369:13–17, 1995.

76. S Morein, AS Andersson, L Rilfors, G Lindblom. *J Biol Chem* 271:6801–6809, 1996.

77. E Selstam, AW Wigge. In C Sundqvist, M Ryberg, Eds. *Chloroplast Pigment–Protein Complexes: Synthesis and Assembly.* Academic Press, 1993, pp. 241–277.

78. E Selstam, J Schelin, T Brain, WP Williams. *Eur J Biochem* 269:2336–2346, 2002.

79. L Rilfors, G Lindblom. *Colloids Surf* 26:112–124, 2002.

80. RP Rand, NL Fuller, SM Gruner, VA Parsegian. *Biochemistry* 29:76–87, 1990.

81. RP Rand, VA Parsegian. In RM Epand, Ed. *Lipid Polymorphism and Membrane Properties* (Current Topics in Membranes, Vol. 44). Academic Press, New York, 1997, chap. 4.

82. SL Keller, SM Bezrukov, SM Gruner, MW Tate, I Vodyanoy, VA Parsegian. *Biophys J* 65:23–27, 1993.

83. JA Szule, NL Fuller, RP Rand. *Biophys J* 83:977–984, 2002.

84. S Leikin, MM Kozlov, NL Fuller, RP Rand. *Biophys J* 71:2623–2632, 1996.

85. N Fuller, RP Rand. *Biophys J* 81:243–254, 2001.

85a. MM Kozlov, W Helfrich. *Langmuir* 8:2792–2797, 1992.

86. RM Epand. *Biochemistry* 24:7092–7095, 1985.

87. DP Siegel, J Banschbach, PL Yeagle. *Biochemistry* 28:5010–5019, 1989.

88. G Basáñez, JL Nieva, E Rivas, A Alsonso, FM Goñi. *Biophys J* 70:2299–2306, 1996.

89. MP Veiga, JLR Arrondo, FM Goñi, A Alonso. *Biophys J* 76:342–350, 1999.

90. MB Ruiz-Argüello, G Basáñez, FM Goñi, A Alonso. *J Biol Chem* 271:26,616–26,621, 1996.

91. RM Epand, RF Epand, N Ahmed, R Chen. *Chem Phys Lipids* 57:75–80, 1991.

92. MJ Darkes, TA Harroun, SM Davies, JP Bradshaw. *Biochim Biophys Acta* 1561:119–128, 2002.

93. PL Yeagle, FT Smith, JE Young, TD Flanagan. *Biochemistry* 33:1820–1827, 1994.

94. N Janes. *Chem Phys Lipids* 81:133–150, 1996.

95. MW Tate, EF Eikenberry, DC Turner, E Shyamsunder, SM Gruner. *Chem Phys Lipids* 57:147–164, 1991.

96. DC Turner, SM Gruner. *Biochemistry* 31:1340–1355, 1992.

97. M Rappolt, A Hickel, F Bringezu, K. Lohner. *Biophys J* 84:3111–3122, 2003.

98. GL Kirk, SM Gruner. *J Physique* 46:761–769, 1985.

99. MW Tate, SM Gruner. *Biochemistry* 26:231–236, 1987.

100. RM Epand, RF Epand, CRD Lancaster. *Biochim Biophys Acta* 945:161–166, 1988.

101. C Valtersson, G van Duyn, AJ Verkleij, T Chojnacki, B de Kruijff, G Dallner. *J Biol Chem* 260:2742–2751, 1985.

102. MJ Knudsen, FA Troy. *Chem Phys Lipids* 51:205–212, 1989.

103. K Boesze–Battaglia, SJ Fliesler, J Li, JE Young, PL Yeagle. *Biochim Biophys Acta* 1111:256–262, 1992.

104. Z Chen, RP Rand. *Biophys J* 74:944–952, 1998.

105. DC Turner, SM Gruner, JS Huang. *Biochemistry* 31:1356–1363, 1992.

106. DM Anderson, SM Gruner, S Leibler. *Proc Natl Acad Sci USA* 85:5364–5368, 1988.

107. US Schwartz, G Gompper. *Langmuir* 17:2084–2096, 2001.

108. G Basáñez, FM Goñi, A Alonso. *Biochemistry* 37:3901–3908, 1998.

109. A Walter, PL Yeagle, DP Siegel. *Biophys J* 66:366–376, 1994.

110. MM Kozlov, LV Chernomordik. *Biophys J* 75:1364–1396, 1998.

111. MM Kozlov, LV Chernomordik. *Traffic* 3:256–267, 2002.

112. Y Kozlovsky, MM Kozlov. *Biophys J* 85:85–96, 2003.

113. EE Kooijman, V Chupin, B de Kruijff, KNJ Burger. *Traffic* 4:162–174, 2003.

114. WJ Brown, K Chambers, A Doody. *Traffic* 4:214–221, 2003.

115. FM Goñi, A Alonso. *Biosci Rep* 20:443–463, 2000.

116. FM Goñi, A Alonso. *Progr Lipid Res* 38:1–48, 1999.

117. JL Nieva, A Alonso, G Basáñez, FM Goñi, A Gulik, R Vargas, V Luzzati. *FEBS Lett* 368:143–147, 1995.

118. JL Nieva, FM Goñi, A Alonso. *Biochemistry* 28:7364–7367, 1989.

119. JL Nieva, FM Goñi, A Alonso. *Biochemistry* 32:1054–1058, 1993.

120. MB Ruiz-Argüello, G Basáñez, FM Goñi, A Alonso. *J Biol Chem* 271:26,616–26,621, 1996.

121. G Basáñez, MB Ruiz-Argüello, A Alonso, FM Goñi, G Karlsson, K Edwards. *Biophys J* 72:2630–2637, 1997.

122. MB Ruiz-Argüello, FM Goñi, A Alonso. *J Biol Chem* 273:22,977–22, 982, 1998.

123. I Martin, JM Ruysschaert, RM Epand. *Adv Drug Delivery Rev* 38:233–255, 1999.

124. LK Tamm, X Han. *Biosci Rep* 20:501–518, 2000.

125. LK Tamm, X Han, Y li, AL Lai. *Biopolymers (Peptide Science)* 66:249–260, 2002.

125a. EI Pécheur, J Sainte-Marie, A Bienvene, D Hoekstra. *J Memb Biol* 167:1–17, 1999.

126. MJM Darkes, SMA Davies, JP Bradshaw. *FEBS Lett* 461:178–182, 1999.

127. TA Harroun, K Balali-Mood, I Gourlay, JP Bradshaw. *Biochim Biophys Acta* 1617:62–68, 2003.

127a. A Colotto, RM Epand. *Biochemistry* 36:7644–7651, 1997.

128. A Colotto, I Martin, JM Ruysschaert, A Sen, SW Hui, RM Epand. *Biochemistry* 35:980–989, 1996.

129. RM Epand, RF Epand, I Martin, JM Ruysschaert. *Biochemistry* 40:8800–8807, 2001.

130. I Martin, RM Epand, JM Ruysschaert. *Biochemistry* 37:17, 030–17,039, 1998.

131. SG Peisajovich, RF Epand, M Pritsker, Y Shai, RM Epand. *Biochemistry* 39:1826–1833, 2000.

132. F Aranda, JA Teruel, A Ortiz. *Biochim Biophys Acta* 1618:51– 58, 2003.

133. A Bertocco, F Formaggio, C Tonino, QB Broxterman, RF Epand, RM Epand. *J Peptide Res* 62:19–26, 2003.

134. RM Epand. *Biosci Rep* 6:647–653, 1986.

135. RM Epand, RF Epamd, CD Richardson, PL Yeagle. *Biochim Biophys Acta* 1152:128–134, 1993.

136. DR Kelsey, TD Flanagan, JE Young, PL Yeagle. *Virology* 182:690–702, 1991.

137. GW Kemble, T Danieli, JM White. *Cell* 76:383–391, 1994.

138. VI Razinkov, GB Melikyan, FS Cohen. 1999. *Biophys J* 77:3144–3151, 1999.

139. RM Markosyan, FS Cohen, GB Melikyan. *Molec Biol Cell* 11:1143–1152, 2000.

140. GB Melikiyan, MG Roth, FS Cohen. *Molec Biol Cell* 10:1821–1836, 1999.

141. RT Armstrong, AS Kushnir, JM White. *J Cell Biol* 151:425–437, 2000.

142. B Schroth-Diez, E Ponimaskin, H Reverey, MF Schmidt, A Hermann. *J Virol* 72:133–142, 1998.

143. DZ Cleverley, J Lenard. *Proc Natl Acad Sci USA* 95:3425–3430, 1998.

144. SM Dennison, N Greenfield, J Lenard, BR Lentz. *Biochemistry* 41:14,925–14,924, 2002.

145. JK Lee, BR Lentz. *Biochemistry* 36:6251–6259, 1997.

146. JA Killian, KU Prasad, DW Urry, B de Kruijff. *Biochim Biophys Acta* 978:341–345, 1998.

147. SL Keller, SM Gruner, K Gawrisch. *Biochim Biophys Acta* 1278:241–246, 1996.

148. EJ Prenner, RN Lewis, KC Neuman, SM Gruner, LH Kondejewski, RS Hodges, RN McElhaney. *Biochemistry* 36:7906–7916, 1997.

149. EJ Prenner, RN Lewis, RN McElhaney. *Biochim Biophys Acta* 1462:201–221, 1999.

150. E Staudegger, EJ Prenner, M Kriechbaum, G Degovics, RN Lewis, RN McElhaney, K Lohner. *Biochim Biophys Acta* 1468:213–230, 2000.

151. MRR de Planque, JA Killian. *Molec Membr Biol* 20:271–284, 2003.

152. JA Killian. *FEBS Lett* 555:134–138, 2003.

153. JA Killian, I Salemink, MRR de Planque, G Lindblom, RE Koeppe II, DV Greathouse. *Biochemistry* 35:1037–1045, 1996.

154. S Morein, E Strandberg, JA Killian, G Arvidson, RE Koeppe II, G Lindblom. *Biophys J* 73:3078–3088, 1997.

155. MRR de Planque, JAW Kruijtzer, RM Liskamp, D Marsh, DV Greathouse, RE Koeppe II, B deKruijff, JA Killian. *J Biol Chem* 274:20,839–20,846, 1999.

156. MRR de Planque, BB Bonev, JAA Demmers, DV Greathouse, RE Koeppe II, F Separovic, A Watts, JA Killian. *Biochemistry* 42:5341–5348, 2003.

157. PCA van der Wel, T Pott, S Morein, DV Greathouse, RE Koeppe II, JA Killian. *Biochemistry* 39:3124–3133, 2000.

158. S Morein, RE Koeppe II, G Lindblom, B de Kruijff, JA Killian. *Biophys J* 78:2475–2485, 2000.

159. MRR de Planque, BB Bonev, JAA Demmers, DV Greathouse, RE Koeppe II, F Separovic, A Watts, JA Killian. *Biochemistry* 42:5341– 5348, 2003.

160. F Liu, RNAH Lewis, RS Hodges, RN McElhaney. *Biochemistry* 40:760–768, 2001.

161. E Strandberg, S Morein, DTS Rijkers, RMJ Liskamp, PCA van der Wel, JA Killian. *Biochemistry* 41:7190–7198, 2002.

162. JF Nagle, S Tristram-Nagle. *Biochim Biophys Acta* 1469:159–195, 2000.

163. MRR de Planque, E Goormaghtigh, DV Greathouse, RE Koeppe II, JAW Kruijtzer, RMJ Liskamp, D de Kruijff, JA Killian. *Biochemistry* 40:5000–5010, 2001.

164. PCA van der Wel, E Strandberg, JA Kiliian, RE Koeppe II. *Biophys J* 83:1479–1488, 2002.

165. JA Szule, RP Rand. *Biophys J* 85:1702–1712, 2003.

166. RH Templer, BJ Khoo, JM Seddon. *Langmuir* 14:7427–7434, 1998.

167. U Schwarz, G Gompper. *Lecture Notes in Physics* 600:107–151, 2002.

168. Y Kozlovsky, LV Chernomordik, MM Kozlov. *Biophys J* 83:2634–2651, 2002.

169. Y Kozlovsky, MM Kozlov. *Biophys J* 82:882–895, 2002.

170. VS Markin, JP Albanesi. *Biophys J* 82:693–712, 2002.

171. S May. *Biophys J* 83:2969–2980, 2002.

172. DC Turner, ZG Wang, SM Gruner, DA Mannock, RN McElhaney. *J Phys II France* 2:2039–2063, 1992.

173. H Chung, M Caffrey. *Nature* 368:224–226, 1994.

174. RH Templer, JM Seddon, JM Warredner. *Biophys Chem* 49:1–12, 1994.

175. RH Templer, DC Turner, P Harper, JM Seddon. *J Phys. II France* 5:1053–1065, 1995.

176. RH Templer, BJ Khoo, JM Seddon. *Langmuir* 14:7427–7434, 1998.

176a. RP Rand, VA Parsegian. *Biochim Biophys Acta* 988:351–376, 1989.

177. MP do Carmo. *Differential Geometry of Curves and Surfaces.* Englewood Cliffs, NJ: Prentice-Hall, 1976, pp. 264–282.

178. MP do Carmo. *Differential Geometry of Curves and Surfaces.* Englewood Cliffs, NJ: Prentice-Hall, 1976, p. 212.

179. PR Cullis, PWM van Dijck, B de Kruijff, J de Gier. *Biochim Biophys Acta* 513:21–30, 1978.

180. CPS Tilcock, MB Bally, SB Farren, PR Cullis. *Biochemistry* 21:4596–4601, 1982.

181. RM Epand, DW Hughes, BG Sayer, N Borochov, D Bach, E Wachtel. *Biochim Biophys Acta* 1616:196–208, 2003.

182. RM Epand, RF Epand, AD Bain, BG Sayer, DW Hughes. *Magnetic Reson Chem* 42:139–147, 2004.

183. Z Chen, RP Rand. *Biophys J* 73:267–276, 1997.

184. N Fuller, RP Rand. *Biophys J* 81:243–254, 2001.

5

Aspects of the Differential Geometry and Topology of Bicontinuous Liquid-Crystalline Phases

ROBERT W. CORKERY

CONTENTS

ABSTRACT

This chapter provides a semiformal, though largely qualitative, look at the mathematics of triply periodic minimal surfaces (TPMS) in relation to bicontinuous and polycontinuous liquid crystals. It is motivated largely by research of Stephen Hyde and others at the Applied Mathematics Department, Australian National University, Canberra, Australia and by the discovery of these TPMS liquid crystal partitions by Kåre Larsson and the late Krister Fontell at Lund University. The article is meant to complement the existing, more rigorous papers and provide an introduction to the more fundamental topics of differential geometry and topology of TPMS.

Bicontinuous liquid crystals contain two mutually interpenetrating labyrinths separated by a hyperbolic partition that is often best described as a triply periodic minimal surface (TPMS). Properties of these surfaces can be broken down into local and global types. Local properties pertain to small, isolated surface patches and include intrinsic measures such as the metric, surface areas, and angles and the Gaussian curvature. The global properties require a look at the surface as a whole, embedded in three-dimensional space. The global measures include parameters such as the Cartesian coordinates, topology, curvature distributions, and overall symmetries. These local and global properties are connected through the Gauss–Bonnet theorem that relates the topology to the Gaussian curvature. For the simplest TPMS, such as the P, D, and G surfaces, the local Gaussian curvature determined from small patches of the surface can give information regarding the stability of bicontinuous liquid crystals.

For some simpler TPMS, the small surface patches are polygonal (e.g., triangles), according to the various symmetries of the surface. These small two-dimensional surface patches can be built up into the entire three-dimensional surface using various three-dimensional construction algorithms based on symmetries, the same way three-dimensional crystals can be built up from unique atoms. Alternatively these patches can be usefully represented in two dimensions by various mappings to the sphere, hyperbolic plane, and complex plane; then built up in those spaces using analogous symmetry operations; and then embedded in three-dimensional space by folding and "gluing" the edges of the two-dimensional representations. The two-dimensional forms of these surfaces offer simple ways of constructing known and novel surfaces, algorithmically, and also explicitly through the Weierstrass parametrization, which gives the exact and explicit (x,y,z) coordinates of the actual minimal surface (not approximates) via the complex plane representation. Armed with this explicit information, modeling the stabilities, phase changes, and other physical processes involving these minimal surfaces is rigorously quantifiable from a geometric perspective. A novel, topology-preserving phase transformation between the P, D, and G cubic phases arises directly from the analysis of rhombohedral and tetrahedral distortions of the P, D, and G surfaces using this method.

Networks decorating TPMS are of interest for generating and analyzing crystal networks. The simplest networks overlay the polygonally decorated surfaces built up of the fundamental patches. Families of networks can be generated for individual TPMS using supergroup–subgroup symmetry relationships between the decorating tesselation and the underlying symmetry elements of the TPMS itself. Of direct relevance to liquid crystalline mesophases is the set of polycontinuous networks that can be generated on various TPMS. These are generated by commensurately decorating two-dimensional surface representations of TPMS with tree networks. With some well-chosen constraints, folding and gluing operations can result in surfaces decorated by entangled thickets of unconnected networks. In turn, these embedded networks can be used to form tri-, quadra-, octa- and other polycontinuous morphologies, with their respective mutually interpenetrating labyrinths separated by a single triply periodic minimal surface.

5.1 INTRODUCTION AND HISTORICAL CONTEXT

Bicontinuous liquid crystals are characterized by a partitioning of contrasting chemical moieties into two intertwined labyrinths. A powerful tool for the description of the corresponding partitioning surfaces is differential geometry, developed initially by Gauss. The basis of modern differential geometry is Riemannian geometry. This deals with surfaces as spaces (Riemann spaces) unto themselves, bearing intrinsic geometric properties, rather than objects embedded in a surrounding space of higher dimension. The main characteristics of Riemann spaces or surfaces are their curvatures. All the information regarding the curvatures of surfaces is found in the Riemann-Tensor.

The mathematics of this is not important for this article. Knowing the tensor is equivalent to knowing the Gaussian curvature for points on the surface through which geodesics pass. Geodesics are the curves forming the shortest path on the surface between two points. This is a straight line on a plane or a great circle on a sphere.

By way of analogy, imagine a bug confined to a surface. Intrinsic properties are those properties of the surface the bug can express using only arc lengths. Besides arc length, these properties are surface areas and angles. The bug uses local coordinates for points $\sigma(u,v)$ on the surface but does not know its (x,y,z) coordinates. A parametrized curve on the surface is then described, i.e., $(u(t),v(t))$. The arc length, $s(t)$, along the curve can then be measured, and this can be differentiated with respect to t. Indeed, the bug can also confirm whether a curve is a geodesic by determining whether it is the shortest route. The bug, in choosing a coordinate system, has defined the metric, and equivalently, what is known in modern differential geometry as the first fundamental form. The extrinsic properties of the surface include the principal curvatures, their directions, and their cartesian (x,y,z) coordinates. Remarkably, as Gauss first showed, the Gaussian curvature is intrinsic to the surface, although it is composed of the product of the principal curvatures, because it can be determined from considering the metric alone.

Many of you reading this volume know that the P-surface (Im3 m spacegroup), a special case of a bicontinuous surface, has a trigonometric approximation often used in modeling and visualization:

$$\cos x + \cos y + \cos z = 0$$

The formula shows that this is a level cut (zero value) through a surface defined in three-dimensional cartesian space (R^3). Despite the profound usefulness of trigonometric approximations in the study of bicontinuous and other morphologies, their use, in some senses, defeats the gains made in understanding the explicit structures of surface morphologies made over 175 years ago by Riemann, Gauss, and others.

5.2 CURVATURE

Since Riemann, we know that small pieces of surfaces, or patches, must either be elliptic (positive Gaussian curvature), parabolic (zero Gaussian curvature), or hyperbolic (negative Gaussian curvature).

More formally, two orthogonal principal curvatures (k_1, k_2) can be defined for points on any two-dimensional surface. From these we can obtain the mean curvature ($H = (k_1 + k_2)/2$) or their Gaussian or intrinsic curvature ($K = k_1 \cdot k_2$) at each point. The local geometry is hyperbolic, that is, saddle shaped, for surfaces whose points have principal curvatures opposite in sign, such that their Gaussian curvature must be negative. Surfaces such as spheres and ellipsoids have principal curvatures of the same sign, so their intrinsic geometries are defined as elliptic. Surfaces with one principal curvature as zero, such as planar sheets, cones, and cylinders, have parabolic geometries. Figure 5.1 shows these simple concepts. "Sewing" together a continuous set of hyperbolic, saddle-shaped patches can lead to the formation of labyrinthine tunnel networks, defined by the closing of pathways embedded in adjacent patches to form loops. Surfaces such as these can have a global property of being bicontinuous, that is, they divide space into two interpenetrating, convoluted subspaces.

5.3 PERIODICITY

Detection and understanding of molecular periodicity in solids and liquid crystals has evolved hand in hand with the development of x-ray, electron and neutron diffraction, optical microscopy, and various spectroscopic techniques. Singly periodic liquid crystals are the nematics phases. Doubly periodic structures include the smectic phases and the columnar phases. An important close structural relative of the smectic phases is the mesh phases. Historically, it is significant that the discovery by Luzzati and coworkers of nonlamellar, columnar, or rod-like liquid crystalline phases in the sodium soap-water system in the late 1950s [2] led directly to the detection of triply periodic liquid crystalline cubic phases in the strontium soaps in the early 1960s [3]. However, Luzzati did not explicitly recognize the bicontinuous partition in these soaps until after Kåre Larsson noted the presence of these in another triply periodic, cubic liquid-crystalline system [4,5].

Since Larsson, we now recognize that triply periodic liquid-crystalline systems can display bicontinuity. The hyperbolic interface in pure, liquid-crystalline reverse cubic strontium soap sits at the interface of the lipid chains, separating the loci of terminal methyl groups from each of the two distinct labyrinths of otherwise separate (chiral enantiomorphic) soap networks. This is synonymous with the Gyroid. In so-called "normal" (as opposed to "reverse") lyotropic liquid-crystalline systems, two distinct, mutually interpenetrating

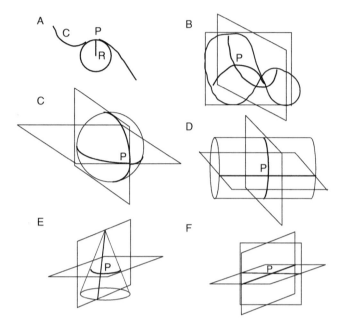

Figure 5.1   This figure illustrates the concept of curvature in a plane, Gaussian (K) curvature, and mean curvature (M). The point P on each of the cartoons (a to f) is a typical point on each of the curves or surfaces (interfaces) shown. In A, we define the curvature in a plane (k) to be scalar and equal to 1/R, where R is the radius of an osculating circle whose local curvature near P is exactly the same as the instantaneous curvature of the curve C at P. The shape in B is easily seen as a saddle; the point P on the saddle is a typical point, with two planar curvatures — one convex and the other concave. We define a convex curve to be positively or negatively curved (depending upon preferred convention), and a concave curve to be the opposite (i.e., negatively or positively curved, respectively). At an arbitrary point on either of these surfaces, we can define two principal orthogonal tangent vectors $k_1\mathbf{e}_1$ and $k_2\mathbf{e}_2$ to the surface at P whose magnitudes are the principle curvatures, k_1 and k_2, where k_1 or k_2 is the maximal or minimal curvature taken from the set of all curves determined in any plane (containing the surface normal at P) passing through P. So we define two types of curvature at a point (P) on a surface (interface) to be a function of the two principal curvatures k_1 and k_2: mean curvature, $M = (k_1 + k_2)/2$, and Gaussian curvature, the product, $K = k_1 \cdot k_2$. In B we can clearly see that at P the mean curvature is the sum of a positive and negative term.

continuous water channels can be separated by a bilayer, the water and headgroups lining the water channels and the midsurface of the bilayer defining the partitioning interface. Many other examples are given in this volume.

5.4 MINIMAL SURFACES

Minimal surfaces are those that have vanishing mean curvature. From a mathematical point of view, every complex analytical function can be used to define a minimal surface. The simplest of these is the plane. Soap films suspended on circular wire loops have vanishing mean curvature due to the surface tension effects pulling the film flat. However, if the circular wire frame is bent so that the rim is no longer planar, the film adopts a saddle shape.

At all points on the saddle surface, excluding the rim, the principal curvatures are as equally convex as concave, i.e., $k_1 = -k_2$. For a surface patch on the saddle-shaped film, mean curvature vanishes and the Gaussian curvature is everywhere negative, except for a few flat points where the curvature becomes zero. Minimal surfaces

Figure 5.1 (Continued) If the principal curvatures were equal and opposite, then we would obtain a mean curvature of zero. The Gaussian curvature for a saddle is always negative, given that k_1 will always be positive if k_2 is negative and vice versa. In the case of the sphere in C the principle curvatures are of equal magnitude and sign, thus the mean curvature $= k_1 = k_2$, in an ellip-soid k_1k_2. In any case the mean curvature can never be zero. In addition, the Gaussian curvature of a point P on a sphere (or ellipsoid) is always positive since both k_1 and k_2 always have the same sign. In D through F we have three apparently distinct surfaces, but a point P on each of these surfaces has zero Gaussian curvature, since either k_1 or k_2 in each plane, cone, or cylinder is zero. The mean curvature of the plane is zero; however, the mean curvatures of the cylinder and cone are nonzero. In short, most points on saddles (hyperbolic surfaces) have negative Gaussian curvature and may have zero mean curvature. All points on spheres and ellipsoids (elliptical surfaces) have positive Gaussian curvature and always have nonzero mean curvature. Most points on planes, cones, or cylinders (parabolic surfaces) have zero Gaussian curvature, and in the case of the plane, zero mean curvature. (From RW Corkery, Artificial Biomineralisation and Metallic Soaps, Ph.D. dissertation, Australian National University, Canberra, Australia, 1998.)

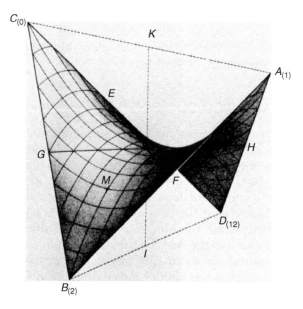

Figure 5.2 Hyperbolic minimal surface showing saddle shape. (From Schwarz original work.)

in E^3 generally are self intersecting, that is, they cut back through themselves. Many of these wrap continuously onto and through themselves and will, if grown infinitely, densely fill space, and are thus non-bicontinuous. Self-intersecting surfaces that have high symmetry are useful structures for describing the nonbicontinuous liquid crystalline systems such as micellar phases in space groups

Figure 5.3 The (a) P, (b) D, and (c) G surfaces are the most well-known TPMS and probably the only surfaces truly proven to exist in liquid-crystalline systems. (Images courtesy of Stephen Hyde and Stuart Ramsden.)

such as Pm3n. These will not be considered further. Other minimal surface patches can be built up into translationally repeating motifs, synonymous with crystalline systems. These are the periodic minimal surfaces. When these are infinitely translatable in the three orthogonal axes of E^3, these are classified as TPMS, or triply periodic minimal surfaces. Another common terminology is "IPMS," or infinite periodic minimal surface.

5.5 TRIPLY PERIODIC MINIMAL SURFACES (TPMS)

TPMS are the most symmetric of the bicontinuous minimal surfaces in E^3. These are characterized by their negative Gaussian curvature, their zero mean curvature, and their ordered distribution of flat points and crystallographic unit cell. These surfaces divide space into two congruent, mutually interpenetrating labyrinths of equal volume. Explicit mathematical parametrization of these surfaces has been reported for simpler surfaces and will be discussed in the next section.

One way to understand the construction of a TPMS is to build it up from small surface patches or Flächenstück. The construction is analogous to building up crystalline unit cells from their respective asymmetric units. An asymmetric unit of an atomic solid is the smallest polyhedral subunit of a crystal that contains the unique atoms. It is smaller than the unit cell and can be used to generate the unit cell via reflection and rotation operations. As its name suggests, the asymmetric unit contains no primary symmetry elements. The rotation, reflection, and inversion symmetry operations are performed on the boundaries of the asymmetric units to build the unit cell. The surface equivalent of an asymmetric unit is the Bashkirov stereohedron. It is also polyhedral and can have curved faces. Analogously, Bashkirov stereohedra, containing the primary surface patch, the Flächenstück, can be combined via rotation and reflection operations to obtain the unit cell of minimal surface, which can then be translated infinitely to give the periodic minimal surface. The Flächenstück of a TPMS perpendicularly intersects the faces of the bounding Bashkirov stereohedron such that the surface is continued across the faces during buildup via reflections and rotations. The intersection of the Flächenstück on the Bashkirov stereohedron yields a closed loop trace characteristic of that particular TPMS under the symmetry constraints of the particular space group.

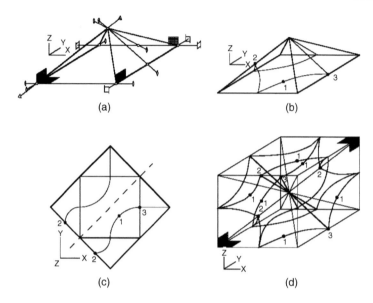

(a)   (b)

(c)   (d)

Figure 5.4 Asymmetric unit for Pm$\underline{3}$n space group with Flächenstück of genus 25 surface. (a) The asymmetric unit is characterized by having mirror planes, rotation axes, and inversion centers on its exterior. (b) Bashkirov stereohedron containing the surface trace of the Flächenstück. This surface trace can be made into a wire frame and dipped in a soap solution (or input into Surface Evolver) to generate its corresponding minimal surface patch or Flächenstück. (c) Unfolded stereohedron and surface trace. Note that the surface generated here has symmetry Im$\underline{3}$m because of the extra mirror plane. The pyramid represents two Im$\underline{3}$m asymmetric units, related by reflection across the dotted line. (d) 1/8 unit cell of the g = 25 Im$\underline{3}$m surface built up via rotations about the three fold axes and inversion through the pyramid apex. (From RW Corkery, Artificial Biomineralisation and Metallic Soaps, Ph.D. dissertation, Australian National University, Canberra, Australia, 1998.)

It is a simple exercise to make a wire frame model (Figure 5.4b) of this trace by bending it over a simple irregular polyhedron (a square pyramid in the case of the Pm$\underline{3}$n space group) and dipping it into a soap solution to generate a film that represents the Flächenstück. These can then be glued at their edges (Figure 5.4d) according to appropriate symmetry operations to give a useful model of the surface unit cell. The Flächenstück of simple TPMS is a simple saddle, bounded by its wire boundaries. When the rotation and reflection operations are used to build up the surface, the saddle troughs become tunnels that, depending on the exact space group, intersect to form the labyrinths characteristic of the TPMS. The troughs on the other side of the saddle do likewise and form the tunnel system on the other side, and thus the bicontinuous network can be constructed. Hyde et al. reported specific ways to derive TPMS via construction of traces on Bashkirov stereohedra according to certain rules. These require a "balanced" trace, and the details can be found in Hyde's PhD thesis [29]. Using combinatorial methods, it is possible to generate all TPMS this way.

5.6 TOPOLOGY OF TPMS

The topology of a surface is a global parameter, as opposed to the intrinsic curvature, which is a local property of a surface patch on the greater surface. The topology describes the morphology of the TPMS through the connectivity of the tunnel networks of the congruent labyrinths and the total curvature of the surface. Topology is more concerned with the warping and stretching of surfaces, and is independent of the local geometric measures. It is denoted by the Euler–Poincaré characteristic, χ, which is equal to the area-weighted integral of the Gaussian curvature over the surface. This result is known as the Gauss–Bonnet theorem, which in another form (conditionally) relates the genus of the surface to the dihedral angles of the surface trace of the Flächenstück of the bounding Bashkirov stereohedron.

The characteristic, χ, is also related to the genus of the (orientable) surface, or the number of handles. So a surface with g handles or holes has a Euler–Poincaré characteristic (χ) of $(2 - 2g)$. For example, a sphere has no handles and has $g = 0$ and $\chi = 2$. A donut has $g = 1$ and $\chi = 0$, a two-holed donut has $g = 2$, $\chi = -2$, a three-holed donut has $g = 3$, $\chi = -4$ and so on. Genus 3 TPMS, such as D, P, and G, also have $\chi = -4$, per unit cell, and are thus (on a unit cell basis) topologically equivalent to the triple donut, that is they have three handles or holes. This is intuitively illustrated for a unit cell of the P-surface. The three tunnel systems are identified as shown in Figure 5.5, and when these are joined up along the guiding lines, give a compactified sphere with three handles. This is more explicitly demonstrated for the P-surface again, this time by using the Euler relationship, $\chi = F - E + V$, where F is the number of faces, E is the number of edges (curved or not) and V is the number of vertices. Take a sphere, divide its surface into, say eight equal spherical triangles. In this case $F = 8$,

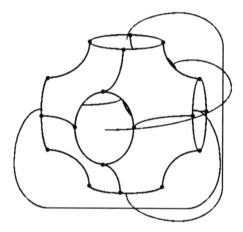

Figure 5.5 Unit cell of the P surface, divided into eight equivalent monkey saddles, and with its tunnels identified.

E = 12, V = 6. Thus χ = 8 – 12 + 6 = 2, as expected for a g = 0 surface. Likewise, take the unit cell of the P-surface, as in Figure 5.5. Divide it into eight equivalent monkey saddles that share edges and vertices. In this case, F = 8, E = 36, V = 24. Thus χ = 8 – 36 + 24 = –4, as expected for a genus 3 (per unit cell) surface (N.B., we could have used other tesselations of these surfaces to demonstrate this).

5.7 PARAMETRIZATION OF SURFACES USING THE GAUSS MAP AND THE COMPLEX PLANE

All minimal surfaces can be expressed in terms of the Weierstrass parametrization. This representation expresses the x, y, and z coordinates of the surface in terms of integrals of a Weierstrass function in the complex plane. This function itself completely specifies the first and second fundamental forms and thus the differential geometry of the surfaces. Thus the problem of mathematically describing a minimal surface reduces to that of determining its corresponding Weierstrass function. This is a simple matter in the case of simple infinite surfaces such as the catenoid and helicoid. More importantly, systematic procedures have been developed to determine the Weierstrass functions corresponding to known TPMS, or indeed to propose new such TPMS using this function as the starting point.

The analytical algorithm is relatively straightforward, based on utilizing the genus and space group of the translational unit of the TPMS, and further building in the functional requirements necessary to match the flat points and their orders, and crystallographic symmetry elements bounding the fundamental unit of the surface. For example, the genus directly specifies the order of the polynomial equation to be solved for the Weierstrass function. Although conceptually simple, as outlined below, the procedure is most manageable for TPMS of relatively low genus and high symmetry. The simplest cases are the TPMS of genus 3, for which the full set of TPMS has been listed using this procedure. This includes the cubic P, D, and G surfaces, their rhombohedral (rPD, rG), tetragonal (tP, tD, tG) and lower symmetry (e.g., orthorhombic, monoclinic) distortions, as well as the tetragonal CLP and hexagonal H surfaces. For all genus 3 TPMS, the Weierstrass representation reduces to elliptic integrals. Other higher-genus TPMS of cubic symmetry have also been parametrized, namely the C(P) (Neovius), I-WP and F-RD, although their integral representations are more complicated. Thus the TPMS that have an exact parametrization still represent only a small fraction of the number proven to exist by analytical or numerical arguments. By the same token, the examples with lowest genus and highest symmetry are those most relevant to bicontinuous structures in liquid crystals and other self-assembled systems.

The initial step in parametrizing a surface is a mapping of the TPMS to the complex plane achieved via two steps. First, normal vectors at each point on the surface are mapped onto the unit sphere via their intersection with the sphere surface. Next, the sphere is stereographically projected onto the complex plane. For simple TPMS, including the P, D, and G surfaces, this composite mapping yields a distribution of the flat points of the minimal surface in the complex plane (see Figure 5.6).

Different types of flat points occur in TPMS, e.g., those at the saddle points of a "regular saddle" and those at the saddle points of so-called "monkey" saddles, as seen in the [111] direction of the Gyroid. Commonly, the flat points of the TPMS lie on the outside of the bounding Bashkirov stereohedra mentioned earlier. In these cases, appropriately chosen fundamental units of the TPMS, i.e., the Flächenstück, can be faithfully represented in the complex plane by mapping only the trace of the surface on its bounding Bashkirov stereohedron. Since the Flächenstück intersects the bounding stereohedron orthogonally, the Gauss map of the surface boundary maps to "generalized spherical geodesic polygons," bounded by great

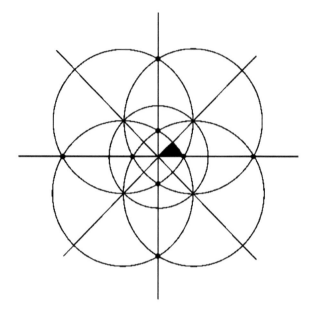

Figure 5.6 Complex plane representation of the fundamental asymmetric patch of the P, D surfaces. This tile is kaleidoscopic (bounded by mirrors) and is essentially the Flächenstück mapped to the complex plane via the Gauss mapping. The (first-order) flat points are represented by dots and decorate the vertices of the asymmetric tiles.

circles on the sphere. Thus, for simple TPMS, the Flächenstück map to simple spherical tiles on the sphere (Figure 5.6). Analogously, building up the translatable unit cell of the TPMS equates to tiling the unit sphere via reflection operations across the bounding great circle segments defining the Gauss map of the Flächenstück. Fogden and Hyde [6] used the set of Schwarz triangular tilings of the sphere to exhaustively determine all branch point distributions within the class of surfaces that includes the P, D, and G surfaces and thus determine all TPMS in this class.

The genus of a surface can be determined by noting that a simple relation exists between the order and multiplicity of branch points and the number of sheets of the Riemann surface on the sphere. If the Riemann surface wraps twice onto the sphere (i.e., in order for a spherical tiling to completely cover the sphere and be superposed onto the identity geodesic polygon — as in the regular class of TPMS), then this is the equivalent of an s = 2 sheeted Riemann surface. The genus is then s + 1 = 3. This is the case for the genus-3 (per unit cell) P, D, and G surfaces.

5.8 HOMOGENEITY OF CURVATURE ON SURFACES

This mathematical concept provides some intuitive understanding of why the P, D, and G cubic surfaces are those most commonly found among the various liquid-crystalline bicontinuous cubic phases. In short, relatively high surface homogeneity equates to relatively low interfacial bending energy. But first we need some definitions. A homogeneous surface is one that is uniformly curved everywhere on the surface. A smooth sphere, for example, has uniform Gaussian and mean curvature everywhere. Thus departure from this uniformity is equitable to inhomogeneity. This can be due to time-invariant distortions or through motions such as harmonic oscillations and thermal fluctuations. The global integrated Gaussian curvature of, for example, the exterior of a swollen micelle remains, however, positive, although it will rarely be spherical. Likewise, an apparently flat lamellar bilayer sheet is subject to various fluctuations that cause local departures from K = 0; however, the global integrated curvature can still vanish. However, in order to determine the global properties, one cannot rely upon the local surface geometry unless the whole surface is adequately homogeneous. Global properties such as surface-to-volume ratios require knowledge of the curvature distribution, which then defines the topology.

It is seemingly esoteric to point to the fact that hyperbolic surfaces with homogeneous curvature cannot exist in three-dimensional Euclidean (E^3) space. So what? Well first, it is necessary to have a qualitative answer to the question of why two-dimensional hyperbolic space (H^2) cannot be embedded in E^3. A proof is not offered here (see Reference 7). But the fact is that the hyperbolic plane (H^2) can contain infinite Euclidean dimensions. H^2 is thus simply more dimensionally expansive than E^3 is. The consequence relevant to us is that in order to accommodate a hyperbolic surface in E^3, we must compromise on homogeneity. That is, we can "fit" a triply periodic minimal surface in E^3 only by introduction of non-constant Gaussian curvature. In H^2, on the other hand, one can realize constant negative Gaussian curvature. In contrast, surfaces of elliptic and parabolic geometries can have constant, homogeneous curvatures. If a surface is sufficiently homogeneous, we can make approximate estimates of the surface's global properties. Such quasi-homogeneous surfaces include the P, D, G and other triply periodic minimal surfaces.

There are also branch points in these surfaces where the curvature vanishes. The location and type of branch points determine the symmetry and geometry of the surface, as we have seen, and also are the locus of the highest concentration of curvature variation on the surface.

5.9 MOLECULAR SHAPE AND THE SURFACTANT PACKING PARAMETER

The bicontinuous liquid-crystalline phases falling between hexagonal and lamellar regions of particular phase diagrams comprise amphiphiles forming bilayers wrapped onto hyperbolic surfaces. The midsurface between the bilayers provides a convenient locus for the bicontinuous description of the structure. In the case of "normal" or type I phases, the midsurface dissects the polar headgroups, while in reverse or type II phases, it dissects the locus of hydrophobic chain ends. Hyde [8–10] has shown that, under the reasonable assumption of constant bilayer thickness, the shape of the resultant parallel polar and hydrophobic interfaces can be quantified in terms of average molecular shape, through a relationship to the Gaussian and mean curvatures and the bilayer thickness. Thus the well known $S = v/al$ average surfactant packing parameter relation can be related to global aggregation geometries, assuming homogeneous interfaces. This assumption implies a relatively small bending energy constraint on the overall aggregate shape.

The useful relationship (e.g., Reference 10)

$$S_i = V/al = 1 \pm Ht_i + 1/3 \cdot Kt_i^2 \qquad (5.1)$$

yields the graphical phase space in Figure 5.7 for different liquid-crystalline forms, but without any constraint on a particular symmetry. Thus bicontinuous intermediate and cubic phases fall under the broad banner of "sponges," which includes the disordered L_3 phases; hexagonal phases fall under the term "rods"; and micellar phases fall under the term "spheres." Figure 5.3 shows the regions corresponding to type-I and -II liquid crystalline phases. Note that bicontinuous monolayers, typical of the interfaces found in microemulsions, appear in this diagram intermediate to spheres of oil and spheres of water. We will not investigate these microemulsion

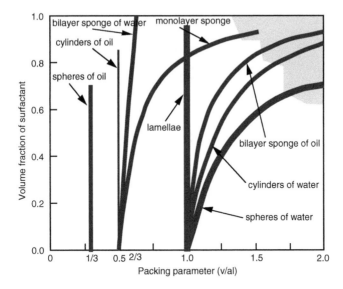

Figure 5.7 Surfactant packing parameter and liquid crystalline forms. (After ST Hyde, *Colloq Phys* C7:209, 1990.)

phases further. Punctured lamellar bilayers, mesh phases, and hexagonal phases with disordered necks connecting neighboring rods all contain strictly hyperbolic interfaces, and these are probably important intermediates during lamellar to sponge and sponge to hexagonal transitions, and possibly for the formation of pores in biomembranes. Details of these intermediate structures are discussed elsewhere [11].

5.10 BENDING ENERGY AND RELATIVE STABILITIES

Low genus and high symmetry TPMS have the highest homogeneity or equivalently the least variation of Gaussian curvature. If a molecule has a preferred average molecular shape defined via the self-assembled aggregate, then any departure from that average will necessarily incur an energetic cost. Further, if the bilayer has relatively high rigidity, as in lamellar phases of ionic amphiphiles, or self-assemblies containing stiff molecules such as fluorinated hydrocarbons, the bending energy can also be a significant if not determining factor on the relative stability of the phase.

Figure 5.8 The hyperbolic plane H² represented as the Poincaré disc. Here H² is decorated with the kaleidoscopic hyperbolic tiles (*246 orbifolds) corresponding to the fundamental patches of the P, D, and G surfaces. These are equivalent to the Flächenstück in E³ and the spherical geodesic triangles of S². The dodecagonal patch is made up of 96 of the *246 tiles. A single unit of the {4,6} tiling is also shown. The {4,6} network is the simplest regular tiling commensurate with the *246 orbifold symmetry and thus represents the simplest network embeddable in the P, D, and G surfaces.

Steric packing models, involving the average surfactant packing parameter or alternatively using cross-sectional area mismatches between the headgroups and chain ends, do not account for the bending energy contributions to bilayer stability. Although the shape parameter diagram in Figure 5.7 is a useful predictor of general phases for given systems, some systems suggest that bilayer rigidity is essential to the stability of some intermediate mesophases at the expense of cubic phases [12–14]. Helfrich introduced the bilayer bending energy model [15], relating the bending energy to the mean and Gaussian curvatures through elastic moduli scalars, and later modified for spontaneously saddle-shaped systems [16]. Under these relationships, the vanishing mean and Gaussian curvatures of static planar bilayers and perfectly homogeneous sponges correlate to zero bending energy contributions to the stability. In reality, bilayers have temporal fluctuations, such as undulations, and thus necessarily incur bending costs. Moreover, quasihomogeneous TPMS in E^3 also necessarily incur a finite but small bending cost.

This is least for (appropriately scaled) Bonnet-related P, D, and G surfaces, whose bending energies are degenerate owing to their equivalence in the hyperbolic plane. These highly homogeneous surfaces can be continuously transformed into each other via direct rhombohedral and tetragonal distortions. In contrast to the Bonnet transformation, the direct, distortion-based transform preserves their topology and zero mean curvature throughout. The related intermediate rPD family of surfaces and the rG, tG, tP, and tD surfaces can attain relatively high degrees of Gaussian curvature homogeneity among known TPMS and so are favored in terms of their bending energetics. These are therefore the lowest-energy forms available to cubic phases undergoing thermal fluctuations and cubic–cubic phase transformations, and should be considered as possibly stable anisotropic intermediates in a number of liquid-crystalline systems where tetragonal and rhombohedral mesh phases may have been previously assigned. In contrast, the two other TPMS of low genus ($g = 3$) and having three- or fourfold symmetry elements, the H and the tCLP family, have relatively higher degrees of Gaussian curvature variation, particularly in comparison with the transformation $G \rightarrow tG \rightarrow tD \rightarrow D \rightarrow rPD \rightarrow P$.

5.11 TPMS SURFACE SYMMETRIES AND PHASE TRANSFORMATIONS

It is obvious to those of you who use x-ray diffraction and polarizing optical techniques for characterization of liquid-crystalline phases that symmetry plays a vital role in understanding these structures. Also, symmetry plays a fundamental role in evenly distributing curvature heterogeneity when bicontinuous structures form and so stabilize the more symmetric cubic phases relative to, say orthorhombic TPMS. Indeed, it is thought that even isolated local patches of hyperbolic bilayers should bear some inherent symmetries, such as mirror planes and rotation axes. L_3 phases are curious in this respect. These are the disordered sponges; they have high topological complexity and display no long-range translational ordering. Their curvature heterogeneity is comparable with high topology (per unit cell) bicontinuous structures, and so given their topological complexity, there is no energetic reason for these to crystallize. Locally, the L_3 structures may display very short-range ordering, similar to patches of the D, P, and G, etc., and so the entropic advantage of having disordered flat point and curvature distributions may dominate the stability considerations. In this respect the L_3 phases can be considered high-temperature molten bicontinuous structures.

The Bonnet transformation maps the quasihomogeneous P, D, and G surfaces in E^3 to each other. The relationship was described by Bonnet in 1853. They all have the same Gauss map representation in the complex plane, C^2, and their fundamental translational unit in H^2 is exactly the same semiregular dodecagon. It is the different ways in which the edges of this surface patch are sewn together that results in the differing forms in E^3. This is similar to choosing different (discrete) values for the association parameter or Bonnet angle $(\theta),(0 - \pi/2)$ in the Weierstrass equations (e.g., $\theta = 0°$ for the D, 38.015° for the G, and 90° for the P). The metrics of these surfaces are identical, and the transformation between these is isometric and conformal, thus preserving lengths and angles. It is a local bending of the surfaces but without stretching. From an energetic point of view, the transform yields a continuous family of surfaces with identical homogeneity, and thus the spectrum of members is energetically degenerate for all values of the association parameter. For this reason the transform offers an energy-free transformation, on the local scale between the different symmetries. However, the transformation does not preserve the global structure, and thus polar or apolar continuous domains must be disrupted in lyotropic bicontinuous cubic phases forced to undergo a Bonnet transformation (in, say, a simulation). This must bear some energetic cost, and thus it has been suggested that the previously mentioned transformations between cubic phases via rhombohedral and tetragonal distortions are more favorable. Either way, the fact still remains that we do not see coexistence of P, D, or G phases with each other, so other factors must be at play in isolating the stability fields of these phases.

5.12 NETWORKS EMBEDDED IN TPMS

The set of possible two-dimensional networks embedded in the hyperbolic plane is far richer than the set of tilings of three-dimensional space. For this reason, embedding of networks in TPMS is a very useful alternative method for generating crystalline networks and for understanding such complex structures as zeolites, clays, and mesoporous inorganics, and more recently complex interpenetrating networks seen in open framework inorganics and coordination compounds [17–19]. The relevance to liquid crystalline phases will be made clear in the following section on polycontinuous networks.

The surface Euler–Poincaré characteristic can also be related to the quantities of an embedded network in the surface through a generalized Euler formula. Thus the average ring size and the connectivity of any network tesselating a surface are coupled to the surface topology and average Gaussian curvature. The equation is:

$$\chi/N = [n\{z + (1 - (z/2)\}]/n, \tag{5.2}$$

where N is the number of vertices in the surface.

This is therefore equivalent to stating:

$$(n - 2)(z - 2) < 4 \text{ for elliptic;}$$

$$(n - 2)(z - 2) = 4 \text{ for flat;} \tag{5.3}$$

$$(n - 2)(z - 2) > 4 \text{ for hyperbolic systems.}$$

So for three-connected sp2 carbon networks, we substitute $(z = 3)$ in Equation (5.3) and immediately deduce that graphite sheets must have an average ring size n = 6 to remain flat. Indeed this is true. Likewise, this is also true for nanotubes, which have the same (zero) average Gaussian curvature. In contrast, Fullerenes such as C_{60} must have some five-rings (or smaller) to be elliptic (i.e., positive Gaussian curvature). Hyperbolic carbons, or Schwartzites, when discovered or synthesized, will contain some 7-rings or 8-rings [20].

Tesselations of TPMS can be generated via commensurate embeddings of fundamental network domains in their respective Flächenstück, prior to building up of the surfaces via symmetry operations. Tesselations can also be constructed using embeddings in the fundamental surface units represented in the hyperbolic plane [21]. Next we will look at hyperbolic crystallography as a way to build up surfaces in E^3.

5.13 A BRIEF INTRODUCTION TO THE HYPERBOLIC CRYSTALLOGRAPHY OF BICONTINUOUS MESOMORPHS AND THEIR USE IN GENERATING SURFACE RETICULATIONS

The angle-preserving (conformal) Poincaré disc represents a mapping of all two-dimensional hyperbolic space (H^2) into the unit circle. Thus the metric is radially symmetric about the center, and is highly nonlinear with respect to the relative Euclidean coordinates that can be read from it. For example, the perimeter is at infinity. Great circles (intersecting the perimeter orthogonally) spanning the disc or touching one side have infinite length, yet arcs of great circles that touch neither side have a measurable and finite length given by the relationship:

$$|ds| = 2|dr|/(1 - r^2), \qquad (5.4)$$

where ds is the hyperbolic arc length, dr is the Euclidean arc length, and r is the radial coordinate.

A fundamental unit used in hyperbolic crystallography is the symmetry group called an orbifold, after Thurston (see Reference 27). This concept is not restricted to two-dimensional tilings but is very useful for elliptic, flat, and hyperbolic tilings nonetheless. The orbifold concept used here is strictly two-dimensional and is analogous to conventional three-dimensional Euclidean space groups, in that tilings of space can be built up using symmetry operations on the fundamental unit.

The orbifold group [28] (or just orbifold) consists of closed polygons with vertex angles denoted by π/a, π/b, π/c.... Here we consider the kaleidoscopic groups, i.e., those orbifolds bounded only by intersecting reflection lines, having n-fold rotational elements at the vertices. The orbifold symbols that represent these kaleidoscopic groups have symbols *abc.... Each orbifold element (in this case just reflections and rotations) has an orbifold characteristic ($c = 2 - \Sigma\zeta$), that is the cost associated with the two-dimensional symmetry elements on its boundary. For reflections (symbol *), the cost is $\zeta = 1$; for rotations, the cost is $\zeta = n - 1/2n$, where $n = a,b,c$.... For example, the *236 triangular orbifold, which is the fundamental surface patch of the {6,3} tiling (graphite)[*] has $c = 2 - (1 + 1/4 + 2/6 + 5/12) = 0$, as expected for a flat net. The *246 triangular orbifold element (the fundamental hyperbolic patch of the P, D, and G surfaces) has a cost, $c = 2 - (1 + 1/4 + 3/8 + 5/12) = -0.04167$, as expected for 1/96th of the genus 3 unit cell. These costs are equivalent to the Euler–Poincaré characteristic and are thus proportional to the integral Gaussian curvature of the equivalent a symmetric domain in two dimensions. So for homogeneous (or constant Gaussian curvature) surfaces, the Gauss–Bonnet theorem implies this (cost) scales with the area of the patch. The identity or fundamental

---

[*] The Schäfli symbol for two-dimensional tilings is {n,z}, where n = ring size and z = vertex connectivity.

triangular orbifold can be built up into larger surface patches via reflections across its own boundaries to build a composite polygon. Sadoc and Chavrolin [22] found that the P, D, and G surfaces have the same disc-like translational patch in E^3 that when "pulled back" into H^2 is an identical dodecagon for each surface. This group is made up of 96 triangular elements obtained via reflections from the identity orbifold.

Robbins et al. [21] and Hyde and Ramsden [23] have shown that reticulations of TPMS can be systematically constructed using this hyperbolic tiling method as a basis. Their method requires that all the subsymmetries of the hyperbolic representation of the TPMS be known. If this is so, then it is possible to generate networks commensurate with these. For example, place a single vertex in each subdomain (whether this is a single fundamental triangle or multiple triangles), join it with an edge if adjacent, and then map it into E^3 using the cover map (the cover map is a unique function that maps the nonunique translational patch in the hyperbolic plane onto the minimal surface).

5.14 POLYCONTINUOUS LIQUID CRYSTALS

Mounting evidence suggests bicontinuous liquid crystals are a restricted subset of a larger set of liquid-crystalline systems that have polycontinuous morphologies [24]. In this section a brief review of the new field of polycontinuous networks in relation to TPMS is given.

In the previous section, tilings and networks of the TPMS were introduced. The basis for constructing the more intricate interwoven polycontinuous networks is the development of embedded tree networks in TPMS, and can be algorithmically derived using embeddings of graphs in the subset of tilings commensurate with the translational surface patch of TPMS. Hyde et al. [24,25] have derived families of these for the relatively simple, low genus TPMS, namely the P, D, G, H, and I-WP.

Regular {n,z} tilings or networks of the hyperbolic plane have edge lengths that are determined solely by the values of n and z. These are related via a hyperbolic trigonometric function.

One way to generate trees on H^2 is via asymptotic triangles. These are hyperbolic triangles made up of three great circles that meet at infinity, i.e., on the disc boundary (infinity). They have zero vertex angles and parallel edges, yet finite area. These can tile H^2 and thus can be described using kaleidoscopic orbifold nomenclature, and a symmetric example has symbol*23∞. Regular trees, having a distinct vertex connectivity (or branching number), z, and edge length, a, can be simply generated by taking the dual of the symmetric asymptotic tiling. Significantly, the duals of asymmetric tilings can give infinite disconnected trees.

All trees T(a,z) have edges that are geodesic arcs of equal length (a) that meet at z-coordinated vertices at equal angles. There is a critical edge length that plays an important role in determining how the tree fills H^2. If the tree edge lengths are relatively short (subcritical), the branching predominates and the tree edges self-intersect and densely fill space. When the branching is not so predominant, the tree branches perfectly merge to form a honeycomb tiling of H^2, i.e., a {n, z} tiling, discussed in the previous section. If the edge lengths are greater than a critical length (supercritical), and certain symmetry criteria are met, a "forest" of trees can decorate the underlying tiling.

Three-, four-, and six-connected forests overlaying given honeycomb tilings of known TPMS have been generated by Hyde et al. (e.g., Reference 24) with these points in mind. These authors have developed an algorithm analogous to making helical windings on cylinders (elliptic geometries) but adapted it for hyperbolic surfaces. Helices scribed on cylinders are easily generated by first unwrapping a cylinder to form a flat sheet. Next, a square grid is drawn on it, and a straight line is marked through the squares to form a set of parallel diagonals. Upon reforming the cylinder via a rolling up of the sheet, the diagonals, if chosen appropriately, will fuse, so that a continuous set of parallel helices is formed. On the surfaces of TPMS, hyperbolic analogues of these helical patterns can be constructed. The analogues of the helices in this case are interpenetrating networks or trees. For example, recall that the P, D, and G surfaces can be tiled with a {4,6} network, made up of *246 orbifolds. Networks of disclinations are inserted at the vertices of the tiles, and lines are drawn between these. The tiles can have tree branches along their edges or cutting diagonally through, depending upon what characteristic edge lengths are desired for the tree limbs. In Figure 5.9, the {4, 6} tile (oversized for clarity) has tree branches on its geodesic edges (edge length is $\cosh^{-1}(3)$). These bifurcate at each vertex. A forest of disjoint trees results when the entire hyperbolic plane is tesselated with these decorated {4, 6} tiles. A related forest of trees with characteristic edge length cosh− 1(5) can be similarly constructed, the difference being that the {4,6} tile is decorated with a tree limb bisecting its diagonal; otherwise, each

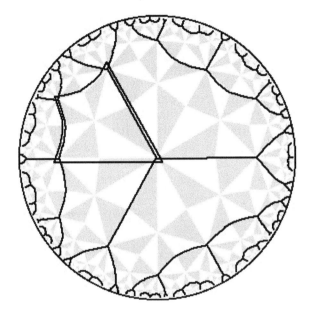

Figure 5.9 A forest of trees commensurate with the {4,6} hyperbolic tiling. For clarity, only a (slightly enlarged) single tetragon of the {4,6} tiling is shown. Note that the tree "limbs" decorate the edges of the tiles (along geodesic arcs), and at every vertex of the tiling, disclinations or branches or bifurcations occur. Each tree extends to the disc edge at infinity, and does not ever meet. When the fundamental dodecagonal patch (made up of eight {4,6} tiles) is wrapped up and its edges "glued" to form a TPMS, polycontinuous networks (defined by the trees) decorate the surface.

vertex still hosts a bifurcation of the network (see Figure 7a and b of Hyde and Oguey [25]).

These parallel trees embedded in H² are then wrapped up, via the identification process discussed previously, to obtain TPMS decorated with polycontinuous networks. Significantly for liquid-crystalline systems, these networks may be considered as the medial axis of channel systems of polycontinuous labyrinths, bearing minimal surface partitions. They are distinguished from other multicontinuous systems reported to date [26] in that they are not made up of nested surfaces, which necessarily have nonzero mean curvature.

Several beautiful polycontinuous structures have been described thus far [24,25]. One such example is the cubic quadra-continuous Gyroid-like structure. It has four interwoven chiral labyrinths of the same handedness, each identical to one of the two enantiomorphic Y graphs defined by the channels of the bicontinuous gyroid labyrinths. It is constructed from the networks shown in Figure 5.10. The algorithm for generating these polycontinuous surfaces from entangled networks will be published soon by Hyde et al. It is impossible to construct triply continuous cubic phases with identical interpenetrating labyrinths, but it is possible in the hexagonal system. Tricontinuous structures have been reported [25] by embedding three and six coordinated forests onto the hexagonal H surface. This set of interpenetrating networks can be used to construct the minimal surface shown in Figure 5.11. Thus it is a very good candidate for the morphology of liquid crystalline systems made up from star triblock copolymers with blocks that are subject to micro-phase separation, analogous to the bicontinuous partitions of diblock copolymers.

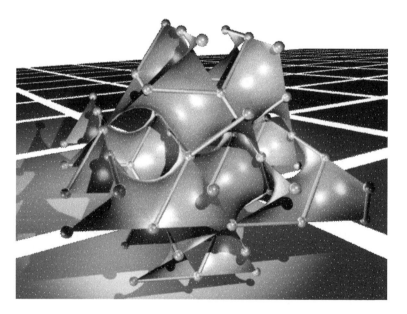

Figure 5.10 Threefold forest, with edge length cosh-1 (3) embedded in the D-surface. The forest yields an entangled set of four independent chiral networks. Image courtesy of Stephen Hyde and Stuart Ramsden.

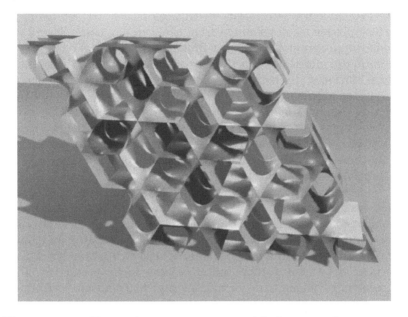

Figure 5.11 Tricontinuous structure with hexagonal symmetry, constructed from forests of trees decorating the H surface. The color-coded labyrinths are identical, interpenetrating, and completely independent. Image courtesy of Stephen Hyde and Stuart Ramsden.

5.15 FINAL REMARKS

The differential geometry and topology of TPMS has a very distinct language compared with that of the liquid-crystalline systems it can be used to describe. This chapter gives an introduction to the mathematics of these structures. The most fundamental concepts discussed here are the curvatures and symmetries intrinsic to two-dimensional fundamental patches of TPMS. The local properties of the surface are determined from these fundamental surface units. The global properties such as topology and curvature variation can be determined once the patches have been built up into larger structures, and these can in turn be related to physical properties of liquid crystals, such as their relative stabilities.

Crystallographic symmetry operations can be used to build up the global TPMS structures in three dimensions. More naturally, the fundamental surface patches can be manipulated and characterized by considering these in two-dimensional spaces such as the surface of the unit sphere and the complex and hyperbolic planes.

The explicit parametrization of surfaces is carried out using the surface properties determined from the complex plane. Using two-dimensional symmetries, larger fundamental patches of TPMS can be built up, and these can then be glued to give the three-dimensional surfaces. In both two-dimensional and three-dimensional spaces, the surface patches can be decorated with network elements prior to symmetry operations that build the unit cells of TPMS. This way commensurate honeycomb and entangled networks can be systematically generated, so that these decorate the surfaces of TPMS. From these, algorithms are now being developed to generate TPMS with polycontinuous morphologies.

In relation to experimental liquid crystal studies, perhaps the most exciting developments reviewed here are the researches on rhombohedral and tetrahedral distortions of the P, D, and G surfaces, and the development of algorithms for generating polycontinuous liquid crystalline morphologies. These two sets of studies are considered by this author and by others to be predictive of yet unrecognized phase transformations and morphologies within the soft-condensed matter field.

ACKNOWLEDGMENTS

Stephen Hyde and Stuart Ramsden kindly supplied images of TPMS and their embedded networks and the tricontinuous hexagonal structure. The author thanks Andrew Fogden for his review of the sections on parametrization.

REFERENCES

1. RW Corkery, Artificial Biomineralisation and Metallic Soaps, Ph.D. dissertation, Australian National University, Canberra, Australia, 1998.

2. V Luzzati, H Mustacchi, A Skoulios, *Nature* 180:600, 1957.

3. V Luzzati and P Spegt, *Nature* 215:701, 1967.

4. K Larsson, K Fontell, N Krog, *Chem Phys Lipids* 27:321, 1980.

5. K Larsson, *Nature* 304:664, 1983.

6. A Fogden, ST Hyde, *Acta Crystallograph* A48:442, 1992.

7. D Hilbert and S Cohn-Vossen, *Geometry and the Imagination,* Chelsea Publishing Co., New York, 1952.

8. ST Hyde, *Colloq Phys* C7:209, 1990.

9. ST Hyde, *Pure Appl Chem* 1992, 64:1617, 1992.

10. ST Hyde, *Curr Opin Solid State Mater Sci* 1:653, 1996.

11. ST Hyde, *Colloids Surf A: Physicochem Eng Aspects* 129–130: 207, 1997.

12. P Kékicheff, GJT Tiddy, *J Phys Chem* 93:2250, 1989.

13. E Blackmore, GJT Tiddy, *J Chem Soc Faraday Trans II,* 84:1115, 1988.

14. A Fogden, M Stenkula, CE Fairhurst, MC Holmes, MS Leaver, *Prog Colloid Polym Sci* 108:129, 1998.

15. W Helfrich, *Z Naturforsch* C28:693, 1973.

16. A Fogden, ST Hyde, G Lundberg, *J Chem Soc Faraday Trans* 87:949, 1991.

17. S Mukhopadhyay, PB Chatterjee, D Mandal, G Mostafa, A Caneschi, J vanSlageren, T Weakley, J R, and M Chaudhury, *Inorg Chem* 43:3413, 2004.

18. CJ Kepert, and MJ Rosseinsky, *Chem Commun* 1:31–32, 1998.

19. B Abrahams, F, S Batten, R, H Hamit, BF Hoskins, and R Robson, *Chem Commun* 11:1313–1314, 1996.

20. ST Hyde, *Phys Chem Miner* 20:190, 1993.

21. V Robins, SJ Ramsden, ST Hyde, 2D Hyperbolic Groups Induce Three-Periodic Euclidean Reticulations, *Eur Phys J B* 39(3): 365–375, 2004.

22. JF Sadoc, J Charvolin, *Acta Crystallograph* A45:10, 1989.

23. ST Hyde, SJ Ramsden, Some Novel Three-Dimensional Euclidean Crystalline Networks Derived from Two-Dimensional Hyperbolic Tilings, *Eur Phys J B* 31(2): 273–284, 2004.

24. ST Hyde, S Ramsden, *Europhysics Lett* 50:135, 2000.

25. ST Hyde, C Oguey, *Eur Phys J B* 16:613, 2000.

26. V Babin, P Garstecki, R. Holyst, *Phys Rev B* 66:1, 2002.

27. J Montesinos, *Classical Tesselations and Three-Manifolds,* Springer-Verlag, Berlin, 1987.

28. JH Conway, *Groups, Combinatorics, and Geometry,* in *Lond. Math. Soc. Lecture Note Series,* Vol. 47, 1992, p. 438.

29. ST Hyde, Infinite Periodic Minimal Surfaces and Crystal Structures, Ph.D. dissertation, Monash University, Melbourne, Australia, 1986.

Section 2

Physical Chemistry and Characterization

6

Novel L₃ Phases and Their Macroscopic Properties

R. BECK AND H. HOFFMANN

CONTENTS

6.1 INTRODUCTION

Due to their amphiphilic properties, surfactant molecules in aqueous solution can form a large variety of different aggregates. This is related to the existence of many options to divide space into polar and apolar regions. The detailed chemical structure of the surfactant molecules determines the area of the polar group and the spontaneous curvature of the surfactant layer.

Roughly speaking, one may distinguish between spherical (micelles), cylindrical (rodlike) and lamellar (disklike) shapes. These supramolecular objects can organize themselves again on a larger scale, building structures with long-range ordering (liquid crystal phases) or only short-range correlation (liquid isotropic phases).

In water, the hydrophobic surfactant chains are located in the interior of the aggregates without a contact to the polar solvent, as they are shielded by the hydrophilic headgroups. The micellar so-called L_1 phase is composed of monolayers of the surfactant, while bilayers build the classical flat lamellar $L_{\alpha h}$ phase or spherical vesicles ($L_{\alpha l}$ phase). The closed spheres of the $L_{\alpha l}$ phase include parts of the solvent in their interior.

One of the most interesting and astonishing structures is the so-called L_3 phase. This phase has been found in a great variety of amphiphilic systems [1,2]. Its structure has been deduced by small-angle neutron scattering (SANS) [3,4], small-angle x-ray scattering (SAXS) [5–9], light scattering [10,11], conductivity [12,13], and self-diffusion NMR [14,15] and visually confirmed by freeze-fracture electron microscopy (FFEM) [16,17] to be a multiply connected bilayer structure that divides a solvent into two interwoven labyrinths. Pictorially, the branched bicontinuous tubes are envisioned as a spongelike fluid membrane. This membrane structure resembles the bicontinuous microemulsions without the presence of oil. The basic structure of bilayers instead of monolayers exists. The L_3 phase is an isotropic disordered solution with a relatively low viscosity. It scatters considerable light and frequently exhibits streaming birefringence. Early studies on the L_3 phase have focused on its unique properties such as the absence of long-range order [18], existence at high dilution [12], Newtonian flow behavior at low shear [19], and streaming birefringence at high shear [20]. Also there exist a number of studies exploring possible applications, for example as model structures for studying intercellular membranes [21] or as drug transport vehicles [22].

In this work a survey of the macroscopic properties of the sponge phase and of a novel L_3 phase with extraordinary properties and phase behavior is presented.

6.2 THE "CLASSIC" SPONGE PHASE

In the long history of research in surfactant science, the L_3 phase has only recently been discovered to be a distinct phase [23]. Without detailed investigation, it can easily be mistaken for an L_1 phase. Since its discovery in 1980 by J. Lang and R. Morgan in the system $C_{10}E_4$/water [24], it has been found in many other systems. It is general knowledge now that L_3 phases are encountered when dilute L_α phases are made more lipophilic by changing a physicochemical parameter of the surfactant system. This is most easily done by increasing the cosurfactant/surfactant ratio in nonionic surfactant systems, by increasing the temperature in alkylpolyglycol systems, or by increasing the salt concentration in mixed surfactant systems where one compound is an ionic surfactant [1,25,26].

These common procedures for the preparation of L_3 phases do not always work. In many known situations, L_3 phases are not observed when a dilute L_α phase is made more lipophilic [27]. Although L_3 phases are generally observed in phase diagrams with zwitterionic surfactant and intermediate-chain alcohols, the L_3 phase does not occur when the hydrocarbon cosurfactants are replaced by perfluorocosurfactants [28]. L_3 phases are also not observed in zwitterionic surfactants, with longer chain cosurfactants, even though the L_α phases are observed in these mixtures. These unexpected differences in the behavior of surfactants with cosurfactants are explained by the size of the normal bending and Gaussian bending constant of the bilayers.

L_3 phases can also be formed from block copolymers such as the Poloxamers [29]. Even though there was much controversy about the structure of the L_3 phase at its discovery, its structure is now well known. Early publications had suggested that the phase consisted of freely rotating disklike micelles in random orientation at the overlap concentration. Although this model is certainly wrong, it still can be of use to estimate macroscopic properties such as the translational and rotational diffusion constants of the surfactant aggregate and the viscosity [30].

Now it is well known that the surfactant and cosurfactant in an L_3 phase are present as a continuous network [31]. The first experimental study proposing such a structure was done in Montpellier by Porte et al. [32], independently of a theoretical proposition by Cates et al. [33]. The phase can therefore be considered as a molten L_α phase in which the aqueous part and the surfactant part are continuous in nature; it is therefore a bicontinuous phase. In spite of this bicontinuous character, it has a low viscosity with a value in the range of several times the value of the water viscosity. What is even more surprising is the fact that the viscosity of the L_3 phase has an extremely low concentration dependence.

In ternary phase diagrams, the L_3 phase can usually be observed over a large concentration range or volume fraction but only over an extremely narrow cosurfactant/surfactant ratio [34,35]. On one side of the existence region is a narrow two-phase L_α/L_3 phase region, while on the other side is a two-phase L_1/L_3 region. In Figure 6.1 the phase diagram of the system $C_{14}DMAO$, n-heptanol, and water is given, representing the typical phase behavior of a ternary system with an L_3 phase. The L_3 phase is well defined from the lamellar $L_{\alpha h}$ phase by a two-phase region. Several kinds of vesicular phases also exist; they are described in detail in literature [17].

The existence of the phase is extremely sensitive to ionic charges. Usually it becomes unstable when in neutral cosurfactant/surfactant systems a few percent of the nonionic surfactant is replaced by an ionic surfactant. The phase also is shear sensitive, and high shear can transform the L_3 phase into a metastable L_α phase that relaxes back into the L_3 phase when the shear is stopped [36–38].

The phase transition L_3/L_α and the interface between these two structures were the subjects of several investigations in recent years. The L_α phase is a stack of parallel layers with an interlamellar distance d_α and a layer thickness ξ_α. The sponge phase is made of infinite multiconnected membranes with a characteristic length d_3 and a layer thickness ξ_d (Figure 6.2). The phase is disordered and isotropic. The L_α and L_3 phases have the same local structure (membranes) but different characteristic lengths, and their interface is characterized by a large anisotropy of the interface tension as a function of the direction of the L_α layers in the interface. An epitaxy phenomenon is reported [39].

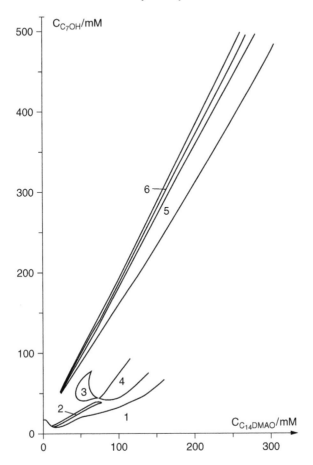

Figure 6.1 Phase diagram of the system $C_{14}DMAO$, *n*-heptanol, and water at 25°C. (From H Hoffmann, C Thunig, U Munkert, HW Meyer, W Richter. *Langmuir* 8:2629, 1992.) 1: micellar L_1 phase; 2 to 4: vesicular L_1^*, L_3^*, and $L_{\alpha l}$ phases; 5: lamellar $L_{\alpha h}$ phase; 6: L_3 phase.

At the interface between the two phases exists a tilt angle of the L_α layers, which corresponds to a geometry of the characteristic lengths d_α and d_3 [40]. The L_3 is connected to the L_α by bridges and passages [41] (Figure 6.3). The interfacial energy is estimated assuming the elastic curvature of the membrane in the vicinity of the interface. Hoffmann et al. have published remarkable freeze-fracture microphotographs of flat interfaces with tilted L_α bilayers in the L_α–L_3 interface (Figure 6.4), which show an abrupt transition from L_α to L_3.

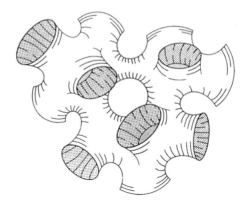

Figure 6.2 Schematic drawing of the spongelike L3 phase. (From M Skouri, J Marignan, J Appell, G Porte. *J Phys II* 1:1121, 1991.)

In 1998 in the group of Hoffmann and Platz, a classic L_3 phase was likewise found in an ionic surfactant system existing of calcium dodecylsulfate/pentanol or hexanol and water [42]. Later this was also observed in the corresponding heptanol and octanol systems [43]. It is quite interesting to find a sponge phase in a charged system and as far as we know it was for the first time.

The L_3 phase in the $Ca(DS)_2$ system is also only existent in a phase region with a narrow cosurfactant/surfactant ratio. It is reached

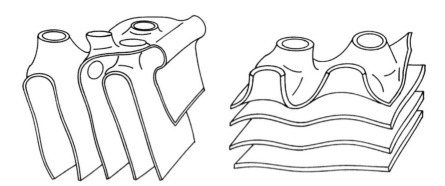

Figure 6.3 Possible interface at the transition between the lamellar $L_{\alpha h}$ and the sponge phase. (From C Blanc, O Sanseau, V Cabuil. *Mol Cryst Liq Cryst* 332:523, 1999.)

Figure 6.4 FF-TEM micrograph of the system 100 mM C$_{14}$DMAO + 185 mM n-heptanol in water. The L$_{\alpha h}$ phase is completely surrounded by an L$_3$ phase. (From H Hoffmann, C Thunig, U Munkert, HW Meyer, W Richter. *Langmuir* 8:2629, 1992).

from the lamellar phase via a small two-phase area. The phase boundaries between L$_\alpha$ and the sponge phase are difficult to make out due to their similarity in refractive index. With polarizers, however, a clear boundary is seen.

6.3 THE NOVEL L₃ PHASE

In this section we describe a novel L$_3$ phase that was observed for the first time in a ternary phase diagram of Ca- or Mg-salts of anionic surfactants, branched cosurfactants, and water [44,45]. As for other ternary cosurfactant/surfactant systems, the L$_3$ phase occurs with increasing cosurfactant/surfactant ratios after the L$_\alpha$ phase. It is a low-viscosity, optically isotropic fluid with a low flow birefringence. The time constants τ of the electric birefringence results scale with $\tau \sim \emptyset^{-3}$. Its conductivity is very much higher than that of the neighboring L_α phase, but there are some marked differences from known L_3 phases. This novel L_3 phase is thermodynamically stable in spite of the ionic structure of the surfactant. No two-phase region was found in the Ca-α-sulfonated alkyl fatty acid methylester system between the L_α and the L_3 phase, and the L_3 phase was stable over a wide cosurfactant/surfactant ratio between one and two. In SANS measurements it shows a broad correlation peak that occurs at about the same position as the sharper peak in the L_α phase. The structure of the L_3 phase is demonstrated by FF-TEM micrographs.

The novel L_3 phase with extraordinary properties was first discovered when we studied the sequence of phases of the system calciumtetradecylmethylester-α-sulfonate (Ca(C$_{14}$-αMES)$_2$) (1) and 2-ethylhexyl-monoglycerinether (EHG) (3) in water. It was also be found in the Ca- and Mg-salts of dodecylsulfates (Ca/Mg(DS)$_2$) (3) with EHG and another heavily branched cosurfactant, 1-phenylpro-pyl-1-amine (PPA) (4).

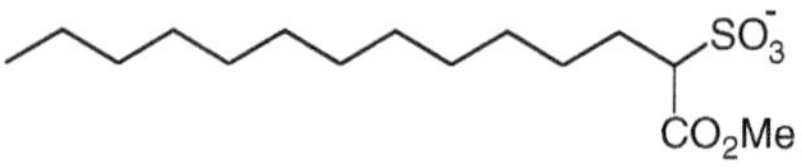

(1)

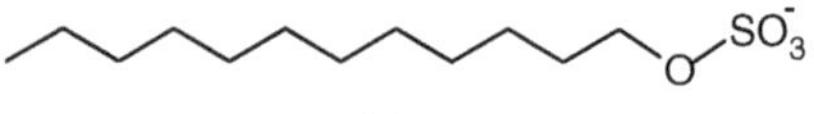

(2)

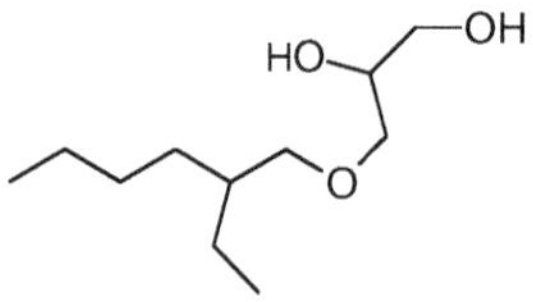

(3)

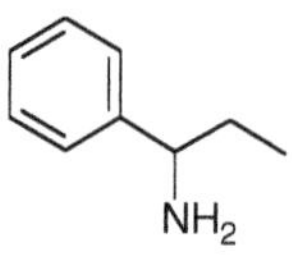

(4)

6.3.1   Phase Behavior and Conductivity

The ionic conductivity is an interesting method to investigate the structure and phase behavior of a surfactant system and especially the sponge phase. In a system with charged particles or ions, the conductivity can probe the obstruction posed by the structure to the mobility of the free ions. In a water-rich sponge phase the conductivity are ions in the solvent, while the membrane is the insulator. The conductivity values of a sponge phase are much higher than those of lamellar $L_{\alpha l}$ or $L_{\alpha h}$ phases. This indicates a continuous water path through the system and relative free mobility of the charges. Nevertheless, it is remarkable that the reduced conductivity (i.e., the conductivity of the L_{α} compared to the conductivity for the same ionic concentration in the L_1 phase) is found to be 2/3 of the value of the pure solvent [3,32]. This was interpreted as a result of the tortuous path through a dielectric medium filled with randomly distributed insulating surface elements.

Taking a closer look at the evolution of conductivity over a wider range of volume fraction of surfactant, it is plausible that for constant ionic concentration the conductivity has to decrease. This could be explained by the higher and higher opposition displayed by the multiconnected membrane to the mobility of the free ions. For the dilute L_3 region, this is correct and was observed in many systems.

For higher concentrated L_3 phases the conductivity shows sometimes a rather different behavior. As Gomati et al. observed [46], the conductivity for more concentrated L_3 phases rises after the decrease in the dilute region with a jump to a very high value and stays there at a constant level. They explained this with a transition from the symmetric sponge to the assymmetric sponge. The symmetry between the two sides of the bilayer is not preserved, and a infinite piece of connected membrane with permitted defects arises [10,47]. This corresponds to a structure where infinite paths are formed in the diluent allowing free passages of the free ions between the two subvolumes and therefore a higher conductivity.

In the following section an overview is given of the phase behavior of the cosurfactant/surfactant system by measuring the conductivity and preparing the samples in respect to these results. The samples were photographed between crossed polarizers to distinguish between the isotropic sponge phase and the anisotropic lamellar phases.

For the visualization of microstructure in complex fluids, such as microemulsions, dilute lamellar phases, or sponge phases, electron microscopy is a useful technique. With TEM one gets a clear picture of the multiconnected spongelike structure of the L_3 phase. With images from electron microscopy, one can observe the structure of membranes, phase transitions in the samples, and even undulations of the bilayer.

It should always be considered that the preparation process of the TEM replica influences the long-range order in the samples, but the characteristic repeat distances are preserved [16]. So one has to discriminate between artifacts that emerge in the preparation process and reproducible pictures that come close to reality.

6.3.1.1 The System Calciumtetradecylmethylester-α-Sulfonate (Ca(C$_{14}$-αMES)$_2$)/2-Ethyl-hexyl-Monoglycerinether (EHG) (3) in Water

Figure 6.5 shows the section of the phase diagrams of the ternary system when for a 50 mM solution of the Ca-salt of an α-sulfonated

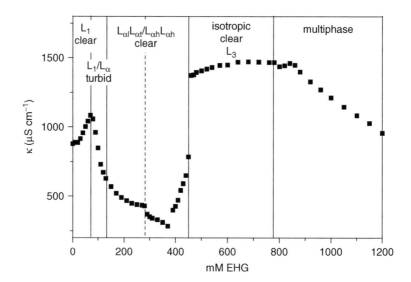

Figure 6.5 Plots of the electric conductivity κ against the EHG concentration for a constant surfactant concentration of 50 mM Ca(C$_{14}$-αMES)$_2$ at 25°C. The sequence of different mesophases is schematically shown in the diagram.

alkylfattyacidmethylester and the cosurfactant 2-ethylhexyl-monoglyceride the cosurfactant concentration is increased. The conductivities of the various phases were observed when they were stirred. For this reason, only one conductivity value is given for two phase regions.

In the first part of the diagram, the conductivity of the low viscosity clear solution increased slightly until 80 mM EHG concentration was reached. In this region micellar structures that are optically isotropic exist. After adding more EHG, the solution became turbid and the conductivity decreased to very low values even when the solution was clear again. This phase is different from the micellar phase. It is highly viscous, and between crossed polarizers there is a strong birefringence. Afterwards, it turned out to be the vesicular $L_{\alpha l}$ phase followed by the lamellar $L_{\alpha h}$ phase.

So far the system shows the normal and expected behavior up to the $L_{\alpha h}$ phase, which is a birefringent phase of low viscosity with the typical birefringence pattern of random domains.

With increasing cosurfactant concentration the sequence L_1, L_1/L_{α}, $L_{\alpha l}$ and $L_{\alpha h}$ was observed. At the end of the L_{α} phase, the conductivity increases and the birefringent L_{α} phase is replaced by an isotropic, highly conductive phase that exists over a wide cosurfactant/surfactant region. A two-phase region between the L_{α} and the isotropic phase could not be found by preparing samples with small variation in their composition around this transition and leaving the samples for long times under temperature-controlled conditions without stirring.

All samples around the phase boundary were either completely birefringent or completely isotropic. Figure 6.6 shows samples between crossed polarizers. In the L_1/L_{α} two-phase region (100 mM EHG), the vesicles settled down as a cloudy precipitate on the bottom of the tube. All birefringent samples are clear and extend to EHG concentrations of up to 450 mM. At this concentration the birefringence suddenly disappears without a two-phase region between the clear phases.

A FF-TEM micrograph of the novel phase is given in Figure 6.7. The micrograph is fuzzier than the micrographs that have been obtained on other L_3 phases [16,17]. However, it still shows the typical features of L_3 phases. In particular the spacings that can be evaluated from the micrograph are in the same range as the spacings expected for L_3 phases of this concentration [48]. The dimensions for the spacing are also consistant with the dimensions that

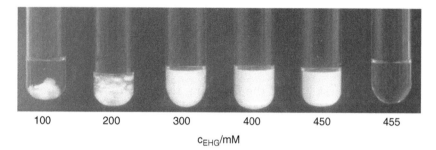

c_{EHG}/mM

Figure 6.6 Samples of a solution of 50 mM Ca(C$_{14}$-αMES)$_2$ and EHG in water between crossed polarizers.

can be extracted from the SANS data. We are thus confident that the newly found phase is indeed an L$_3$ phase, even though some of its properties are somewhat different from those of the L$_3$ phases that have been found earlier.

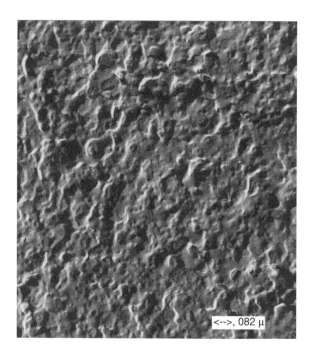

Figure 6.7 FF-TEM micrograph of the systems 50 mM Ca(C$_{14}$-αMES)$_2$ and 600 mM EHG.

6.3.1.2 The System Calciumdodecylsulfates $(Ca(DS)_2)$/ 1-Phenylpropyl-1-Amine (PPA) in Water

In Figure 6.8 one can see the conductivity and phase diagram of another Ca-salt with a different cosurfactant. It is the system calciumdodecylsulfate and 1-phenylpropyl-1-amine (PPA). $Ca(DS)_2$ has a similar phase behavior as the corresponding α-sulfonated alkylfattyacidmethylester salt with EHG. The EHG was replaced by a different branched cosurfactant to verify our assumption that a sterical large group is crucial for the novel L_3 phase. The $Ca(DS)_2$ has a much higher Krafft temperature (about 55°C) compared with the corresponding α-sulfonated alkylfattyacidmethylester salt. A precipitate region exists therefore until the concentration of 50 mM PPA is reached, followed by an turbid vesicular $L_{\alpha l}$ phase. It is interesting that in the vesicle phase the conductivity increases with rising cosurfactant concentration and levels off when the classical lamellar phase is reached. In the $L_{\alpha h}$ region the conductivity stays constant and even decreases somewhat. When the phase border of the novel L_3 phase is reached, the conductivity ascends abruptly and, as in the α-sulfonated alkylfattyacidmethylester system, the values are constant up to the beginning of the multiphase region. It is a striking fact that in this system the novel phase did not stretch across such a wide cosurfactant/surfactant

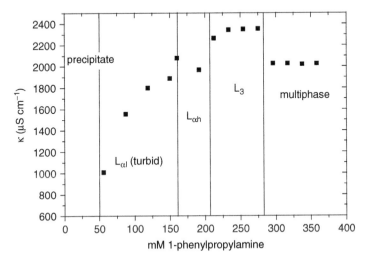

Figure 6.8 Plots of the electric conductivity κ against the PPA concentration for a constant surfactant concentration of 50 mM $Ca(DS)_2$ at 25°C.

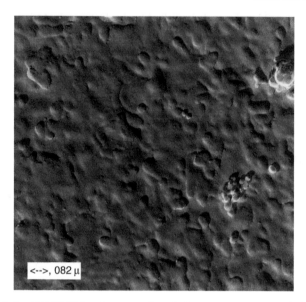

Figure 6.9   FF-TEM micrograph of the systems 50 mM Ca(DS)$_2$ and 270 mM PPA.

ratio as in the previous system. Furthermore, at this surfactant concentration a two-phase region existed between the lamellar and the sponge phase. But nevertheless the conductivity and phase behavior are quite similar.

In Figure 6.9, the TEM micrograph shows the similarity between the EHG system and the PPA system and at the same time on this picture one can notice tubes and holes in the membrane.

6.3.1.3   Magnesiumdodecylsulfates (Mg(DS)$_2$)/ EHG/Water in Mixtures with Na(DS)$_2$ and C$_{12}$DMAO

Contrary to Ca(DS)$_2$, the analogous Mg-salt possesses a much lower Krafft temperature of 31°C. The conductivity/phase diagram of Mg(DS)$_2$ with EHG has the same shape as the phase diagram with Ca salts, and the phase sequences and conductivity behavior are similar. Figure 6.10 shows the schematic phase behavior between the 25 and the 100 mM Mg(DS)$_2$. As expected, with higher surfactant concentration the phase boundaries shift to higher cosurfactant concentrations and are more extended. Over the whole investigated concentration

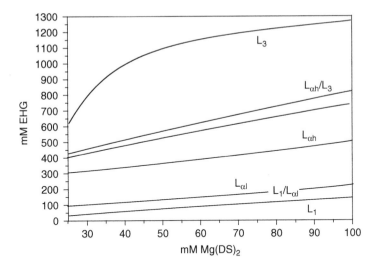

Figure 6.10   Phase diagram of the system $Mg(DS)_2 + EHG$ between surfactant concentrations of 25 and 100 mM $Mg(DS)_2$.

range exists a two-phase $L_{\alpha h}/L_3$ region. In opposition to the L_1 and L_α regions, the L_3 phase has no linear dependence to the surfactant concentration in this part of the phase diagram. This means that the novel L_3 phase cannot be diluted with water arbitrarily. At low concentration the sponge region shrinks dramatically. This is also different from the phase behavior of classic sponge phases (Figure 6.1).

Samples between crossed polarizers of the 50 mM $Mg(DS)_2$/ EHG in water solution are given in Figure 6.11. The $L_{\alpha l}$ and $L_{\alpha h}$ feature

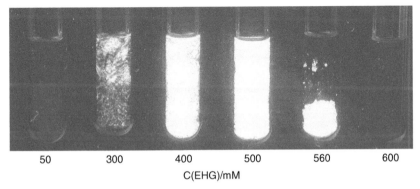

Figure 6.11   Samples of a solution of 50 mM $Mg(DS)_2$ and EHG in water between crossed polarizers.

the typical birefringence patterns between the polarizers. There is a two-phase region in the sample 50 mM MG(DS)$_2$ + 560 mM EHG. The lamellar phase is on the bottom of the test tube and the L$_3$ phase on the top.

For a better understanding of this novel phase in charged surfactants, the phase behavior of the Mg-salt was investigated when the Mg^{2+} ions were replaced by Na$^+$ ions in sodium dodecylsulfate (NaDS) (surfactant solution 100 mM). In Figure 6.12, the surprising results are shown. In spite of the higher charge of the bilayers caused by the higher dissociation of the sodium ions, the L$_3$ phase does not disappear and to the contrary gets even broader.

This is in good agreement with Figure 6.13. The charge of the bilayers was lowered by adding the zwitterionic uncharged dode-cyldimethylamineoxid (C$_{12}$DMAO). With rising content of the zwit-terionic surfactant, the sponge phase gets smaller and smaller, as the lamellar and micellar phases do. In pure C$_{12}$DMAO, most phases disappear, and the C$_{12}$DMAO/EHG/ water system exceeds the solu-bilization border with EHG concentration of 30 mM. Only with excess salt can a diverse phase behavior be induced in the system C$_{12}$DMAO/EHG/water.

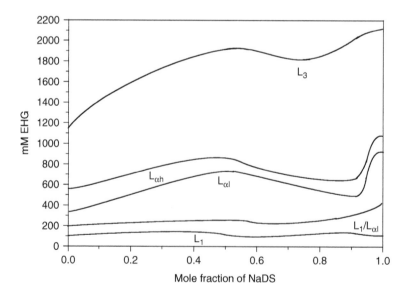

Figure 6.12 Phase diagram of the charged system Mg(DS)$_2$/Na(DS) + EHG. The Na/(Na + 2Mg) ratio varies while the surfactant concen-tration (DS)$^-$ is constant at 100 mM.

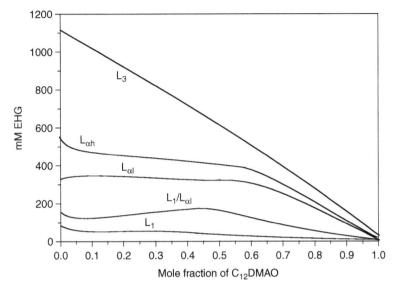

Figure 6.13 Phase diagram of the system $Mg(DS)_2/C_{12}$ DMAO + EHG. The ionic/zwitterionic surfactant composition varies, while the total surfactant concentration is 100 mM.

This is strong evidence that the used cosurfactant requires a system with free charges inside and that the stiff bilayer of charged systems is no handicap for the formation of the novel sponge phase because the steric-demanding cosurfactant can shift the charged surfactant layers apart and change in this manner the curvature of the membrane. These results require specific consideration, and therefore it is necessary to examine in detail the structure and behavior of the bilayer.

6.3.1.4 Discussion

A surfactant bilayer may be modeled as a continuous film of incompressible, two-dimensional fluid. The configuration of bilayer-forming surfactants in solution consists of self- and mutually avoiding random surfaces. The surfactant layer divides the solution into a "Inside" (I) and "Outside" (O). Once a point has been chosen as "Inside," the classification of all other points is determined. If the two sides of the fluid film are equivalent, then the Hamiltonian of the entire system is unaffected by the interchange I/O.

To describe the statistical behavior of the flexible membrane it is necessary to introduce the elasticity of the surface. The total bending free energy E of a surface configuration is given by the integral of the Hamiltonian H over the total surface A. [49,50]:

$$E = \int H \, dA = \int \frac{\kappa}{2} \left(\frac{1}{R_1} + \frac{1}{R_2} - H_0 \right)^2 + \frac{\bar{\kappa}}{R_1 R_2} \, dA \qquad (6.1)$$

with R_i the principal radii of curvature of the surface and κ and $\bar{\kappa}$ the average and the Gaussian bending constants, respectively. H_0 is the spontaneous radius of curvature and is zero by symmetry for bilayers but can be different from zero for films.

Taking a closer look at the effects of thermal fluctuation on the membranes, κ has to be compared to $k_B T$ (k_B is the Boltzmann constant and T the temperature). When κ is much larger than $k_B T$, the membrane will be rigid and thermal fluctuation will have only little effect on the shape of the membrane, but when κ is in the same magnitude as the thermal energy, the effects become important. The thermal fluctuations cause undulation on the membrane, which leads to a renormalization of the elastic constants κ and $\bar{\kappa}$, which is again a function of membrane size [52,53]. The consequence of this is for example that in the case of the decrease of the elastic constant with the correlation length on the membrane plane over a certain distance, the membrane will lose its orientational order. The length at which the bending constant becomes zero is called the persistence length of the membrane. For length larger than the persistence length, the membrane starts to crumble. This means that the temperature influences the ordering behavior of the membrane. With rising temperature, the bending constant gets close to zero and even changes its algebraic sign. This causes the temperature-dependent transition from a lamellar to a sponge phase.

But also for smaller length than the persistence length a thermal effect can happen. Helfrich [54] has shown that long-range repulsive interactions are caused from the restriction of the thermal undulation of stacked membranes (as in lamellar phases) resulting from steric exclusion by neighbors in the stack. When the distance d between the membranes is smaller than the persistence length, the entropy loss per unit area could be written as:

$$\frac{F}{A} = \frac{3\pi^2}{128} \frac{(k_B T)^2}{\kappa} \frac{1}{d^2} \qquad (6.2)$$

Because of the long-range repulsive interaction, lamellar systems can be swollen by a large factor by adding solvent. The repeat distance d can vary between about 5 and 1000 nm [55,56].

The stability of the lamellar phase over a wide range of concentration is therefore among others the result of undulation interactions. On dilution this phase has to melt in an isotropic liquid phase like the micellar L_1. Like the lamellar phase, the sponge phase is stabilized by steric interactions resulting out of the thermal undulation (Equation 6.2).

Back to Equation (6.1), it can be summarized that for stability reasons κ has to be positive (here in the magnitude of the thermal energy) and $\bar{\kappa}$ negative with an absolute value twice as large as κ. By adding small amounts of amphiphilic impurities such as alcohols into the bilayer or changing the bulk salinity, the changing conditions can be large enough to make $\bar{\kappa}$ less negative. In this case the sponge phase with many handles becomes energetically possible.

In addition, a higher salinity in the solution can also compensate for a disruptive effect that appears in solutions of charged surfactants. In ionic systems, there exist besides the steric interactions strong electrostatic interactions. The decay of these repulsive forces occurs exponentially, according to the Debye–Hückel equation:

$$V_x = V_0 e^{-\frac{x}{L_D}} \tag{6.3}$$

where x is the distance from an isolated bilayer of surface potential V_0 and L_D is the Debye length, which is proportional to the reciprocal root mean square of the ionic strength. L_D is therefore the characteristic decay length of the electrostatic interactions. The salinity causes a shielding of the charges on the membrane. The electrostatic interactions are reduced, and the rigidity of the membrane declines. This leads to a higher flexibility of the bilayer and with this to a possible structural change from a lamellar to a sponge phase.

There are thus three different options to change the flexibility of the bending constant of the membrane, which is important for the creation of an L_3 phase:

- The temperature change, which causes undulation in the case of a compatible κ
- Addition of impurities such as cosurfactants, which are assembled in the bilayer and can also remodel the bending constant

- Excess salt, which shields the charge of the bilayer and leads to a more flexible membrane

As described earlier, membranes divide space into two sections — inside and outside. This division in the sponge phase can happen in two ways. The first is the symmetric phase, in which both parts posses an equal volume. Consequently, the average curvature of the surfactant film given in Equation (6.1) is exactly equal to zero, respecting the symmetry of the Hamiltonian [47]. By changing the temperature or concentration of one of the components, the symmetry can be broken up and a phase transition to an asymmetric phase with a averaged curvature not equal to zero occurs within the L_3 phase. This phase transition could be continuous or discontinuous [57].

This symmetric to asymmetric sponge phase transition was theoretically predicted [10,45], and there exist several theoretical models to understand the behavior of the sponge phase. Here we only want to mention these models. A detailed discussion would go beyond the scope of this work.

There exist beyond others a microscopic (lattice) model [33,58] and a phenomenological model based on a Landau–Ginzburg description of the free energy [59]. With this model, many properties of the sponge phase can be explained, such as viscosity [19,60], dynamic properties [61,62] and transport properties [13,46].

6.3.2 Phase Transition L_α to L_3

A Landau theory can also be used to describe the transition from the symmetric sponge to a lamellar state. [63]. The theoretical analysis with the Landau–Ginzburg formalism was first performed by Marcelja and Radic [64] for a second-order process and extended by de Gennes [65] to a smectic transition. The first-order transition has been considered by Poniewierski and Sluckin [66]. So not only the second-order nematic-smectic but also the first-order isotropic-smectic transition is described by the Landau theory.

But nevertheless there are also signs that the phase transition between sponge and lamellae could be a second-order transition in certain circumstances. To proclaim this transition generally as a first-order transition is a simplification of real features of this structural change. Of course there is a drastic change from an ordered phase to an isotropic phase. Anyway, this transition requires a continuous description, especially for the border regions. Golubovic and

Lubensky have proposed such a treatment and have included the steric repulsion in the sponge free energy.

Interesting investigations about the nature of the transition layer at the interface between lamellar L_α and the L_3 phase in swollen surfactants were performed by Antelmi [67], Lavrentovich [41], and Blanc [40]. Of course, the L_α and the L_3 phases in equilibrium do not have the same composition, and therefore it is obviously a first-order transition as in any multicomponent systems. The surfactant volume fraction Φ shows a jump $\Phi_3 - \Phi_\alpha$ when going from the L_α to the L_3 phase. But of course in the experimental situation on the phase border this cannot happen. The surfactant bilayer must vary continuously through the transition layer (if there is any) for energetical and topological reasons. Also, the solvent volume fraction has to vary continuously from its value Φ_α in the bulk of L_α to its value Φ_3 in the bulk of L_3.

The transition between the L_α and the L_3 phase can not only be induced by varying the surfactant/cosurfactant ratio (and with this the volume fraction Φ) [8,32,68], but also by varying the temperature [1,69,70] or changing the salt [23,71]. These transitions take place between equilibria of thermodynamically stable phases. But in the last 10 years, the possibility of transforming the sponge phase to L_α just by the act of shearing arose. This is not uncommon because it is one of the main features of the isotropic sponge phase that it become birefringent under flow. Many studies of this phenomenon have been completed.

The most interesting part of this phenomenon is that there can occur a change of the order of the transition. Detailed theories suggested the important contribution to the free energy from membrane fluctuation modes of the lamellar phase that causes the transition to be of first order, so that indeed a L_α/L_3 coexistence may be anticipated [53,72]. But submitting the system to a symmetry-breaking shear field suppresses a number of fluctuation modes associated with the degeneracy of the possible orientations of the ordered state [72]. Detailed experiments by Yamamoto and Tanaka [38] reveal that upon shearing the system, the coexistence region shrinks and eventually disappears altogether. This is a strong point for a shear-induced change in the order of the dynamical phase transition. There appears to be a continuous transition that occurs when the transition is induced by a strong shear.

The lamellar–sponge transition is driven by at least two major energetic factors. The first is the entropy-driven order-disorder phase transition [33] and the second is the Gaussian-curvature–driven topological transition [8,11,61,73]. Without considering the contribution of the Gaussian curvature, the transition is described purely by the fluctuation-induced first-order transition [63,74]. The elastic energy associated with the Gaussian curvature may modify the character of the transition since the sponge phase also has some order.

These results suggest that the observed transition in the studied systems is a second-order phase transition. This is in sharp contrast to previously studied L_α/L_3 phase transitions, which were always found to be first order and consequently had a two-phase region in between the separate single-phase regions L_α and L_3. Furthermore, it is surprising that the new phase has such a wide cosurfactant/surfactant region over which it is stable. In view of results on previously studied systems, it is furthermore surprising that a L_3 phase exists in this phase diagram at all, because the present micellar structures are charged and there is no excess salt to shield the charge density. The absolute conductivities in the L_1 phase indicate that the dissociation degree of the Ca^{2+} ions from the surfactants in the L_1 phase is about 5%, and that the charge density increases in the novel phase to about 10%. This increase is probably due to the decrease of the charge density of the surfactant aggregates by the incorporation of the cosurfactant into the surfactant assemblies.

6.3.2.1 Temperature Dependence of Conductivity

As mentioned above, the transition can also be observed at constant composition but varying temperature. The reason for this is a change of the bending constant caused by thermal undulations. Especially in the system $Ca(C_{14}\text{-}\alpha MES)_2/EHG/water$, where the transition seems to be a second-order transition, this effect should show a dramatic change of the physical properties. The transition could be observed by the change of conductivity. Again, only a birefringent phase or an isotropic phase was observed. The conductivities varied continuously from the low value of the L_α phase to the high value of the isotropic phase. The results are shown in Figure 6.14. The temperature was changed for one sample by small steps of 0.2 degrees, followed by a delay to allow equilibrium to be reached.

6.3.2.2 Small-Angle Neutron Scattering

As in the lamellar phase, two characteristic lengths can be deduced from neutron scattering data: the thickness ξ of the bilayer on a local

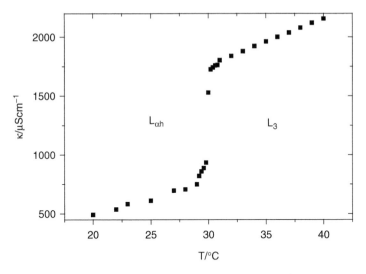

Figure 6.14 Plots of the conductivity κ against the temperature for a 50 mM Ca(C$_{14}$-αMES)$_2$ solution and 440 mM EHG.

scale and the average distance d between the bilayers characterizing the cell size of the solvent in the sponge. The scattering intensity can be divided into two major regimes: a large q-regime with a long decreasing tail at higher q and a small q regime where correlation between the bilayers leads to a broad peak. The large q regime obeys a $1/q^2$ behavior for $q < 2\pi/\zeta$ and a $1/q^4$ behavior at larger q. ζ is the bilayer thickness, which can be calculated in this way.

The position of the broad peak due to the bilayer–bilayer correlation also gives structural information. In lamellar phases, the peak due to the long-range order of the bilayers can be exactly located in q-space as $q = 2\pi/d$ with $d = \zeta/\Phi$. For a random distribution of connected bilayers as in sponge phases, $d = \alpha\zeta/\Phi$ with α larger than 1. The peak position for the sponge phase indeed shows a $1/\Phi$ dependence with $\alpha = 1,4$ to $1,6$.

In Figure 6.15 some SANS data are shown for some L$_\alpha$ and novel phases. For the L$_\alpha$ phase we see the typical correlation peak, which represents the interlamellar distance, and an indication of a shoulder at double the scattering vector of the main correlation peak. The novel phase shows a much broader peak which occurs, in contrast to results on normal L$_3$ phases, at the same position as for the L$_\alpha$ phase. Normally the L$_3$-peak is expected to shift by about 1/3 to a lower q-value.

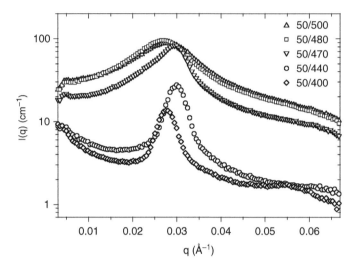

Figure 6.15 Radially averaged SANS scattering intensity I as a function of the scattering vector q. The composition of the samples is given in the inset, where the first number is the concentration of $Ca(C_{14}\text{-}\alpha MES)_2$ in millimolar and the second the concentration of EHG in millimolar.

6.3.2.3 NMR Characterization from an Ordered to a Disordered State

The shape of a NMR line is dominated by the interaction of an electric quadrupol moment of a given ^2H nucleus with its local electric field gradient originating in molecules from the charge distribution in the corresponding chemical bond. The resulting spectrum contains information on the distribution of the orientation of the electric field gradient tensor with respect to the external magnetic field [75]. Molecular dynamics has a strong impact on the ^2H NMR spectrum because the NMR frequency depends on orientation. In isotropic phases like the L_3, the fast molecular motions average the quadrupol interactions to zero, and a single sharp peak is observed in the spectrum. In liquid-crystalline phases the quadrupol interaction is not averaged to zero, and a residual contribution is retained. In a sufficiently fluid system, the director is oriented in the magnetic field, and a doublet is observed. The quadrupol splitting depends on the order parameter of the deuterated molecules and the molecular motion in the individual phase [76].

In the following ^{2}H NMR spectra, the transformation of the birefringent lamellar phase into the sponge phase was studied by recording the spectra of the sample at different temperatures.

Figure 6.16 shows the spectra at four different temperatures of the system 135 mM Ca(DS)$_2$/634 mM EHG/water. All ^{2}H NMR spectra were obtained by applying a solid echo pulse sequence with $\pi/2$ pulse length of 2,6 µsec and a 40 µsec interpulse delay in a magnetic field of 9,3 T.

At 275 K a clear doublet is obtained which indicates the ordered lamellar L$_{\alpha h}$ phase. At higher temperatures the transition starts, and by passing some transitional states at 283 and 285 K a sharp single peak for the isotropic sponge phase is received.

6.3.3   Electric Birefringence and Scaling Laws

Many properties of the L$_3$ phase can be described by power laws with specific exponents. These exponents can now be well described by a theoretical model [8]. On the theoretical side, it is well known that the sign and the size of the Gaussian bending constant plays a crucial role for the understanding of the L$_3$ phase. It is a phase of bilayer membranes with a mean curvature of zero and where the Gaussian modul is positive [77]. According to Porte, the transition from lamellar to L$_3$ occurs when the Gaussian bending modulus changes from negative to positive [78]. While these features are generally accepted, there are still some controversial details about some theoretical models [51,79,80].

If we look at the extension of the sponge phase from the water-rich corner of the phase diagram to the cosurfactant corner, which is classically the region of existence of the inverse micellar phase, the physical properties of the sponge phase must be discussed in terms of continuous structural transformation of the multiconnected bilayer, especially because the morphology transformations of the aggregates are not systematically accompanied by phase change. With the decrease of swelling, the volume fraction of the membrane becomes higher and simultaneously the cell size of the solvent decreases [32].

The swelling of phases obeys simple power laws. Huse and Leibler [47] derived the scaling law for the restricted case of swollen cubic phases, and later Porte et al. [7,8,32] generalized it based on a simple scaling argument without respect for short- or long-range order.

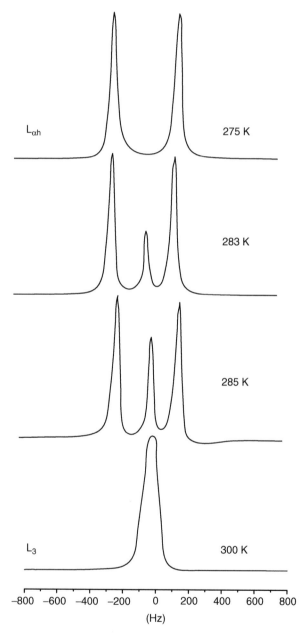

Figure 6.16 ^{2}H NMR spectrum of the system 135 mM Ca(DS)$_2$ + 634 mM EHG in D$_2$O at varying temperatures.

Therefore, this scaling law is applicable to all cases where infinite fluid membranes interact through self avoidance only.

Scaling laws are often the result of the invariance of some characteristic quantity with respect to a set of spatial transformations. In the case of fluid elastic membranes, the invariant quantity is the elastic Hamiltonian H from Equation (6.1), and the spatial transformation is the set of isotropic dilations, where isotropic means the change in scale in all three directions of space. This means for Equation (6.1) that a dilation of ratio λ will transform dA into λ^2 dA and each R_i to λR_i so that H remains identical. This relation could be used also in structures with no well-defined measurable characteristic distance.

Another important assumption has to be made. It is the absence of small thermal ripples on the membrane that can influence the phase behavior and complicates the calculation model. The isotropic dilated system corresponds with the nondilated system by the ratio λ. Both systems have the same configuration, the same elastic energy, and the same statistical weight and therefore bring the same contribution to the free energy of each system. This means that the free energy of the membranes is also scale invariant like the elastic energy. For infinite membranes (not for vesicles) the free energy [7] is therefore:

$$F = \mu_A A + B_A \frac{A^3}{V^2} \tag{6.4}$$

B_A is a function of $\kappa, \bar{\kappa}$, and T; μ_A is the standard chemical potential, and V is the total volume. The second part of the equation is extensive and scale invariant. A/V is proportional to Φ, and so this equation can be written as the free energy per unit volume of the system:

$$\frac{F}{V} = \mu_\Phi \Phi + B_\Phi \Phi^3 \tag{6.5}$$

The linear term does not affect the stability and physical properties of the phase, while the second part reflects the Φ^3-dependence scale invariance of the statistics of the membrane and is important for its properties. Before presenting an example of a scaling law derived from this relation, a closer look has to be taken at the assumption we made.

The main point is the invariance of the Hamiltonian through the dilation. Therefore, it is important that the membranes only interact through self avoidance. From the same assumption of scale invariance comes the d^{-2} dependence of the Helfrich steric interaction [54] in swollen lamellar phases.

Another important idealization is the assumption that the membrane just has curvature degrees of freedom. In reality, fluid membranes form spontaneously from surfactants when mixed with cosurfactants. In this case, the cosurfactants cause an additional degree of freedom through the local composition of the membrane, which can be coupled to its local curvature. The coupling contribution to the overall Hamiltonian of the system could influence its invariance in systems with cosurfactants, though this is not always the case.

In an electric birefringence experiment, the sample is subjected to an electric field E. In bilayers, a local anisotropy exists along their local normal n. These bilayers possess a tendency to orient with respect to the electric field. The initially isotropic sponge phase becomes anisotropic under the field and shows an optical birefringence Δn.

$$\Delta n = B_K E^2 \lambda_0 \qquad (6.6)$$

where λ_0 is the wavelength of the light, and the proportionality constant B_K is the so-called Kerr constant.

After the electrical field is switched off, the structure will progressively relax back to the isotropic initial state in an exponential way (Figure 6.17).

In this case we can write:

$$T\Delta S \sim \eta_0 V \left(\frac{\partial \beta}{\partial t} \right)^2 \qquad (6.7)$$

It could be assumed that all dissipation of energy is due to viscous flows of the solvent inside the cells and passages of the sponge. In Equation (6.7), η_0 is the viscosity of the solvent and V the volume segment $\partial \beta / \partial t$ is the fixed deformation rate, which is proportional to the velocity gradients that appear in the relaxation process. In a dilation process, the velocities of corresponding points are proportional to the dilation ratio, but their gradients remain invariant. So we can conclude that the entropy is independent of the dilution Φ. For short times, the relaxation of the system is adiabatic without work exchange. This means:

$$\Delta U = 0 = \Delta F + T\Delta S \qquad (6.8)$$

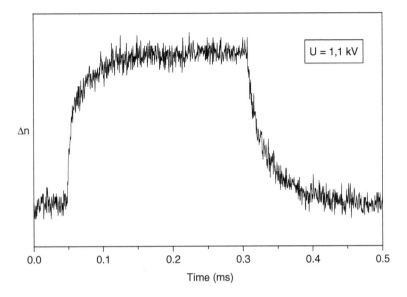

Figure 6.17 Representative electric birefringence signal. In the first part Δn rises to a constant level and decrease after switching off the electric field.

$T\Delta S$ is independent of Φ, and the free energy depends on the dilution with Φ^3 [7], so we determine immediately that the relaxation time τ_R must scale as:

$$\tau_R \sim \Phi^{-3} \tag{6.9}$$

Figure 6.17 shows the electric birefringence signal of a phase and Figure 6.18 the concentration dependence of the rotation time, which was determined from these signals. Both the absolute values of the time constant and their concentration dependences are the same as for previously studied L_3 phases. In particular the $\tau \sim \varnothing^{-3}$ dependence was found, which can be explained on the basis of a theoretical model [61].

6.3.4 Rheological Properties

As mentioned before, the dilution of an L_3 structure acts like a simple dilation with the conservation of the total membrane areas and a change of the characteristic structural length d of the L_3, which is proportional to Φ^{-1}. What does this mean for viscosity?

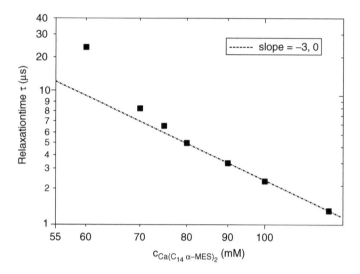

Figure 6.18 Relaxation time τ_R versus the concentration of $Ca(C_{18}\text{-}\alpha MES)_2$ for the L_3 phase; double logarithmic.

In L_3 phases the viscosity remains relatively low, i.e., 10 times that of water. It is remarkable to observe such a low viscosity for a surfactant system with relatively high concentrations, and it is also hard to imagine that a highly connected bicontinuous structure such as the sponge phase behaves as a Newtonian fluid. The Newtonian behavior means that viscosity follows Newton's law:

$$\tau = \eta\dot{\gamma} \tag{6.10}$$

where $\dot{\gamma}$ is the shear rate. This suggests that no structural relaxation time is relevant for the viscosity.

In Figure 6.19, the typical rheogram for the L_3 phase in $Ca(DS)_2$, EHG, and water is shown. The storage (G') and loss modulus (G'') together with the complex viscosity (η^*) as a function of frequency are given. The complex viscosity is constant for the whole frequency range. This is typical for a Newtonian fluid.

The viscosity η is a function of the volume fraction Φ of the membranes. For rising volume fractions, the viscosity increases in two different ways.

In dilute classical L_3 phases, there is an linear dependence between viscosity and volume fraction, which was investigated by Snabre and Porte [19]. The scaling law for this relation can be

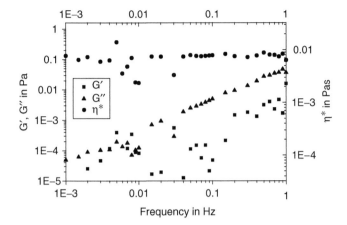

Figure 6.19 Storage modulus G′, loss modulus G″, and complex viscosity η^* as a function of frequency ω for a solution of 50 mM Ca(DS)₂ + 280 mM EHG in water.

deduced in the same manner as the one of the birefringence relaxation time and comes to the following result:

$$\eta(\Phi) = A\eta_0(1-\Phi) + B\eta_s\Phi \qquad (6.11)$$

where A and B are prefactors, which include the ratios between the velocity gradients of the solvent in the cells and passages of the sponge (and also of the surfactant in the membrane) and the shear rate. η_0 is the zero-viscosity of the solution and η_S is the viscosity of the solvent. For extrapolations to volume fractions $\Phi = 0$, the function extrapolated not to η_0 but to a viscosity that was confirmed by Snabre [19] and Gomati [82] with the value of $\eta(0) = 3\eta_0$, which means that $A = 3$. The viscosity value changes between 5 and 10 mPa.

In the concentrated classic L₃ phases, the viscosity can rise to 20 mPa (nevertheless very low in relation to the structurally related but high viscosity cubic phase). But it is more important that in these regions the viscosity increase with the volume fraction following an expotential law of the form:

$$\eta(\Phi) = Ce^{D_\Phi} \qquad (6.12)$$

where C and D are constants. This exponential increase of the L₃ phase viscosity is similar to that for microemulsions [83], which also indicates the structural similarity.

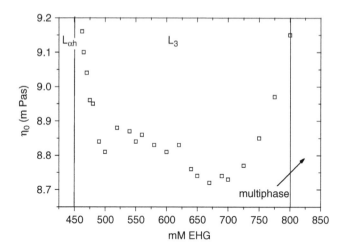

Figure 6.20 Zero-shear viscosity η_0 against the EHG concentration in the L_3 phase. The $Ca(C_{14}\text{-}\alpha MES)_2$ concentration was 50 mM.

In Figure 6.20, the viscosities of the L_3 phases are shown as a function of the composition of the phases. The viscosities are remarkably constant over the whole composition region, but exact measurements show subtle but characteristic variations. This result is again very surprising. It means that while the novel observed phase is indeed an L_3 phase, its dynamic behavior remains about constant even when the cosurfactant/surfactant ratio is varied by a factor of two. Normally one would expect that the flexibility and bending constant of the phase decrease with increasing cosurfactant/ surfactant ratio. It is conceivable that this effect is counterbalanced by the increased degree of dissociation of the phase, which should lead to a stiffening of the bilayers.

6.4 OUTLOOK AND CONCLUSION

The L_3 phase with its fascinating structure shows extraordinary physical properties, which have been studied in recent years. The low viscosity of the isotropic surfactant solution in spite of the relatively high concentrations arouses interest for technical applications. As yet the adaptability of these systems is limited by the narrow temperature range; respectively, the narrow range of cosurfactant/surfactant ratio in which the sponge phases exist. Nevertheless the whole phase exists over a wide surfactant concentration area.

By introducing branched cosurfactants, it is possible to expand the existing range of the L_3 phase without losing their major benefits. This novel sponge has been found to exist in ionic Ca salts of surfactants and features all the properties that are typical for a classic sponge phase, such as the scaling laws in electric birefringence, the flow birefringence, the low viscosity, the neutron scattering and NMR behavior, and the missing order, which is normal for an isotropic fluid. In face of these similarities, there are some striking differences besides the wide cosurfactant/surfactant existing range. For example, the conductivity is much higher than in ordinary L_3 phases. This could be related to the different structure seen in the FF-TEM micrographs.

It is also noteworthy that the branched cosurfactant EHG sometimes even requires ionic charges in the surfactant solution for the formation of all the expected phases. This can be achieved by ionic surfactants or alternately by charging the nonionic surfactant with ionic surfactants or excess salt.

In the future, attention should certainly be drawn to the thermodynamics of this novel sponge phase, especially to its existence only in ionic charged systems and the fact that it is induced only by branched cosurfactants, which do not show any interesting phase behavior in uncharged surfactant systems. These two points are the most remarkable facts that should be investigated in view of theoretical models. It is obvious that both effects are crucial for this novel phase, and it is also plausible that each effect influences the other.

The thermodynamics that cause the seemingly second-order phase transition from $L_{\alpha h}$ to L_3 in the system 50 mM Ca(C$_{14}$-αMES)$_2$/ EHG in water are also of special interest. Neither the physical reasons for nor the mechanisms of the phase transition were well known until now. There are some indications of the possibility of a second-order phase transition between lamellar and sponge phases, but nevertheless this behavior is still quite ambiguous and controversial.

As mentioned before, some physical properties of the novel phase are rather different from those of the classical sponge phase. The possible use of this special behavior for technical applications is of interest. In particular, the wide existing range of this phase combined with such advantages of the L_3 phase as low viscosity, isotropy, or high conductivity have several possibilities for applications.

The investigation of the transition from the classical sponge to the novel sponge is another major challenge that should be helpful for the understanding of these phases. In this connection, mixtures between the two different L_3 phases should be studied. These mixtures are composed of course of two totally different surfactant systems, One with an ionic surfactant and a branched cosurfactant and the other made of an nonionic (or zwitterionic) or charged shielded (with brine) ionic surfactant and for example a n-alkyl alcohol as cosurfactant. These multicomponent systems are very complicated, and finding theoretical models for their behavior seems to be more difficult than the examination of the macroscopic properties.

REFERENCES

1. DJ Mitchell, GJD Tiddy, L Waring, T Bostock, MP McDonald. *J Chem Soc Faraday Trans I* 79:975, 1983.

2. G Gompper, M Schick. *Self-Assembling Amphiphilic Systems*, Academic Press, San Diego, 1994.

3. D Gazeau, AM Bellocq, D Roux, T Zemb. *Europhys Lett* 9:447, 1989.

4. R Strey, J Winkler, L Magid. *J Phys Chem* 95:7502, 1991.

5. A Maldonado, W Urbach, R Ober, D Langevin. *Phys Rev E* 54:1774, 1996.

6. N Lei, CR Safinya, D Roux, KS Liang. *Phys Rev E* 56:608, 1997.

7. G Porte, J Appell, P Bassereau, J Marignan, M Skouri, I Billard, M Delsanti. *Physica A* 176:168, 1991.

8. G Porte, J Appell, P Bassereau, J Marignan. *J Phys France* 50:1335, 1989.

9. R Strey, R Schomäcker, D Roux, F Nallet, U Olsson. *J Chem Soc Faraday Trans* 86:2253, 1990.

10. D Roux, ME Cates, U Olsson, RC Ball, F Nallet, AM Bellocq. *Europhys Lett* 11:229, 1990.

11. M Skouri, J Marignan, J Appell, G Porte. *J Phys II* 1:1121, 1991.

12. R Strey, R Schomäcker, D Roux, F Nallet, U Olsson. *J Comput Phys* 86:2253, 1990.

13. C Vinches, C Coulon, D Roux. *J Phys II* 4:1165, 1994.

14. B Balinov, U Olsson, O Södermann. *J Phys Chem* 95:5931, 1991.

15. K Fukuda, U Olsson, U Würz. *Langmuir* 10:3222, 1994.

16. R Strey, W Jahn, G Porte, P Bassereau. *Langmuir* 6:1635, 1990.

17. H Hoffmann, C Thunig, U Munkert, HW Meyer, W Richter. *Langmuir* 8:2629, 1992.

18. DA Antelmi, P Kelicheff, P Richetti. *J Phys II* 5:103, 1995.

19. P Snabre, G Porte. *Europhys Lett* 13:641, 1990.

20. H Pleiner, HR Brand. *Europhys Lett* 15:393, 1991.

21. G Lindblom, L Rilfors. *Biochim Biophys Acta* 988:221, 1988.

22. K Alfons, S Engstrom. *J Pharm Sci* 87:1527, 1998.

23. CD Adam, JA Durrant, MR Lowry, GJT Tiddy. *J Chem Soc Faraday Trans I* 80:789, 1984.

24. J Lang, RD Morgan. *J Chem Phys* 73:5849, 1980.

25. M Jonströmer, R Strey. *J Phys Chem* 96:5993, 1992.

26. K. Fontell, in *Colloidal Dispersions and Micellar Behavior,* ACS Symposium Series 9, American Chemical Society, Washington, DC, 1975, p. 270.

27. CA Miller, M Gradzielski, H Hoffmann, U Krämer, C Thunig. *Colloid Polym Sci* 268:1066, 1990.

28. M Bergmeier, H Hoffmann, F Witte, S Zourab. *J Coll Interface Sci* 203:1, 1998.

29. E Hecht, K Mortensen, H Hoffmann. *Macromolecules* 28:5465, 1995.

30. CA Miller, M Gradzielski, H Hoffmann, U Krämer, C Thunig. *Progr Colloid Polym Sci* 48:243, 1991.

31. C Coulon, D Roux, AM Bellocq. *Phys Rev Lett* 66:1709, 1991.

32. G Porte, P Bassereau, J Marignan, J May. *J Phys France* 49:511, 1988.

33. M Cates, D Roux, D Andelman, SC Milner, S Safran. *Europhys Lett* 5:733, 1988 [Erratum: 7:94, 1988].

34. J Marignan, F Gauthier-Fournier, J Appell, F Akoum, J Lang. *J Phys Chem* 92:440, 1988.

35. R Gomati, J Appell, P Bassereau, J Marignan, G Porte. *J Phys Chem* 91:6203, 1987.

36. H Hoffmann, S Hofmann, A Rauscher, J Kalus. *Progr Colloid Polym Sci* 84:24, 1991.

37. HF Mahjoub, KM McGrath, M Kléman. *Langmuir* 12:3131, 1996.

38. J Yamamoto, H Tanaka. *Phys Rev Lett* 77:4390, 1996.

39. C Quilliet, C Blanc, M Kleman. *Phys Rev Lett* 77:522, 1996.

40. C Blanc, O Sanseau, V Cabuil. *Mol Cryst Liq Cryst* 332:523, 1999.

41. OD Lavrentovich, C Quilliet, M Kleman. *J Phys Chem* 101:420, 1997.

42. U Hornfeck, M Gradzielski, K Mortensen, C Thunig, G Platz. *Langmuir* 14:2958, 1998.

43. A Zapf, U Hornfeck, G Platz, H Hoffmann. *Langmuir* 17:6113, 2001.

44. R Beck, H Hoffmann. *Phys Chem Chem Phys* 3:5438, 2001.

45. R Beck, Y Abe, T Terabayashi, H Hoffmann. *J Phys Chem B* 106:3335, 2002.

46. R Gomati, M Daoud, A Gharbi. *Physica B* 239:405, 1997.

47. DA Huse, S Leibler. *Phys Rev Lett* 66:437, 1997.

48. DH Huse, S Leibler. *J Phys France* 49:605, 1988.

49. RG Laughlin. In *Advances in Liquid Crystals*. GH Brown, Ed. Academic Press, New York. 1978, pp. 44 and 99.

50. PG Nilsonn, B Lindman. *J Phys Chem* 88:4764, 1984.

51. D Roux. *Physica A* 213:168, 1995.

52. W Helfrich. *J Phys* 46:1263, 1985.

53. L Peliti, S Leibler. *Phys Rev Lett* 54:1690, 1986.

54. W Helfrich. *Z Naturforschung* 33a:305, 1978.

55. F Larche, J Appell, G Porte, P Bassereau, J Marignan. *Phys Rev Lett* 56:1700, 1986.

56. P Bassereau, J Marignan, G Porte. *J Physique* 48:673, 1987.

57. D Roux, C Coulon, M Cates. *J Phys Chem* 96:4174, 1992.

58. L Golubovich, T Lubensky. *Europhys Lett* 10:513, 1987.

59. G Gompper, M Schick. *Phys Rev Lett* 65:116, 1990.

60. C Vinches, C Coulon, D Roux. *J Phys France* 2:453, 1992.

61. G Porte, M Delsanti, I Billard, M Skouri, J Appell, J Marignan, F Debeauvais, *J Phys II France* 1:1101, 1991.

62. SC Milner, M Cates, D Roux. *J Phys France* 51:2629, 1990.

63. S Brazovskii. *Sov Phys JETP* 41:85, 1975.

64. S Marcelja, N Radic. *Chem Phys Lett* 42:129, 1976.

65. PG de Gennes. *Langmuir* 6:1448, 1990.

66. A Poniewierski, T Slucklin. *J Liq Cryst* 2:281, 1987.

67. DA Antelmi, P Kelicheff, P Richetti. *Langmuir* 15:7774, 1999.

68. R Gomati, J Appell, G Porte, P Bassereau, J Marignan. *J Phys Chem* 91:6203, 1987.

69. C Quilliet, M Kleman, M Benillouche, F Kalb. *C R Acad Sci Ser II* 319:1469, 1994.

70. Y Nastichin, E Lambert, P Boltenhagen. *C R Acad Sci Ser II* 321:205, 1995.

71. CA Miller, O Gosh. *Langmuir* 2:312, 1986.

72. R Bruinsma, Y Rabin. *Phys Rev A* 45:994, 1992.

73. DC Morse. *Phys Rev E* 50:R2423, 1994.

74. M Cates, SC Milner. *Phys Rev Lett* 62:1856, 1989.

75. G Briganti, AL Segre, D Capitani, C Casieri, C La Mesa. *J Phys Chem* 103:825, 1999.

76. D Capitani, C Casieri, G Briganti, C La Mesa, AL Segre. *J Phys Chem* 103:6088, 1999.

77. DA Anderson, HT Davis, LE Scriven, *J. Chem Phys* 91:3246, 1989.

78. G Porte, J Appell, J Marignan, M Skouri. *J Phys II France* 1:1121, 1991.

79. J Daicic, U Olsson, H Wennerström, G Jerke, P Schurtenberger, *J Phys II France* 5:199, 1995.

80. L Golubovic. *Phys Rev E* 50:R2419, 1994.

81. R Lipowski, S Leibler. *Phys Rev Lett* 54:1690, 1986.

82. R Gomati, N Bouguerra, A Gharbi. *Physica B* 322:262, 2002.

83. EG Richardson. *Kolloid-Z* 65:32, 1932.

7

Bicontinuous Cubic Phases of Lipids with Entrapped Proteins: Structural Features and Bioanalytical Applications

VALDEMARAS RAZUMAS

CONTENTS

7.1   INTRODUCTION

Lipid/protein interactions and the diversified aggregates of these compounds play significant parts in many biological systems and biochemical processes. The digestion of fats, transport of lipids, and, of course, biomembranes should be considered as the prominent examples of the above statement.

In biomembranes, lipids and proteins are the dominant constituents. On the average, proteins account for about half of the membrane mass in eukaryotic cells. In specific cases, as, for instance, in the inner mitochondrial membrane, the protein-to-lipid mass ratio may run as high as 3.2 [1]. Membrane proteins perform important biological functions; they catalyze chemical reactions, mediate the flow of nutrients and wastes, transmit the extracellular signals to the inside of the cell, etc. Thus, not surprisingly, the efforts to explore the unique features of biomembranes have stimulated a considerable body of work performed on biomimetic lipid/protein systems. In this respect, there is no doubt that the reversed bicontinuous cubic phases of lipids are among the most elegant models of the lipid bilayer. Moreover, as pointed out in several studies (see, e.g., Refs. 2–5), a number of cubic mesophases have been observed in conditions close to those prevailing in living organisms. For this reason, these three-dimensional structures seem to be involved in different biological processes (e.g., fat digestion [5,6], membrane fusion [7]), and might occur in cellular membranes (formation of so-called cubic membranes; for review, see Ref. 8). The possible biological implication of the bicontinuous cubic phases is an added reason for utilizing these meso-phases with entrapped proteins as the bilayer-based biomembrane models. Besides, as illustrated in other chapters of this book, as well as in the last section of this chapter, the protein (enzyme)-containing cubic phases are of interest in many practical applications.

Three main topics are reviewed in this chapter:

1. Structural features of the protein-containing bicontinuous cubic phases of lipids. The focus is on the reversed (Q_{II}) phases—their stabilities and crystallographically determined parameters.
2. Molecular features of the bicontinuous lipid/protein/water Q_{II} phases as revealed by spectroscopic methods.
3. Electrochemical and bioanalytical applications of the bicontinuous Q_{II} phases with entrapped proteins.

Other related subjects, such as protein crystallization within the lipidic cubic phases, protein delivery systems based on the cubic particles, and protein- and polypeptide-induced cubic structures during membrane fusion, are presented in the corresponding chapters of this book.

7.2 STRUCTURAL FEATURES OF THE PROTEIN-CONTAINING BICONTINUOUS CUBIC PHASES OF LIPIDS

To the best of the author's knowledge, the first observation of the protein-containing cubic phase was made in 1969 by Gulik-Krzywicki and co-workers [9] in studies of the aqueous mixture of phosphatidylinositol and lysozyme (most likely from hen egg white). In addition to the lamellar and hexagonal phases, the authors monitored a few samples with three-dimensional periodicity in this ternary system. At the time, however, the x-ray reflections of these phases had not been interpreted with confidence, and the types (direct or reversed) and classes (bicontinuous or micellar) of those lysozyme-containing three-dimensional structures remain unknown. It was not until about 13 years later that the studies of lipidic cubic phases with entrapped proteins were resumed. In 1982, Larsson and Lindblom [10] reported the entrapment of a very hydrophobic wheat protein fraction, A-gliadin, into the bicontinuous Q_{II} phase, presumably corresponding to the body-centered lattice, formed by aqueous 1-monooleoylglycerol (monoolein, MO). Subsequently, the possibility of obtaining the bicontinuous Q_{II} phases of lipids with entrapped proteins was unambiguously proved for several proteins; these aqueous lipid/protein systems are listed in Table 7.1.

In parallel with the mixtures indicated in Table 7.1, the capacity to form cubic phases has been ascribed to a number of other lipid/protein compositions: MO/α-lactalbumin, soybean trypsin inhibitor, myoglobin, pepsin, or conalbumin [11,12]; 1-monopalmitoleylglycerol

Table 7.1 Aqueous Lipid/Protein Systems Forming Reversed Bicontinuous Cubic Phases

System Components	Cubic Phase	Space Group	Ref.
MO[a]/wheat A-gliadin	Q^{230} or Q^{229}	*Ia3d* or *Im3m*	10
MO/lysozyme (hen egg	Q^{229}	*Im3m*	11–13[c]
white)[b]	Q^{224}	*Pn3m*	12
MO/bovine serum	Q^{230}	*Ia3d*	12
albumin			
MO + alginate[d]/BSA-	Q^{230}	*Ia3d*	14
FITC[e]			
MO/bovine milk casein	Q^{230}	*Ia3d*	15
MO/horse heart cytochrome c	Q^{229}	*Im3m*	2
	Q^{224}	*Pn3m*	2, 16
MO/bacteriorhodopsin	Q^{230}	*Ia3d*	17, 18
(*Halobacterium halobium*)	Q^{224}	*Pn3m*	
MO + PC[f]/glucose oxidase	Q^{230}	*Ia3d*	19
(*Aspergillus niger*)			
MO/bovine hemoglobin	Q^{230}	*Ia3d*	20
	Q^{224}	*Pn3m*	
MO/glucose oxidase	Q^{230}	*Ia3d*	21
(*Aspergillus niger*)	Q^{224}	*Pn3m*	
MO/MP-11[g] (from horse heart	Q^{230}	*Ia3d*	22
cytochrome c)	Q^{224}	*Pn3m*	
MO/cholesterol oxidase	Q^{224}	*Pn3m*	23
(*Rhodococcus*)			

[a] 1-Monooleoylglycerol. In Refs. 13, 16, 19, 21, and 22, the monooleoylglycerol-based commercial preparations of mono- and diglycerides.

[b] In Refs. 11 and 12, the source of the enzyme was not indicated.

[c] In Refs. 11 and 12, the phase was improperly indexed as the Q^{230} phase of space group type *Ia3d*.

[d] Sodium salt of alginic acid.

[e] Bovine serum albumin tagged with fluorescein isothiocyanate.

[f] Phosphatidylcholine preparation from soya beans containing 97.5% phosphatidylcholine.

[g] Heme undecapeptide, microperoxidase-11, which contains residues 11 to 21 of cytochrome c and exhibits the peroxidase-type enzymatic activity.

or palmitoyl-lysophosphatidylcholine/bacteriorhodopsin [17]; MO/ceruloplasmin [19]; MO or 1-palmitoyl-2-hydroxy-*sn*-glycero-3-phosphocholine (PLPC)/α-chymotrypsin [24]; PLPC/bacteriorhodopsin [24,25] or melittin [25]; MO/lactate oxidase, urease, or creatinine deiminase [26]; distearoylphosphatidylglycerol + MO/cytochrome c [27]; (MO + 1-palmitoyl-2-oleoyl-3-phosphatidylserine) or (dielaidoylphosphatidylethanolamine + peptide antibiotic alamethicin)/protein kinase C [28]; MO/insulin [29]. However, in all the above instances, the authors did not provide compelling experimental evidence (e.g., x-ray diffraction data) for the types and classes of the protein-containing phases. As a rule, it has been expected that the intrinsic structural features of lipid phases would be unaffected by the entrapped proteins. As shown below, this is not always the case.

We now turn to the systems summarized in Table 7.1. On the one hand, it is evident from Table 7.1 that quite large globular proteins (the molecular mass of the largest protein, glucose oxidase from *Aspergillus niger*, is ca. 160 kDa) can be accommodated in the bicontinuous Q_{II} phases of lipids; on the other hand, all three reversed bicontinuous cubic structures, Q^{230}, Q^{229}, and Q^{224}, are found in the lipid/protein/water systems. It can also be seen from Table 7.1 that 1-monooleoylglycerol, which is capable of forming the Q^{230} and Q^{224} phases *per se* [30], enters into the composition of all systems. Nevertheless, among them, the aqueous mixture of MO and lysozyme (an enzyme that destroys bacterial cell walls; molecular mass, 14.6 kDa [31]) deserves particular attention since, as distinct from other systems, it has been characterized by a ternary lipid/protein/water phase diagram [11].

7.2.1   MO/Lysozyme/Water System

Figure 7.1 represents the MO/lysozyme/water phase diagram at 40°C. The unexamined region in Figure 7.1 is determined by the high viscosity of an aqueous solution containing above 45 wt% of protein, as well as by extremely slow phase equilibration at the MO concentrations above 80 wt%. The diagram in Figure 7.1 shows that the ternary cubic phase region extends up to 34 wt% of lysozyme. On the other hand, when compared to the binary MO/water system [30], the protein-containing cubic phase swells to higher water content (up to ca. 52 wt% of water). In the cubic phase region, the Q^{229} and Q^{224} phases were identified. (In Refs. 11 and 12, the Q^{229} phase of space group type *Im3m* was improperly indexed as the Q^{230} phase of space group type *Ia3d*; the x-ray diffraction measurements were revised in [15]). For example, the MO/lysozyme/water sample of

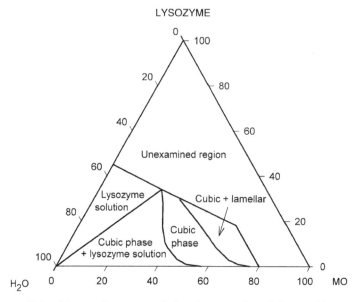

Figure 7.1 Phase diagram of the 1-monooleoylglycerol/lysozyme/ water system at 40°C. (From B Ericsson, K Larsson, K Fontell. *Biochim Biophys Acta* 729: 23–27, 1983.)

42.6:11.5:45.9 wt% composition forms the Q^{229} phase characterized by the unit cell axis $a_o = 130.9$ Å (ca. 10 protein molecules per unit cell), whereas the mixture of 31.6:34.2:34.2 wt% composition leads to the Q^{224} phase of $a_o = 166.1$ Å (ca. 64 protein molecules per unit cell) [12,15]. Since the lipid-to-water ratios in these samples are similar (ca. 0.91), it might be concluded that the increased amount of protein determines the Q^{229} Q^{224} phase transition. Indeed, the structural features of the latter phase make possible the entrapment of larger protein amounts.

Thus, when the centers of the lipid bilayer in the Q^{229} and Q^{224} phases are modeled by the "Schwarz's primitive"– (P) and "diamond"-type (D) infinite periodic minimal surfaces (IPMS), respectively, also assuming that all other molecularly defined surfaces lie parallel to IPMS, a quite simple relationship exists between a_o and the surface averaged radius of water channel, $\langle R_w \rangle$, [32,33]:

$$\langle R_w \rangle = (-A_o/2\pi\chi)^{1/2}a_o - ls \qquad (7.1)$$

where A_o is the dimensionless surface area of the minimal surface in the unit cell of unity lattice parameter (2.3451 and 1.9189 for the Q^{229} and Q^{224} phases, respectively), χ is the Euler characteristic per unit cell (–4 and –2 for the Q^{229} and Q^{224} phases, respectively), and l is the lipid monolayer thickness. For the pure cubic phases of 1-monooleoylglycerol, the values of l can be approximated by the following relation [33]:

$$l = 18 \exp (-0.0019 \, T) \qquad (7.2)$$

where T and l have measurement units of °C and Å, respectively.

Thus, at T = 40°C, l equals to ca. 17 Å, and, according to Equation (7.1), the values of $\langle R_w \rangle$ for the above Q^{229} and Q^{224} phases of the aqueous MO/lysozyme system should be close to 23 and 48 Å, respectively. Considering the dimensions of the hen egg white lysozyme ($30 \times 30 \times 45$ Å [31]), it is evident that approximately two layers of the native protein molecules fit into the water channel of the Q^{224} phase. By contrast, each aqueous labyrinth of the protein-containing Q^{229} phase is capable of accommodating no more than one monomolecular "layer" of protein.

Here, it should be noted that the formation of the Q^{229} phase in the aqueous MO/lysozyme system has been also demonstrated at 25°C [13]. In the latter study, an increase of the lattice constant a_o from 123.4 (± 0.4) to 128.3 (± 1.0) and 130.1 (± 0.9) Å was observed when the protein content in the mixture was increased respectively from 5 to 7 and 8 wt%, yet maintaining the lipid-to-water weight ratio constant and equal to ca. 1.4. Thus, it seems that the protein content is also capable of tuning to some extent the dimensions of the phase.

7.2.2   MO/Cytochrome c/Water System

Subsequently to the description of the MO/lysozyme system, Mariani et al. [2] published a partial phase diagram of the MO/cytochrome c/water system. The diagram was constructed at the constant protein-to-lipid weight ratio of 0.083. As is evident from Figure 7.2, in parallel with the Q^{229} and Q^{224} phases, over the temperature range from 10 to 35°C and at lower water content, this ternary mixture also forms the reversed Q^{212} phase of space group type $P4_332$. It has been proposed that the latter phase is composed of a three-dimensional network of rods and a set of identical micelles, each of which entraps one molecule of cytochrome c (molecular mass, 12.4 kDa; prolated spheroid, $30 \times 34 \times 34$ Å [34]). Figure 7.2 shows that the Q^{224}-phase region extends over the water content range from 23 to 44 wt% and up to ca. 16°C. We also determined the formation of this cubic phase at 22°C for the aqueous MO/cytochrome c mixtures, which contained 0.3 to 5 wt% of protein and 37.7 to 41 wt% of water [16]. Further, the phase diagram in Figure 7.2 indicates the Q^{224} to Q^{229} phase transition upon increasing the water content and/or temperature.

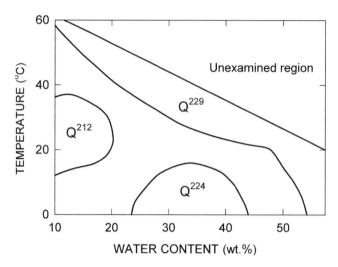

Figure 7.2 Partial phase diagram of the monoolein/cytochrome c/water system. The protein-to-lipid weight ratio equals 0.083. (From P Mariani, V Luzzati, H Delacroix. *J Mol Biol* 204: 165–189, 1988.)

At low water content, the ternary Q^{229} phase samples are stable only at significantly elevated temperatures, whereas the highly hydrated Q^{229} phase exists at considerably lower temperatures.

Compared to the studies of the MO/lysozyme/water system, the investigations of cytochrome c within the bicontinuous Q_{II} phases of MO [2,16] are characterized by considerably lower amount of entrapped protein (no more than four cytochrome molecules per unit cell). Thus, at 10°C, the Q^{224} phase of the MO/cytochrome c mixture containing 32 wt% of water exhibits a_o = 90.4 Å, indicating only ca. two protein molecules per unit cell [2]. Another discrepancy of greater concern is that Mariani and coworkers observed the Q^{224} to Q^{229} phase transition as the content of entrapped cytochrome c increases [2], i.e., the opposite effect when compared to the MO/ lysozyme/water system. In this respect, the following study of hemoglobin encapsulation in the bicontinuous Q_{II} phases of aqueous MO leads to even deeper confusion.

7.2.3 MO/Hemoglobin/Water System

Using the x-ray diffraction measurements, Leslie and coworkers [20] investigated the phase behavior of aqueous MO in the presence of a tetrameric protein, bovine hemoglobin (molecular mass, 64.5 kDa;

dimensions, $64 \times 55 \times 50$ Å [35]). The authors determined the entire phase diagram of the MO/hemoglobin/water system; however, it was not provided in the paper. The formation of the ternary bicontinuous Q_{II} phases was investigated at hemoglobin contents between 1 and 10 wt%; in so doing, the MO concentration of 50 wt% was kept constant, whereas the aqueous solution was phosphate-buffered saline (exact concentrations of the components unspecified; pH most likely equals 7.4 at 25°C).

As can be seen from Figure 7.3, the pure ternary Q^{224} phase forms at hemoglobin contents below 2.5 wt%, the Q^{224} and Q^{230} phases coexist at 2.5 to 5 wt% of the protein, and the pure ternary Q^{230} phase is observed over the protein content range of 5 to 10 wt%. Thus, one can envision that the entrapment of hemoglobin, or in other words, an increase in the protein concentration, induces the $Q^{224} \rightarrow Q^{230}$ phase transition. Figure 7.3 also shows dependence between the phase unit cell axis and protein content in the phase. It is evident from the lower plot in the figure that the a_o value of the Q^{224} phase initially

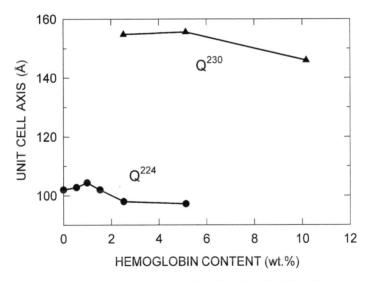

Figure 7.3 Dependence of the unit cell axis of cubic phases on the content of entrapped bovine hemoglobin in the MO/PBS samples containing 50 wt% of MO at 25°C. PBS, phosphate buffered saline from Sigma (exact concentrations of the components unspecified; pH most likely equals 7.4 at 25°C). (Adapted from SB Leslie, S Puvvada, BR Ratna, AS Rudolph. *Biochim Biophys Acta* 1285: 246–254, 1996.)

slightly increases with increasing protein concentration. However, upon reaching the maximum value of ca. 104 Å at the protein content of 1 wt%, the unit cell axis decreases with further increase of the entrapped protein content. In the two-phase coexistence region, as was to be expected, the lattice parameter of each phase remains practically constant and insensitive to the sample composition. Further, the a_o value of the pure ternary Q^{230} phase decreases with increasing the protein content up to 10 wt% (a_o equals ca. 146 Å, indicating about three protein molecules per unit cell). Here, it will be recalled that the protein content in the samples has been increased at the expense of water. Thus, since the structural parameter of all pure phases decreases with sample dehydration, the results of Figure 7.3 might be caused by the interaction of hemoglobin with the phase matrix as well as by the hydration degree of MO. We will return to this problem in Section 7.3.

In Section 7.2, it remains to discuss the MO/glucose oxidase/water system, a partial phase diagram of which was published by Barauskas et al. [21].

7.2.4 MO/Glucose Oxidase/Water System

As already noted above, glucose oxidase (GOD) from *Aspergillus niger* is the largest enzyme entrapped in the bicontinuous Q_{II} phases of lipids. This enzyme is a homodimeric glycoprotein of molecular mass ca. 160 kDa (583 amino acid residues, up to about 20% mannose-type carbohydrate, and two tightly bound flavin adenine dinucleotide cofactors) with approximate dimensions $60 \times 52 \times 77$ Å [36].

Figure 7.4 presents a partial phase diagram of the pseudoternary MO/GOD/water system at 25°C. The mixture was called "pseudoternary" since a monooleoylglycerol-based commercial preparation of mono- and diglycerides was used instead of pure 1-monooleoylglycerol. Besides, in the study under discussion, from the practical point of view (see the last section of this article) the typical commercial preparation of GOD was also applied without additional purification.

As seen from Figure 7.4, when compared to the aqueous 1-monooleoylglycerol system [30,33,37], the pseudobinary MO/water mixture displays a similar sequence of the bicontinuous Q_{II} phases. However, the phase boundaries are shifted towards lower hydration because of admixed diglycerides and polyunsaturated monoglycerides in the MO preparation. Thus, the pseudobinary Q^{230} phase exists up to 30 wt% of water, whereas the Q^{224} phase is observed over the water content range from 30 to 37 wt%. At higher hydration, the phase of space group type *Pn3m* coexists in equilibrium with excess water.

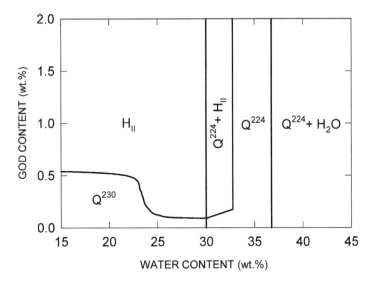

Figure 7.4 Partial phase diagram of the pseudoternary MO/glucose oxidase (GOD)/water system at 25°C. (Adapted from J Barauskas, V Razumas, T Nylander. *Prog Colloid Polym Sci* 116: 16–20, 2000.)

However, even small amounts of GOD preparation induce drastic changes in the phase behavior of MO. The phase diagram also shows that the cubic phases exhibit different sensitivity to the addition of enzyme preparation. At low hydration, the Q^{230} phase accommodates up to 0.5 wt% of enzyme, whereas the swollen phase accepts only up to 0.1 wt% of GOD preparation. At higher enzyme content, the cubic phase of space group type *Ia3d* transforms to the reversed hexagonal phase, H_{II}. The Q^{224} phase of MO is much less sensitive to the entrapment of GOD preparation. At a water content higher than 34 wt%, this phase accommodates up to 6 wt% of enzyme (not shown in Figure 7.4; see Ref. 21). Interestingly, these pseudo-ternary cubic phases also exhibit different thermotropic behavior. Thus, the onset of the H_{II} phase formation in the Q^{230} and Q^{224} phases of the pseudo-binary MO/water system, containing 20 wt% of water, was observed at ca. 55 and 48°C, respectively. In the case of the pseudoternary Q^{230} phase, even 0.4 wt% of GOD reduces the $Q^{230} \rightarrow H_{II}$ phase transition temperature by almost 25°C. In contrast, the entrapment of up to 1.5 wt% of GOD in the Q^{224} phase leaves the phase transition temperature practically unaltered.

The x-ray diffraction data on the above cubic MO/GOD/water phases have indicated that the entrapped enzyme preparation has no effect on the structural parameters of the phases. Considering the dimensions of GOD, this is a rather unexpected result. Moreover, analysis of the experimental unit cell dimensions of the phases rules out the possibility that the native enzyme dimers of about 40 Å hydrodynamic radius [36] can be positioned exclusively in the water channels of the cubic phases. Thus, the Q^{230} phase, containing respectively 0.5 and 20 wt% of GOD and water, is characterized by $a_o = 114.9$ Å, whereas for the sample containing 0.1 and 26 wt% of the enzyme preparation and water, the value of a_o is 127.4 Å (J Barauskas, personal communication, 2000). These two values indicate that, on the one hand, the two samples specified above accommodate a maximum of one molecule of GOD per about 35 and 128 unit cells, respectively; on the other hand, in these cubic phases, the calculated $\langle R_w \rangle$ values (Equation (7.1); $l = 16$ Å [21], $A_o = 3.0910$ and $\chi = -8$ [32,33]) are equal respectively to 12.5 and 15.6 Å. In the case of the MO/GOD/water Q^{224} phase of 56:6:38 wt% composition, the a_o value of 88.1 Å (J Barauskas, personal communication, 2000) indicates the entrapment of maximum one GOD molecule per ca. six unit cells. Here too, the water channel of calculated $\langle R_w \rangle = 18.4$ Å cannot accommodate the native dimer of GOD. However, Li and coworkers [38] have demonstrated that glucose oxidase is capable of penetrating into the monolayers of neutral glycolipids; therefore, it might be hypothesized that a similar molecular event takes place in the GOD-containing cubic phases of aqueous MO.

Still, considering the above structural analysis of the MO/GOD/water cubic phases, grounds exist to believe that not only do the protein molecules determine the phase behavior of this system, but that other components of the commercial GOD preparation (e.g., traces of amylase, maltase, glycogenase, invertase, galactose oxidase, balance phosphate buffer, and sodium chloride) also exert some influence on the phase stability. Nevertheless, as will be demonstrated in Section 7.4 of this chapter, the phase diagram shown in Figure 7.4 is essential to the development of glucose-sensitive bioanalytical devices.

7.2.5 Concluding Remarks

The above MO/GOD/water system is the last of four aqueous lipid/protein systems forming bicontinuous Q_{II} phases and characterized in greater detail by phase diagrams. Clearly, four studies are inadequate to give a full understanding of the physical principles that control protein-induced phase transitions, effects of these biopolymers on the structural parameters of the bicontinuous Q_{II} phases, and other features of such complex mesomorphic structures. As already mentioned above, even opposite-phase transformations might be observed under changes of the protein content in the systems containing different proteins. Thus, it is evident that studies at a molecular level, revealing the inter- and intramolecular interactions in these protein-containing three-dimensional structures, are of paramount importance in an effort to account for many aspects of the behavior of such complex systems. Information along these lines is available, for instance, from the spectroscopic experiment, and the next section aims to summarize our knowledge in this area of research.

7.3 MOLECULAR FEATURES OF THE BICONTINUOUS Q_{II} PHASES OF LIPIDS WITH ENTRAPPED PROTEINS — SPECTROSCOPIC STUDIES

Only two techniques of vibrational spectroscopy, Raman scattering and Fourier-Transform Infrared (FT-IR), have been applied fairly successfully in revealing the molecular aspects of organization of the bicontinuous Q_{II} phases with entrapped proteins. Unfortunately, as with the above-discussed phase behavior studies, the spectroscopic investigations of the protein-containing bicontinuous Q_{II} phases are few in number. Among the systems studied are the MO/lysozyme Q^{229} phase [13], MO/cytochrome c Q^{224} phase [16], and MO/hemoglobin Q^{230} and Q^{224} phases [20]. Two other protein-containing systems, distearoylphosphatidylglycerol + MO/cytochrome c [27] and MO/ bacteriorhodopsin [39], which presumably form bicontinuous cubic phases, have been also characterized by FT-IR [27,39] and Raman [39] spectroscopies; however, the phase types and classes of the investigated samples remain unidentified.

Here, mention may be made of the UV/vis and circular dichroism (CD) spectroscopic studies of the presumably micellar Q^{223} phases of space group type *Pm3n* with entrapped α-chymotrypsin [24], bacteriorhodopsin [24,25], and melittin [25]. Although these three-dimensional structures are unrelated to the subject of the present chapter, in the study by Portman and coworkers [24], the CD spectrum of a 1-monoleoylglycerol/α-chymotrypsin/buffer system was also recorded. The sample was prepared by mixing molten MO (65 wt%) and 195 μ*M* enzyme solution in 17 m*M* phosphate buffer, pH 6.0; therefore, there are grounds to believe that the bicontinuous Q_{II} phase could finally be formed. Notably, in the wavelength region of aromatic chromophores (250 to 330 nm), the CD spectra of α-chymotrypsin in the buffer solution and in the aqueous MO matrix were very similar, indicating that the aromatic amino acid residues of this serine protease (source not indicated; molecular mass of the enzyme from bovine pancreas equals ca. 21.6 kDa [40]) did not undergo conformational changes upon entrapment.

Now, let us discuss the Q^{230}, Q^{229}, and Q^{224} phases with entrapped proteins, which have already been listed at the beginning of this section. Here again, monooleoylglycerol constitutes the major weight portion of all these ternary or pseudoternary mixtures. Unfortunately, shortage of the spectroscopic information is also typical for the binary or pseudobinary MO/water system. So far, there have been only two publications, by Larsson and Rand [41] and by Razumas and coworkers [13], where Raman spectroscopy is used to characterize the bicontinuous MO/water Q_{II} phases. Larsson and Rand [41] pioneered in applying this spectroscopic technique to the liquid-crystalline phases of aqueous MO, including the Q^{230} phase with 25 wt% of water. However, in that study, only a narrow spectroscopic range of the C–H stretching vibrations (2800 to 3000 cm^{-1}) was examined. The MO/water Q^{224} phase of 61:39 wt% composition was investigated more comprehensively over the wavenumber range of 600 to 3050 cm^{-1} [13]. Figure 7.5 presents the FT-Raman spectrum of this cubic phase, and Table 7.2 summarizes the Raman frequencies and assignments of the vibrations. The table also contains spectroscopic features of solid and melted MO for comparison. A number of the characteristics of these different samples were used in the Raman spectroscopic study of lysozyme entrapped in the MO-based Q^{229} phase [13].

7.3.1   Raman Spectroscopy of the MO/Lysozyme/Water Q^{229} Phase

Figure 7.6 shows the Raman spectrum of the MO/lysozyme/water Q^{229} phase of 54:8:38 wt% composition over the wavenumber range 600 to 1800 cm^{-1} at 22°C. It is evident from the figure and Table 7.2 that only four bands in the spectrum can be associated with the protein molecule:

1. At 756 cm^{-1} — symmetric benzene/pyrrole in-phase breathing in tryptophan (Trp) and symmetric ring breathing in tyrosine (Tyr) and phenylalanine (Phe)
2. At 1006 cm^{-1} — symmetric benzene/pyrrole out-of-phase breathing in Trp and symmetric ring stretch in Phe

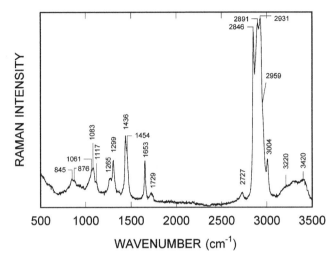

Figure 7.5 FT-Raman spectrum of the MO/water Q^{224} phase of 61:39 wt% composition at 22°C. The monooleoylglycerol-based mixture of mono- and diglycerides (about 25:1 by weight) was produced by Danissco Ingredients (Brabrand, Denmark) with the following fatty acid composition (batch TS-ED 173): 90 wt% oleic acid, 5 wt% linoleic acid, 2.7 wt% stearic acid, 1 wt% palmitic acid, 0.3 wt% linolenic acid, and 1 wt% other fatty acids. Excitation wavelength, 1064 nm. Laser power at the sample, 0.3 W. The spectrum was obtained by averaging 50 scans on a Spectrum GX NIR FT-Raman spectrometer (Perkin Elmer).

    3. At 1358 cm^{-1} — $\delta(CH_2)$ and pyrrole ring vibration in Trp
    4. At 1548 cm^{-1} — symmetric phenyl ring mode in Trp

When compared to the protein aqueous solution, the bands of the aromatic amino acid residues of the entrapped protein occur at about the same peak-frequencies, which is to say that, most likely, the native structure of lysozyme does not undergo significant alterations in the cubic phase. However, the protein Amide III (for the protein solution, at 1236, 1254, and 1263 cm^{-1} [13]) and Amide I (for the protein solution, at 1656 cm^{-1} [13]) bands of the entrapped lysozyme overlap with the strong $\delta(=C-H)$ and $\nu(C=C)$ bands of MO, respectively. For this reason, in our study [13], we have not attempted to determine the effect of lipid matrix on these spectral features of protein. Clearly, this fact prevents additional validation of the above conclusion on the entrapped protein structure.

Table 7.2   Raman Frequencies (cm^{-1})[a] and Vibrational Assignments[b] of Solid and Melted MO[c] and of the MO/Water Q^{224} Phase of 61:39 wt% Composition

Solid MO	Melted MO	MO/Water Q^{224} Phase	Assignment
850 m	852 m	845 m	$\delta(CH_2, CH_3, b)$
884 m	867 m	876 m	$\delta(CH_2, CH_3, b)$
1060 s	1060 m	1061 m	$\nu(C–C)$
1096 s	1081 s	1083 s	$\nu(C–C)$
1121 s	1116 m	1117 m	$\nu(C–C)$
1262 s	1263 s	1265 m	$\delta(=C–H, b)$
1297 vs	1302 vs	1299 s	$\delta(CH_2, tw)$
1438 vs	1437 vs	1436 vs	$\delta(CH_2, b)$
1459 vs	1455 vs, sh	1454 vs, sh	$\delta(CH_2, b)$
1657 vs	1654 vs	1653 vs	$\nu(C=C)$
1731 w	1732 w	1729 w	$\nu(C=O)$
?	2718 w	2727 w	Fermi resonance between $\delta(CH_2, b)$ and $\nu^s(CH_2)$
2846 s	2846 vs	2846 vs	$\nu^s(CH_2)$
2883 vs	2892 vs	2891 vs	$\nu^{as}(CH_2)$
2928 m, sh	2929 s, sh	2931 vs	$\nu^s(CH_3)$
2962 m, sh	2958 m, sh	2959 m, sh	$\nu^{as}(CH_3)$
3003 w	3002 w	3004 m	$\nu(=C–H)$
3350 w	?	3220 m	$\nu(O–H)$; hydrogen bonded
3485 w	?	3420 m	Fermi resonance between $\nu(O–H)$ and $\delta(H_2O)$

[a] The spectra of solid and melted MO were excited at 488 nm, argon laser power ca. 40 mW [13]. The FT-Raman spectrum of the cubic phase was excited at 1064 nm, Nd:YAG (an yttrium aluminum garnet crystal doped with triply ionized neodymium) laser power ca. 0.3 W (recorded for this chapter; see Figure 7.5). vs, very strong; s, strong; m, medium; w, weak; vw, very weak; sh, shoulder.

[b] ν, stretching mode; δ, deformation mode; tw, twisting; b, bending; s, symmetric; as, asymmetric.

[c] The monooleoylglycerol-based mixture of mono- and diglycerides (about 25:1 by weight) produced by Danissco Ingredients (Brabrand, Denmark) with the following fatty acid composition (batch TS-ED 173): 90 wt% oleic acid, 5 wt% linoleic acid, 2.7 wt% stearic acid, 1 wt% palmitic acid, 0.3 wt% linolenic acid, and 1 wt% other fatty acids.

Sources: V Razumas, Z Talaikyte, J Barauskas, K Larsson, Y Miezis, T Nylander. *Chem Phys Lipids* 84: 123–138, 1996; K Larsson, RP Rand. *Biochim Biophys Acta* 326: 245–255, 1973.

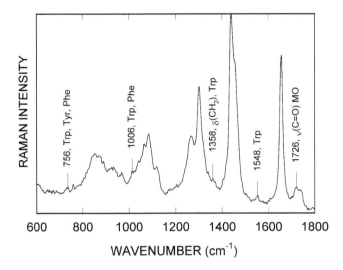

Figure 7.6   Raman spectrum of the MO/lysozyme/water Q^{229} phase of 54:8:38 wt% composition at 22°C. Spectrum excited at 488 nm, argon laser power ca. 40 mW. (From V Razumas, Z Talaikyte, J Barauskas, K Larsson, Y Miezis, T Nylander. *Chem Phys Lipids* 84: 123–138, 1996.)

We have also concluded in Ref. 13 that low content (up to 8 wt%) of lysozyme in the Q^{229} phase has practically no effect on the degree of *gauche/trans*-isomerization and mobility of the MO acyl chains. This was evident from the analysis of the C–C (1000 to 1200 cm^{-1}) and C–H stretching (2800 to 3000 cm^{-1}) vibrations of MO, respectively. Thus, the conformation-sensitive intensity ratio of the C–C stretching modes at 1083 and 1117 cm^{-1} were practically identical in the pseudobinary and pseudoternary cubic phases. Similarly, the entrapment of lysozyme left the intensity ratio of the asymmetric methylene stretching band (2891 cm^{-1}) to that of the symmetric methylene stretching feature (2931 cm^{-1}) unaltered, indicating invariant mobility of the MO acyl chains for the MO/water and MO/lysozyme/water samples. However, in the same study, we also observed that the MO C=O stretching mode for the hydrogen-bonded group in the MO/lysozyme/water Q^{229} phase was shifted downwards by ca. 3 cm^{-1} (Figure 7.6) when compared to the pseudobinary system (Figure 7.5). This effect suggests that the protein somewhat increases the degree of hydrogen bonding of the MO carbonyl group, although the relatively low intensity of this feature in the Raman spectra of lipids complicates interpretation. In this respect, the FT-IR spectroscopy is more suitable.

7.3.2 Fourier-Transform Infrared Spectroscopy of the MO/Cytochrome c/Water Q^{224} Phase

Figure 7.7 presents the FT-IR spectrum of the MO/water Q^{224} phase of 61:39 wt% composition. The absorption spectrum closely resembles the FT-Raman spectrum of the same sample, even though a more detailed comparison reveals a few differences. For example,

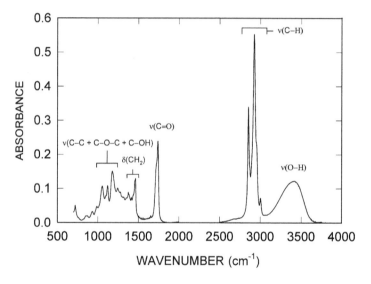

Figure 7.7   FT-IR spectrum of the MO/water Q^{224} phase of 61:39 wt% composition at 22 °C. Some of the bands or their groups are assigned as indicated. Greek characters show the type of vibrational mode (v, stretching; δ, deformation). The monooleoylglycerol-based mixture of mono- and diglycerides (about 25:1 by weight) was produced by Danissco Ingredients (Brabrand, Denmark) with the following fatty acid composition (batch TS-ED 173): 90 wt% oleic acid, 5 wt% linoleic acid, 2.7 wt% stearic acid, 1 wt% palmitic acid, 0.3 wt% linolenic acid, and 1 wt% other fatty acids. The cubic phase sample was assembled between ZnSe windows. A pair of ZnSe windows was used as reference. The spectrum was obtained on a Spectrum GX FT-IR System spectrometer (Perkin Elmer).

the C=O stretching modes are much more pronounced when employing the IR spectroscopic technique, whereas the intensive ν(C=C) mode at 1653 cm^{-1}, observed by Raman scattering (Figure 7.5), is lacking in the FT-IR spectrum. It should be also noted that Nilsson and coworkers [42–44] have performed a detailed FT-IR study of the MO/water system.

In the FT-IR spectroscopic study of the MO/cytochrome c/water Q^{224} phase [16], several vibrational features of MO have been exploited in an effort to elucidate the details of the molecular organization of this pseudoternary phase. The methylene wagging modes of MO were primarily analyzed in that study.

The CH$_2$ wagging modes of lipids are observed in the spectral range from 1330 to 1390 cm^{-1} and provide information on the conformational state of the hydrophobic acyl chain. For the pseudobinary and pseudoternary MO-based Q^{224} phases, three features in this wavenumber range were examined:

1. The double-*gauche* band at 1354 cm^{-1}
2. The band of *gauche-trans-gauche* conformation (kink) at 1367 cm^{-1}
3. The feature of the end-CH$_3$ group symmetric bending (umbrella mode) at 1378 cm^{-1}

The latter band has been used as an internal standard of the system since this mode is insensitive to the acyl chain length and conformation. Thus, the absorbance ratios I_{1354}/I_{1378} and I_{1367}/I_{1378} were measured for four Q^{224} samples. Table 7.3 shows the results, which can

Table 7.3 Absorbance Ratios of the Double-*Gauche* (1354 cm^{-1}) and *Gauche-Trans-Gauche* (1367 cm^{-1}) Modes Relative to the Umbrella Band of the End-CH$_3$ Group (1378 cm^{-1}) Determined from the FT-IR Spectra of the Aqueous[a] Q^{224} Phases[b]

Phase Composition	I_{1354}/I_{1378}	I_{1367}/I_{1378}
MO/water (61:39 wt%)	0.32	0.35
MO/cytochrome c/water (61.6:0.6:37.8 wt%)	0.31	0.31
MO/cytochrome c/water (54:5:41 wt%)	0.26	0.30
MO/buffer (61:39 wt%)	0.19	0.10
MO/cytochrome c/buffer (59.85:0.29:39.86 wt%)	0.19	0.12

[a] Buffer used was 20 mM Na-phosphate buffer (pH 7.0), containing 0.1 M NaClO$_4$.
[b] Standard deviation of the ratios not higher than ±0.01.
Source: V Razumas, K Larsson, Y Miezis, T Nylander. *J Phys Chem* 100: 11766–11774, 1996.

be summarized as follows. First, in the case of pure water-based phases, the entrapped cytochrome c decreases the number of the double-*gauche* and *gauche-trans-gauche* conformers, indicating increased conformational order of the MO acyl chain. Second, the components of the buffer solution exhibit qualitatively identical properties as does the protein in the pure water. However, when buffer is used for the preparation of the cytochrome-containing phase, the low content of entrapped protein leaves the chain conformation unchanged or even slightly increases disorder since the I_{1367}/I_{1378} ratio is higher for the pseudoternary cubic phase (Table 7.3).

For the MO bilayer head-group region, four stretching modes were analyzed in Ref. 16:

1. The $\nu(C_3-OH)$ band at ca. 1050 cm^{-1}
2. The $\nu(C_2-OH)$ band at ca. 1120 cm^{-1}
3. The mode of the ester bond at ca. 1180 cm^{-1}
4. The $\nu(C=O)$ band at ca. 1720 to 1740 cm^{-1}

In the case of the MO C–OH and C=O stretching modes, no reliable shift of the peak-frequencies was observed upon the entrapment of protein. By contrast, the absorption maximum of the ester band at 1182 cm^{-1} for the MO/water Q^{224} phase of 61:39 wt% composition shifts downwards by about 3 cm^{-1} for the MO/cytochrome c/water Q^{224} phase of 54:5:41 wt% composition. From the latter observation it has been concluded that cytochrome c slightly increases the deviation from the dihedral angle of 180° in the C–C–C(O)–O–C segment of MO.

Altogether, the above FT-IR spectroscopic findings show that the entrapment of cytochrome c in the Q^{224} phase of aqueous MO results in a slightly increased conformational order of the MO acyl chains as well as in a moderate rearrangement of the interfacial head-group region.

Here, attention should be drawn to the fact that the MO/DSPG/ cytochrome c/water mixture of 41:7:5:48 wt% composition, where DSPG stands for distearoylphosphatidylglycerol, was also investigated by FT-IR spectroscopy [27]. However, the sample was not characterized by x-ray diffraction, and the purity, type, and class of the phase were not determined experimentally. Nevertheless, as in the case of the MO/cytochrome c/water Q^{224} phase, the FT-IR investigation of the CH$_2$ wagging region also indicated conformational rearrangements along the acyl chains of MO/DSPG mix upon the entrapment of cytochrome c. In the protein-containing pseudoquarternary sample, somewhat higher mobility of the lipid OH groups, reduced hydrogen bonding of the C=O functionalities, and inhibition of the PO$_2^-$ group hydration were also indicated. What is more important, it has been demonstrated that the molecules of cytochrome c probably retain their native secondary structure in the matrix of the aqueous MO/DSPG mixture. Thus, on the one hand, as in the case of the protein aqueous solution, the Amide II band of cytochrome was located at 1551 cm^{-1}; on the other, the fitting procedure over the spectral range of 1481 to 1796 cm^{-1}, wherein five vibrational modes (Amide II, water bending, Amide I, and two carbonyl) overlapped, resolved the Amide I band at 1654 cm^{-1}, which is also typical of the native cytochrome c.

The following FT-IR investigation of the MO/hemoglobin/ water Q^{230} and Q^{224} phases supports the above conclusion that the secondary structure of proteins might be retained within the bicontinuous Q_{II} phases of lipids.

7.3.3 Fourier-Transform Infrared Spectroscopy of the MO/Hemoglobin/Water Q^{230} and Q^{224} Phases

In the FT-IR study of the MO/hemoglobin/water Q_{II} phases by Leslie and coworkers [20], the $\nu(C=O)$, $\nu^{as}(C(O)-O-C)$, Amide I, and $\nu(C-H)$ modes have been monitored. Three MO/hemoglobin/water samples of 50:2.5:47.5, 50:5:45 and 50:10:30 wt% composition were examined. As seen from Figure 7.3, the first two preparations are mixtures of the Q^{230} and Q^{224} phases, whereas the sample with 10 wt% of protein represents the pure Q^{230} phase. Besides, it should be remembered that the aqueous solution used by these authors was phosphate-buffered saline (exact concentrations of components unspecified; pH most likely equals 7.4 at 25°C).

In the spectral range of the C=O vibrational modes (1700 to 1750 cm^{-1}), hemoglobin had no effect on the peak-frequencies of the features. However, a slight decrease in the bandwidth of the hydrogen-bonded C=O group at 1722 cm^{-1} and a slight increase in the bandwidth of the free C=O group mode at 1740 cm^{-1} were observed for the samples containing 5 and 10 wt% of protein. These effects indicate that, in the samples with higher amounts of hemoglobin, water penetrates the lipid interface to a lesser degree, whereas the protein increases hydrogen bonding. According to the authors, the shift of the water association band from 2092 cm^{-1} in the protein-free sample (presumably containing 40 wt% of water, although unspecified in the paper) to 2130 cm^{-1} in the presence of hemoglobin supports the above conclusions.

The stretching mode of the MO ester group was only slightly sensitive to the entrapment of hemoglobin. That is, the peak-frequency shifted downwards from 1183 to 1181 cm^{-1} with increasing the protein content from 2.5 to 10 wt%. By contrast, when compared to the binary sample containing 60 wt% of MO, for the ternary mixture with 10 wt% of protein, the peak-frequency of the symmetric v(C–H) mode was shifted upwards. Figure 7.8 additionally illustrates the variation of peak-frequency of this mode with temperature. The authors also observed broadening of the asymmetric C–H stretching mode at around 2925 cm^{-1} for the MO/hemoglobin/PBS sample of 50:5:45 wt% composition, although other protein-containing phases did not exhibit that effect. The upward shift of the v^s(C–H) mode for the protein-containing system shows a decrease in the MO acyl chain order. However, it should be remembered that the MO/PBS sample containing 40 wt% of PBS most likely forms the Q^{224} phase (not indicated in Ref. 20), whereas the Q^{230} phase is observed for the ternary system with entrapped protein (see Figure 7.3).

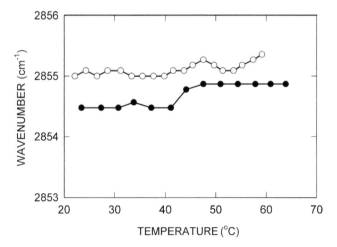

Figure 7.8 Peak-frequency variation of the symmetric C–H stretching mode with temperature for the MO/PBS sample of 60:40 wt% composition (filled circles) and MO/hemoglobin/PBS sample of 50:10:40 wt% composition (open circles). PBS, phosphate buffered saline from Sigma (exact concentrations of the components unspecified; pH most likely equals 7.4 at 25°C). (Adapted from SB Leslie, S Puvvada, BR Ratna, AS Rudolph. *Biochim Biophys Acta* 1285: 246–254, 1996.)

Finally, in Ref. 20, it has been demonstrated that the correlation coefficient of the protein Amide I band at about 1650 cm^{-1} from the spectra of native and entrapped hemoglobin (the cubic phase composition unspecified) equals 0.95 after 2 h incubation of the cubic phase sample at 25°C. Although the parameter decreases to 0.92 after 24 h incubation, the results obtained indicate that the cubic phase matrix does not significantly alter the secondary structure of hemoglobin.

7.3.4 Concluding Remarks

Summarizing the foregoing spectroscopic studies, it can be concluded that, as one might expect, the entrapped proteins can induce subtle changes in the molecular organization of the lipid bilayer of the bicontinuous Q_{II} phases. The effects can appear in the polar head-group region as well as in the interior of the hydrophobic layer of acyl chains. Clearly, the degree to which these biopolymers exert an influence on the bilayer's molecular features depends on several factors, e.g., the nature of the protein and lipid, the lipid/protein ratio, and the ionic composition of the aqueous solution. The structural alterations reveal themselves to a greater or lesser extent depending upon these factors. For example, the above Raman study of the aqueous MO and MO/lysozyme Q_{II} phases indicated unchanged acyl chain order upon the protein entrapment [13]. By contrast, it has been demonstrated in Ref. 16 that cytochrome c increases conformational order of the MO acyl chains in the Q^{224} phase, whereas the opposite effect has been observed for hemoglobin in the MO/PBS Q^{230} phase [20]. An important point is that these vibrational spectroscopy techniques were capable of monitoring the unaltered secondary structure of hemoglobin and probably native features of lysozyme within the MO cubic phases. However, the published Raman and FT-IR studies do not provide unassailable proof concerning the tertiary and, for example, in the case of hemoglobin, quaternary structure of proteins.

Here, it is worthwhile to mention a few publications in which differential scanning calorimetry (DSC) has been applied in parallel with the Raman and FT-IR techniques. It was expected that such combination of the methods would provide additional information on the native qualities of entrapped proteins.

Thus, in the studies of lysozyme within the MO Q^{229} and Q^{224} phases [11,13], the DSC experiments indicated that, regardless of the phase composition (up to 34 wt% of lysozyme), the protein denaturation occurs at about 61 to 64°C (temperature of denaturation, T_D), involving an enthalpy of the process (ΔH_{cal}) of about 370 kJ/mol. Since the determined calorimetric parameters of the entrapped protein are in good agreement with the thermal characteristics of the native lysozyme in water, it could be concluded that the enzyme in the cubic phases of MO maintains native qualities of the tertiary structure regarding temperature and enthalpy of denaturation. This conclusion is also consistent with the results of Raman investigation [13], which were discussed in Section 7.3.1 of this chapter.

As distinct from lysozyme, the T_D and ΔH_{cal} values for the cytochrome c molecules in the MO Q^{224} phase were decreased by ca. 5°C and 39 kJ/mol, respectively, when compared to the protein molecules in water [16]. However, the DSC measurements also indicated that the entrapped protein triggered a significant decrease in temperature of the $Q^{224} \rightarrow H_{II} \rightarrow$ (L_2, reversed micellar) phase transitions, and the unfolding of cytochrome c took place simultaneously with the $Q^{224} \rightarrow H_{II}$ phase transition. Because of this, it cannot be concluded from the DSC measurements that the cubic phase matrix exerts any adverse effect on the tertiary structure of cytochrome c. In such a situation, only direct monitoring of the protein function, i.e., its redox activity, can provide the answer to this structural problem. As shown below, the electrochemical techniques offer interesting possibilities in this respect, and this is also true for the aqueous MO/cytochrome c system. Thus, it will be shown in Section 7.4.1 of this chapter that the amperometric investigations indicate conclusively native characteristics of cytochrome c in the cubic phase of MO.

7.4 ELECTROCHEMICAL STUDIES AND BIOANALYTICAL APPLICATIONS OF THE PROTEIN-CONTAINING BICONTINUOUS Q_{II} PHASES OF LIPIDS

In 1994, we demonstrated for the first time that electrochemical techniques can give a simple way of monitoring activity of enzymes and proteins within the MO matrix [26]. Actually, the results of that study provided support for the view that even the quaternary structure of the entrapped protein can be preserved. At this point, for completeness of the information on this subject, it should be remembered that Portmann and coworkers [24], as well as Wallin and coworkers [45] have demonstrated applicability of spectrophotometry for the same purpose. However, in Ref. 24, the micellar Q^{223}

phase (space group type *Pm3n*) of 1-palmitoyl-2-hydroxy-*sn*-glycero-3-phosphocholine served as a matrix for the entrapment of α-chymotrypsin, whereas in the study of immobilized glucose oxidase [45], the enzyme activity was measured in the presumably bicontinuous Q_{II} phase of ethoxylated fatty alcohol $C_{16-18}(OCH_2CH_2)_{80}OH$.

Table 7.4 summarizes the electrochemical systems reported thus far in which the protein-containing bicontinuous Q_{II} phases of lipids have been applied. In actual fact, the cubic phase class and type were unambiguously determined only for the aqueous MO + PC/glucose oxidase (Q^{230} [19]), MO/cytochrome c (Q^{224} [16]), MO/cholesterol oxidase (Q^{224} [23]) and MO/MP-11 (Q^{224} [46]) systems. However, since MO has been used in all instances, there are grounds to believe that these lipid/protein mixtures might transform to the maximum swollen Q^{224} phase under the conditions of electrochemical experiments. Here, for example, recall that the MO Q^{224} phase can accommodate up to 6 wt% of the commercial preparation of GOD (see Section 7.2.4 of this chapter), whereas the aqueous MO/GOD mixture used in the

Table 7.4 Bioelectrodes and Electrochemical Biosensors Based on the Protein-Containing Bicontinuous Q_{II} Phases of Lipids

Lipid/Protein System	Substrate	Technique	Ref.
MO[a]/glucose oxidase (*Aspergillus niger*)	β-D-Glucose	Amperometry	26
MO/L-lactate oxidase (*Pediococcus*)	L-Lactate	Amperometry	26
MO/jack bean urease	Urea	Potentiometry	26
MO/Creatinine deiminase (*Corynebacterium lilium*)	Creatinine	Potentiometry	26
MO (+ PC)[b]/glucose oxidase (*A. niger*)	β-D-Glucose	Amperometry	19
MO + PC/porcine ceruloplasmin	Fe^{2+}	Amperometry	19
MO/horse heart cytochrome c	—	Amperometry	16
MO/cholesterol oxidase (*Rhodococcus*)	Cholesterol	Amperometry	23
MO/MP-11[c] (from horse heart cytochrome c)	H_2O_2	Amperometry	46

[a] Monooleoylglycerol-based commercial preparations of mono- and diglycerides. In Ref. 23, 1-monooleoylglycerol.

[b] Without or with a phosphatidylcholine preparation from soya beans containing 97.5% phosphatidylcholine.

[c] Heme undecapeptide, microperoxidase-11, which contains residues 11 to 21 of cytochrome c and exhibits the peroxidase-type enzymatic activity.

electrochemical study [26] contained only 70 ng of this enzyme preparation per gram of MO in the matrix. This fact strengthens the assurance that this bicontinuous Q_{II} phase of MO operates in all electrochemical investigations.

As seen from Table 7.4, two electrochemical techniques, amperometry and potentiometry, have been used for the study of bioelectrodes based on the protein-containing cubic phases. Typical three- and single-electrode cell arrangements for the current and potential measurements are shown in Figure 7.9A and Figure 7.9B, respectively. In both cases, a thin layer (ca. 300 to 350 μm) of the cubic phase with entrapped protein is applied to the surface of the working electrode. The cubic phase layer, as a rule, is subsequently covered with a dialysis membrane to prevent leakage of the protein into the bulk electrolyte solution.

At the beginning, let us consider in greater detail the amperometric systems.

7.4.1 Amperometric Bioelectrodes Based on the Protein-Containing Bicontinuous Q_{II} Phases of Lipids

As can be seen from Table 7.4, with the exception of the MO/cytochrome c system, the amperometric bioelectrodes have been constructed basically using oxidoreductases, which catalyze the oxidation–reduction reactions, or their models (MP-11 represents a model of peroxidase).

The catalytic oxidation of substrates (see Table 7.4) by the entrapped glucose oxidase (GOD), L-lactate oxidase (LO), or cholesterol oxidase (COD) proceeds according to the following equation:

$$S + O_2 \rightarrow P + H_2O_2 \tag{7.3}$$

where P represents gluconolactone (in water, spontaneously transforms to gluconic acid), pyruvate, and 4-cholesten-3-one, respectively.

Reaction (7.3) can be coupled with the electrochemical oxidation of hydrogen peroxide at the appropriate transducer surface (Pt was used for the GOD and LO electrodes [19,26], Au and glassy carbon—for the COD electrode [23]), resulting in the regeneration of oxygen:

$$H_2O_2 \rightarrow O_2 + 2H^+ + 2e^- \tag{7.4}$$

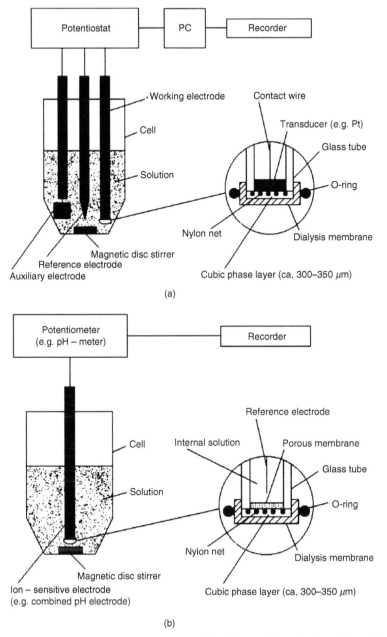

Figure 7.9 Typical equipment used in electrochemical studies of the protein-containing bicontinuous Q_{II} phases of lipids. (a) Three-electrode system for amperometry. (b) Single-electrode system for potentiometry.

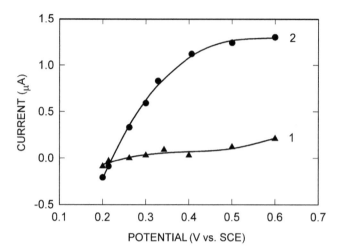

Figure 7.10 Dependence of the residual current (1) and steady-state current of β-D-glucose oxidation (2) on the potential of the Pt working electrode coated with the MO cubic phase with entrapped GOD (70 ng of the commercial GOD preparation per gram of MO). For experiment (2), the D-glucose concentration was 10 mM in 10 mM phosphate buffer (pH 7.0), containing 0.1 M KCl and 1 mM EDTA, 25°C. Disc stirrer rotation speed was ca. 300 rpm; geometric surface area of the working electrode was 0.3 cm^2. (From V Razumas, J Kanapieniene, T Nylander, S Engström, K Larsson. *Anal Chim Acta* 289: 155–162, 1994.)

Figure 7.10 illustrates the amperometric response of the Pt electrode covered with the layer of the GOD-containing MO cubic phase. The catalytic oxidation of β-D-glucose proceeds at electrode potentials (E) higher than 0.2 V (vs. saturated calomel electrode, SCE). In the presence of substrate, the steady-state voltammogram exhibits practically constant catalytic current (I_{cat}) of oxidation at E > 0.45 V (curve 2, Figure 7.10) with a fairly rapid response time (90% of I_{cat} is reached in ca. 2 min). Since the LO-based system demonstrated an identical relationship between I and E, in Ref. 26, a constant potential of 0.6 V was chosen to study different characteristics of these two bioelectrodes. For example, Figure 7.11 shows the dependences of I_{cat} of the GOD- and LO-based electrodes on the bulk concentration of substrate.

It is obvious from Figure 7.11 that the electrodes based on the MO cubic phase with entrapped GOD or LO can be considered as

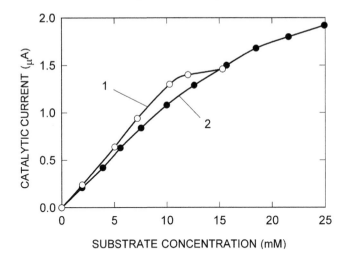

Figure 7.11 Catalytic steady-state current calibration curves for the LO- (1) and GOD-based (2) bioelectrodes at 0.6 V vs. SCE. Other conditions are the same as in Figure 7.10. (From V Razumas, J Kanapieniene, T Nylander, S Engström, K Larsson. *Anal Chim Acta* 289: 155–162, 1994.)

glucose- and lactate-sensitive amperometric biosensors. As may be seen from Figure 7.11, the electrodes are characterized by practically linear I_{cat} responses up to 12 and 10 mM concentrations of D-glucose and L-lactate, respectively. From the analytical point of view, these concentration ranges are clinically relevant since, for example, blood plasma contains 3.6 to 6.1 mM of glucose and 0.4 to 1.8 mM of lactate [47]. Here, it is also important to note that the responses of both sensors exhibited minimal sensitivity to changes in pH (over the pH range from 5 to 8) and temperature (over the range from 15 to 35°C).

As illustrated in Figure 7.12, when compared to the glucose-sensitive electrode, the lactate-sensitive biosensor demonstrated considerably higher long-term stability. Thus, I_{cat} of the latter electrode retained more than 85% of the initial value after 2 months (not shown in Figure 7.12) of storage in the standard buffer solution (for the composition, see the legend to Figure 7.10) at room temperature, whereas the response of the GOD-based sensor was decreased to ca. 18% after 2 weeks. In Ref. 26, it has been proposed that faster deterioration of the glucose-sensitive biosensor might be determined by the higher molecular mass of GOD, which is about

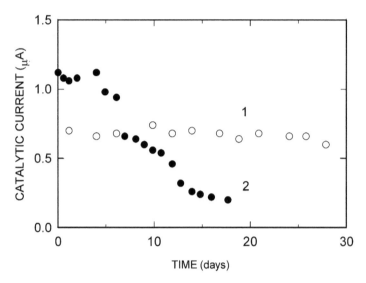

Figure 7.12 Long-term stability of the L-lactate oxidase- (1) and glucose oxidase-based (2) biosensors at room temperature. Other conditions are the same as in Figure 7.10. (From V Razumas, J Kanapieniene, T Nylander, S Engström, K Larsson. *Anal Chim Acta* 289: 155–162, 1994.)

double that of LO (160 and 80 kDa, respectively). In other words, larger enzyme molecules facilitate disintegration of the MO phase matrix, which in turn loses its capacity to fix and to stabilize the native structure of enzyme. However, the study by Nylander and co-workers [19] indicated one more possible reason for the results of Figure 7.12.

In Ref. 19, the bioactive layer on the Pt electrode surface was formed from the MO/PC/GOD/buffer mixture of 48:12:0.044:39.956 wt% composition, where the PC is a phosphatidylcholine preparation from soya beans containing 97.5% phosphatidylcholine. As demonstrated by x-ray diffraction measurements, the mixture initially forms the Q^{230} phase of $a_o = 161.7$ Å. Interestingly, this glucose-sensitive electrode still exhibited about 30% of its initial activity even after 80 days. It is most likely that the impressive long-term stability of this electrode is determined to a large extent by the increased content of the entrapped enzyme. In these conditions, the electrode response might be solely controlled by the diffusional mass transport of substrate(s) across the cubic phase layer (for some details, see Section 7.4.3 of this chapter). For a kinetic mode of operation (e.g., at low content or activity of the entrapped enzyme), the response depends on the kinetic parameters of the biocatalyst, which is to say that the biosensor becomes more sensitive to the nonnatural conditions of enzyme functioning.

In parallel with the glucose-sensitive electrode, Nylander and coworkers have also investigated biocatalytic oxidation of Fe^{2+} ions on the Pt electrode surface modified by the MO/PC/ceruloplasmin/buffer mixture of 49:12.3:0.12:38.58 wt% composition [19]. The cathodic amperometric response of this electrode results from the following enzymatic (7.5) and electrode (7.6) reactions:

$$4Fe^{2+} + O_2 + 4H^+ \rightarrow 4Fe^{3+} + 2H_2O \qquad (7.5)$$

$$Fe^{3+} + e^- \rightarrow Fe^{2+} \qquad (7.6)$$

At -0.3 V vs. SCE, the ceruloplasmin-based electrode showed a linear steady-state response up to a 0.5 mM concentration of Fe^{2+}. However, the current of this biosensor was reduced by ca. 86% after 24 days of its storage at room temperature in $0.05\ M$ piperazine buffer (pH 4.5), containing $0.1\ M$ NaCl.

When compared to the foregoing amperometric biosensors, the cholesterol-sensitive electrode based on the aqueous MO Q^{224} phase with entrapped COD was investigated to a lesser degree [23]. Using cyclic voltammetry in $0.05\ M$ phosphate buffer (pH 7.0) with $0.1\ M$ KCl, the authors demonstrated that the processes shown in Equation (7.3) and Equation (7.4), where S = cholesterol, could be monitored on the Au surface at E > 0.75 V vs. SCE (anodic peak-potential equals 0.9 V). Clearly, since the oxidation of H_2O_2, as shown in Equation (7.4), proceeds at lower overpotentials on Pt (see, for example, Figure 7.10), the application of Au in the construction of the cholesterol-sensitive electrode is of questionable value for the practical use of the system. In the final stage of their study, Ropers and coworkers [23] substituted Au and MO for glassy carbon and synthetic fluorinated surfactant (FS) $C_6F_{13}C_2H_4SC_2H_4(OC_2H_4)_2OH$, 11-perfluorohexyl-9-thio-3,6-dioxaundecanol, respectively. It has been shown that FS also forms the Q^{230} and Q^{224} phases, the latter of which with 60 to 65 wt% of surfactant accommodates up to 0.58 mg/ml of COD [23]. For the glassy carbon electrode modified by the COD-containing FS Q^{224} phase, instead of an anodic process (4), the authors used a cathodic reduction of H_2O_2 at E < -0.3 V vs. SCE (peak-potential at ca. -0.5 V):

$$H_2O_2 + 2H^+ + 2e^- \rightarrow 2H_2O \qquad (7.7)$$

In these experimental conditions, the amperometric response of the biosensor was linear up to 5 mM of cholesterol (the content in blood plasma is about 4.6 to 5.4 mM [47]).

Reaction (7.7) represents the direct reduction of hydrogen peroxide on the electrode surface. We have developed the bioelectrode for the electrocatalytic reduction of this compounds on the Au electrode modified by the MO Q^{224} phase with entrapped heme undecapeptide, microperoxidase-11 (MP-11) [46]. MP-11 is produced by enzymatic digestion of cytochrome c (that is why the product contains amino acid residues 11 to 21 of this protein with covalently linked heme at positions 14 and 17), and it exhibits pronounced peroxidase activity [48]. For the construction of the MP-based bioelectrode, the MO/MP-11/ CH_3OH/water mixture of 62.7:0.3:6.1:30.9 wt% composition was used as it formed the Q^{224} phase of $a_o = 93.4$ Å. The methanol-containing aqueous solution was applied for the phase preparation in an effort to prevent association of the MP-11 molecules. Interestingly, the reduction of H_2O_2 on the Au electrode surface covered by the MP-containing MO cubic phase started already at E < 0.4 V vs. SCE, i.e., the overpotential of the process was decreased by ca. 0.7 V when compared, for example, to the cholesterol-sensitive electrode described in Ref. 23. It has been suggested that the reaction sequence on the MP-based electrode is expressible in terms of the following scheme, which includes one biocatalytic process and two electrochemical reactions, respectively:

$$H_2O_2 + \text{ferri-MP-11} \rightarrow H_2O + \text{Compound I} \qquad (7.8)$$

$$\text{Compound I} + H^+ + e^- \rightarrow \text{Compound II} \qquad (7.9)$$

$$\text{Compound II} + H^+ + e^- \rightarrow \text{ferri-MP-11} + H_2O \qquad (7.10)$$

where ferri-MP-11 indicates heme iron in the oxidation state of +III, Compound I is the high-valent heme-oxygen complex (iron in the formal oxidation state of +V), and Compound II is the MP-11 complex with the formal iron oxidation state of +IV.

Unfortunately, the MP-based electrode displayed one very important drawback. Due to the leakage of relatively low- molecular-mass MP-11 (ca. 1.9 kDa) from the cubic phase layer into the bulk electrolyte solution, the electrode completely deteriorated within about 2 h of operation.

One more point about this system should be mentioned. To ensure the electrocatalytic efficiency of this H_2O_2-sensitive electrode, the Au electrode needs to be pretreated with 4,4'-dithiodipyridine (DTDPy), which forms on the metal surface a self-assembled monolayer of ca. 9 Å thickness [49]. This procedure considerably increases the rate of MP-11 redox conversion on the electrode surface (for more details, see Ref. 49). The same procedure of Au modification was used in studies of the redox conversion of cytochrome c entrapped in the MO Q^{224} phase [16]. This amperometric system is the last to be discussed in this section.

In study [16], the surface of the DTDPy-modified Au electrode was coated by the Q^{224} MO/cytochrome c/water phase of 54:5:41 wt% composition. The cyclic voltammograms of this bioelectrode in 20 mM Na-phosphate buffer (pH 7.5), containing 0.1 M NaClO$_4$, exhibited well-defined anodic and cathodic peaks of the one-electron redox couple ferri-/ferro-cytochrome c. Based on these amperometric curves, the formal redox potential of the entrapped protein $E^{o'} = 15$ mV vs. SCE was calculated. Since the $E^{o'}$ value determined coincides with the accepted formal redox potential of the native cytochrome c [34], this electrochemical result strongly suggests that the protein preserves its tertiary structure and native functional characteristics within the cubic phase matrix. Thus, the problem, which could not be solved by the FT-IR and DSC methods, was finally clarified by the application of electrochemistry. However, one electrochemical result relative to the cytochrome-based system deserves additional attention.

The points in Figure 7.13 show the chronoamperometric response of the cytochrome-containing bioelectrode. In this experiment, considering the results of cyclic voltammetry, the E step from 0.3 to −0.25 V vs. SCE was used to ensure the diffusion-limited reduction of ferricytochrome c to its ferro-form. The response is presented in the coordinates of the Cottrell equation, which holds for the semiinfinite diffusion [50]:

$$i = nF(D_{app})^{1/2}c_0/(\pi t)^{1/2} \tag{7.11}$$

where i is the current density, the number of electrons transferred n = 1, F is the Faraday constant, D_{app} is the apparent diffusion coefficient, c_0 is the bulk concentration of redox active compound, and t represents time.

As evident from Figure 7.13, the experimental plot is nonlinear in the coordinates of Equation (7.11). The shape of the plot suggests that the finite diffusion must be taken into account in the case of this bioelectrode response. For the chronoamperometric technique, the following finite diffusion relationship has been derived [51]:

$$i = \frac{nF(D_{app})^{1/2}c_0}{(\pi t)^{1/2}}\left[1 + 2\sum_{m=1}^{\infty}(-1)^m \exp\left(-\frac{m^2 d^2}{D_{app}t}\right)\right] \tag{7.12}$$

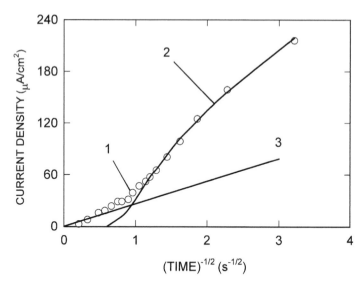

Figure 7.13 Experimental (1) and fitted (2,3) chronoamperometric responses of the 4,4′-dithiodipyridine-modified Au electrode
coated by ca. 300 to 350 μm layer of the MO/cytochrome c/water
Q^{224} phase of 54:5:41 wt% composition. For the experimental points
(1), the potential step was from 0.3 to −0.25 V vs. SCE in 20 mM
Na-phosphate buffer (pH 7.0), containing 0.1 M NaClO$_4$. Curve 2
was calculated according to Equation (7.12), assuming $(D_{app})^{1/2}c_0 =$
1.25×10^{-9} mol/(cm^2 sec$^{1/2}$) and d = 4.3 μm. Plot 3 represents fitting
of the experimental results at t > 2 sec to the Cottrell equation,
Equation (7.11). (From V Razumas, K Larsson, Y Miezis, T Nylander.
J Phys Chem 100: 11,766–11,774, 1996.)

where d is the thickness of the layer, in which diffusion appears as
finite.

Curve 2 in Figure 7.13 represents the best-fitted calculated
data from Equation (7.12), provided that $(D_{app})^{1/2}c_0 = 1.25 \times 10^{-9}$
mol/(cm^2 sec$^{1/2}$) and d = 4.3 μm. Although the calculated $(D_{app})^{1/2}c_0$
parameter agreed with the result of cyclic voltammetry measure-
ments (for more details, see Ref. 16), the fitted d value was about 70
times smaller than the actual thickness of the cubic phase layer (ca.
300 to 350 μm). Moreover, the deviation of the experimental points and
curve 2 is evident at t > 0.7 sec. Considering these two latter observations, it has been concluded that the modified Au electrode initially

(within ca. 0.7 sec) "probes" a thin layer of somewhat transformed cubic phase, whereas the bulk of the protein-containing phase becomes involved in the process at a later time. In practice, as shown by plot 3 in Figure 7.13, the experimental data at $t > 2$ sec could be fitted to the Cottrell equation, Equation (7.11), resulting in the $(D_{app})^{1/2}c_o$ value of 4.81×10^{-10} mol/(cm^2 sec$^{1/2}$). Under the assumption that cytochrome c is evenly distributed in the MO cubic phase and that the phase density is close to unity, the value of c_o in the MO/cytochrome c/water mixture of 54:5:4 wt% composition should be equal to about 4 mM. In this case, the fitted plot 3 leads to the D_{app} value of 1.45×10^{-8} cm^2/sec, which is about 69 times lower when compared to the parameter in the buffer solution [16].

On the one hand, the above discussion of the cytochrome-containing bioelectrode illustrates a highly restricted mobility of protein molecules in the bicontinuous Q_{II} phase; and on the other, the results of Figure 7.13 indicate conclusively that the electrochemical investigations of the lipidic cubic phases are not so straightforward as might appear at first glance.

7.4.2 Potentiometric Biosensors Based on the Enzyme-Containing Bicontinuous Q_{II} Phases of Lipids

When compared to the amperometric bioelectrodes just discussed, potentiometric systems, which function on a basis of the lipidic cubic phases with entrapped proteins, are few in number. As indicated in Table 7.4, only two potentiometric biosensors, developed for the determination of urea and creatinine, have hitherto been reported [26]. Both substances are important metabolites. The amount of urea excreted (20 to 35 g/day) with the urine (0.5 to 2 L/day) is directly related to the metabolized amount of proteins, whereas urine creatinine (1.0 to 1.5 g/day) is an important reference product of muscle metabolism [52].

The urea- and creatinine-sensitive systems were constructed using a flat pH-glass electrode coated by the MO cubic phase layer with the entrapped hydrolytic enzymes urease and creatinine deiminase, respectively (for the equipment details, see Figure 7.9B). Here, as with a number of amperometric systems, it has been assumed that the MO/urease/water and MO/creatinine deiminase/water mixtures form on the electrode surface the maximum swollen Q^{224} phase, although this assumption has not been proved experimentally.

The biosensors' response, expressible by relation $E(mV) = -59.16 \times \Delta$ pH, is based on the consumption of protons according to the corresponding biocatalytic reactions:

$$CO(NH_2)_2 + 2H_2O + H^+ \rightarrow HCO_3^- + 2NH_4^+ \tag{7.13}$$

$$\text{Creatinine} + H_2O + H^+ \rightarrow \text{N-Methylhydantoin} + NH_4^+ \tag{7.14}$$

The calibration graphs for two urea-sensitive electrodes, containing 3.5 and 1.75 mg of urease per gram of MO in the matrix, are shown in Figure 7.14. As evident from Figure 7.14, the response is proportional to the amount of entrapped urease. In this respect, the creatinine-sensitive pH electrodes behaved in analogous fashion. Thus, the biosensors, containing 0.7, 1.4, and 8 mg of creatinine deiminase per gram of MO, gave linear responses between 0.05 and 2 mM concentrations of creatinine in 0.1 M phosphate buffer (pH 8.7),

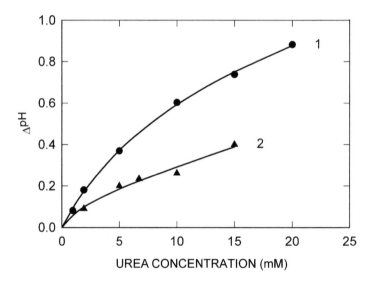

Figure 7.14 Dependence of the urease-based pH electrode response on urea concentration in 10 mM phosphate buffer (pH 7.4), containing 0.1 M KCl and 1 mM EDTA, 25°C. Enzyme content in the MO cubic phase (mg per gram of MO): 3.5 (1) and 1.75 (2). Response was determined at t = 3 min after injection of the urea sample. Disc stirrer rotation speed was ca. 300 rpm. (Adapted from V Razumas, J Kanapieniene, T Nylander, S Engström, K Larsson. *Anal Chim Acta* 289: 155–162, 1994.)

containing 0.1 M KCl and 1 mM EDTA. The results could be expressed by Equation (7.15), Equation (7.16), and Equation (7.17).

$$\Delta pH = 0.022(\pm 0.001)[\text{Creatinine}] \ (mM) \qquad (7.15)$$

$$\Delta pH = 0.044(\pm 0.002)[\text{Creatinine}] \ (mM) \qquad (7.16)$$

$$\Delta pH = 0.068(\pm 0.002)[\text{Creatinine}] \ (mM) \qquad (7.17)$$

However, when compared to the amperometric sensors, the potentiometric electrodes exhibited two very important drawbacks:

1. They showed extremely slow response time, reaching 90% of the steady-state pH only in 10 to 15 min. The problem most likely occurs because of the penetration of the cubic phase and/or its components into the H^+-sensitive membrane of the pH electrode.
2. The second problem is due to the response dependence on the buffer capacity of the electrolyte solution (β). Considering Equation (7.13) and Equation (7.14), the electrode signal must diminish as β increases. This was demonstrated in Ref. 26 for the urea-sensitive electrode, containing 1.75 mg of urease per gram of MO in the matrix. Thus, at the 5 mM concentration of urea, the result could be expressed by the following equation:

$$\Delta pH = 0.391(\pm 0.014) - 0.043(\pm 0.003) \ \beta \ (mM) \qquad (7.18)$$

In parallel to the limitations mentioned above, the urea-sensitive electrodes were not nearly as stable as the amperometric electrodes for the determination of glucose and lactate. Thus, the pH electrode coated by the phase layer, containing 3.5 mg of urease per gram of MO, was completely inactive after two days of operation. By contrast, the creatinine-sensitive pH electrode of the highest activity (8 mg of enzyme per gram of MO) retained about 23% of the initial response after 3 weeks of storage at room temperature in the standard buffer solution (0.1 M phosphate buffer of pH 8.7 with 0.1 M KCl and 1 mM EDTA). It has been suggested in Ref. 26 that the extremely high molecular mass of urease (about 590 kDa for a hexamer) might be responsible for the low stability of the urea-sensitive system. On the other hand, since the molecular mass of creatinine deiminase (ca. 200 kDa) is close to that of glucose oxidase, the creatinine-sensitive potentiometric sensor exhibited better characteristics of long-term stability. However, the effects of these proteins on the structural features of the MO Q^{224} phase remain to be elucidated.

7.4.3   Concluding Remarks

The foregoing presentation of the potentiometric biosensors accomplished a survey of the electrochemical systems based on the lipidic cubic phases with entrapped proteins. In my opinion, both amperometric and potentiometric bioelectrodes discussed in Section 7.4 provide an interesting example of the practical application of the bicontinuous Q_{II} phases of lipids. From the bioelectroanalytical point of view, the merits of these mesophases are their nontoxic character, simple reproduction of the matrix parameters, capacity to preserve activity of relatively large enzymes, and also high viscosity, which enables easy formation of layers on the electrode surface. Besides, as has been demonstrated in our recent study [53], the three-dimensional network of the water channels and lipid bilayer in the bicontinuous Q_{II} phases makes possible detection of the electrochemical activity of entrapped redox molecules. By contrast, the diffusional anisotropy within, for example, a two-dimensional structure of the H_{II} phase completely suppresses the charge transfer process in this matrix.

It should be remembered that, in practice, electrochemical systems based on the lipidic cubic phases are usually quite complex regarding their function. One example is the above-discussed bioelectrode with entrapped cytochrome c. However, the macrokinetic interpretation of the biosensors' action is much more complicated. The latter problem is not extensively discussed in this article since several fundamental studies have been published on this subject (see, e.g., Refs. 54, 55). Nevertheless, to provide a rough idea of the complexity of such theoretical treatment as well as to account for certain functional features of the above-mentioned bioelectrodes, let us briefly outline the simplest case of amperometric biosensor.

It might be assumed that the enzymatic reaction is composed of two elementary processes in which the substrate, S, forms with the enzyme, E, the substrate–enzyme complex, ES, that subsequently transforms to the enzyme and electrochemically active product, P. The overall rate, V, of this biocatalytic reaction in the cubic phase layer of thickness d can be expressed by the well-known Michaelis–Menten equation [56]:

$$V = V_{max}[S]/(K_M + [S]) \tag{7.19}$$

where V_{max} is the maximal velocity at high substrate concentrations, [S], and K_M is the Michaelis constant.

Further, since we are concerned with the steady-state responses, the substrate and product concentration profiles across the cubic phase layer can be obtained by solving the following differential expressions:

$$D_S(d^2[S]/dx^2) = V \tag{7.20}$$

$$D_P(d^2[P]/dx^2) = -V \tag{7.21}$$

where D_S and D_P are the diffusion coefficients of S and P in the cubic phase matrix, respectively, and x is the distance (x = 0 at the electrode surface, x = d at the cubic phase/solution interface).

The solution of Equation (7.20) and Equation (7.21) considerably simplifies by applying two limiting cases for Equation (7.19), where either $[S] \ll K_M$ or $[S] \gg K_M$. Besides, for the amperometric sensor, the following boundary conditions can be applied: (i) at x = 0, d[S]/dx = 0 (the substrate is neither consumed nor generated) and [P] = 0 (the product is amperometrically consumed), and (ii) at x = d, $[S] = [S]_0$ and [P] = 0.

Under the condition $[S] \ll K_M$ and also expressing i_{cat} by $nFD_P(d[P]/dx)_{x\,=\,0}$, Equation (7.20) and Equation (7.21) yield the following expression of the catalytic current density:

$$i_{cat} = \left(\frac{nFD_S[S]_0}{d}\right)\left[1 - \frac{1}{\cosh(\alpha\delta)}\right] \tag{7.22}$$

where $\alpha = [V_{max}/(K_M D_S)]^{1/2}$.

In the case of $[S] \gg K_M$, differential Equation (7.20) and Equation (7.21) lead to the maximum current density:

$$i_{max} = nFV_{max}d/2 \tag{7.23}$$

Now, at $\alpha d \ll 1$, i.e., when the overall process is limited by the rate of enzymatic reactions, Equation (7.22) and Equation (7.23) give a kinetic current density of the electrode:

$$i_k = nFV_{max}[S]_0d/(2K_M) = i_{max}[S]_0/K_M \tag{7.24}$$

By contrast, when $\alpha d \gg 1$, the catalytic current density of the biosensor is controlled by the diffusion of S:

$$i_d = nFD_S[S]_0/d \tag{7.25}$$

Thus, taking into account fairly long linear calibration ranges of the glucose- and lactate-sensitive electrodes (Figure 7.11) as well as their long-term stability (Figure 7.12) and minimal sensitivity to changes in pH and temperature, it is believed that the responses of these biosensors are controlled by substrate diffusion. In other words, the current of the electrodes is governed by Equation (7.25) over a certain range of [S]. By contrast, since similar theoretical treatment is available for the potentiometric biosensors [54,55], it might be concluded that the urea- and creatinine-sensitive electrodes operate under the kinetic conditions insofar as the responses of these bio-sensors are proportional to the concentration of enzyme (i.e., depends on V_{max}; see, for example, Figure 7.14).

The above analysis hopefully demonstrates once again that, in applied work such as biosensor construction, proper use of theory is beneficial for planning experiments as well as making possible the obtaining of the desired result.

Finally, in closing our discussion of the bicontinuous Q_{II} phases of lipids with entrapped proteins, it should be remarked that we are not yet at the point where we can completely understand structural and functional features of these complex systems. Nevertheless, there is good reason to believe that more examples of protein-containing cubic phases will be reported in the immediate future. Specifically, the modification of the bicontinuous Q_{II} phases with functionally active membrane proteins presents a challenging task for the progress in this area of research.

ACKNOWLEDGMENTS

I gratefully acknowledge research support from the Lithuanian State Research and Studies Foundation and the Royal Swedish Academy of Sciences. I also wish to thank my colleagues at the Department of Bioelectrochemistry and Biospectroscopy at the Institute of Biochemistry, as well as at the Departments of Food Technology and Physical Chemistry 1 of Lund University (Sweden) for their diverse contributions to this area of research.

REFERENCES

1. D Voet, JG Voet, CW Pratt. *Fundamentals of Biochemistry*. New York: Wiley, 1999, pp. 239–278.

2. P Mariani, V Luzzati, H Delacroix. *J Mol Biol* 204: 165–189, 1988.

3. G Lindblom, L Rilfors. *Biochim Biophys Acta* 988: 221–256, 1989.

4. K Larsson. *J Phys Chem* 93: 7304–7314, 1989.

5. V Luzzati. *Curr Opin Struct Biol* 7: 661–668, 1997.

6. JS Patton, MC Carey. *Science* 204: 145–148, 1979.

7. G Basañez. *Cell Mol Life Sci* 59: 1478–1490, 2002.

8. S Hyde, S Andersson, K Larsson, Z Blum, T Landh, S Lidin, BW Ninham. *The Language of Shape. The Role of Curvature in Condensed Matter: Physics, Chemistry and Biology.* Amsterdam: Elsevier, 1997, pp. 257–338.

9. T Gulik-Krzywicki, E Shechter, V Luzzati, M Faure. *Nature* 223: 1116–1121, 1969.

10. K Larsson, G Lindblom. *J Disp Sci Technol* 3: 61–66, 1982.

11. B Ericsson, K Larsson, K Fontell. *Biochim Biophys Acta* 729: 23–27, 1983.

12. B Ericsson. Interactions Between Globular Proteins and Polar Lipids. PhD dissertation, Lund University, Lund, Sweden, 1986.

13. V Razumas, Z Talaikyte, J Barauskas, K Larsson, Y Miezis, T Nylander. *Chem Phys Lipids* 84: 123–138, 1996.

14. S Puvvada, SB Qadri, J Naciri, BR Ratna. *J Phys Chem* 97: 11103–11107, 1993.

15. W Buchheim, K Larsson. *J Colloid Interface Sci* 117: 582–583, 1987.

16. V Razumas, K Larsson, Y Miezis, T Nylander. *J Phys Chem* 100: 11766–11774, 1996.

17. EM Landau, JP Rosenbusch. *Proc Natl Acad Sci USA* 93: 14532–14535, 1996.

18. P Nollert, H Qiu, M Caffrey, JP Rosenbusch, EM Landau. *FEBS Lett* 504: 179–186, 2001.

19. T Nylander, C Mattisson, V Razumas, Y Miezis, B Håkansson. *Colloids Surf A: Physicochem Engin Asp* 114: 311–320, 1996.

20. SB Leslie, S Puvvada, BR Ratna, AS Rudolph. *Biochim Biophys Acta* 1285: 246–254, 1996.

21. J Barauskas, V Razumas, T Nylander. *Prog Colloid Polym Sci* 116: 16–20, 2000.

22. J Barauskas, V Razumas, T Nylander. *Biologija Suppl* 1: 5–7, 1998.

23. M-H Ropers, R Bilewicz, M-J Stebe, A Hamidi, A Miclo, E Rogalska. *Phys Chem Chem Phys* 3: 240–245, 2001.

24. M Portmann, EM Landau, PL Luisi. *J Phys Chem* 95: 8437–8440, 1991.

25. EM Landau, PL Luisi. *J Am Chem Soc* 115: 2102–2106, 1993.

26. V Razumas, J Kanapieniene, T Nylander, S Engström, K Larsson. *Anal Chim Acta* 289: 155–162, 1994.

27. V Razumas, Z Talaikyte, J Barauskas, Y Miezis, T Nylander. *Vibr Spectrosc* 15: 91–101, 1997.

28. JR Giorgione, Z Huang, RM Epand. *Biochemistry* 37: 2384–2392, 1998.

29. Y Sadhale, JC Shah. *Int J Pharm* 191: 51–64, 1999.

30. ST Hyde, S Andersson, B Ericsson, K Larsson. *Z Kristallogr* 168: 213–219, 1984.

31. D Voet, JG Voet, CW Pratt. *Fundamentals of Biochemistry*. New York: Wiley, 1999, pp. 300–307.

32. J Engblom, ST Hyde. *J Phys II France* 5: 171–190, 1995.

33. J Briggs, H Chung, M Caffrey. *J Phys II France* 6: 723–751, 1996.

34. HA Harbury, RHL Marks. In: GL Eichhorn, Ed. *Inorganic Biochemistry*. Amsterdam: Elsevier, 1973, pp. 902–954.

35. D Voet, JG Voet, CW Pratt. *Fundamentals of Biochemistry*. New York: Wiley, 1999, pp 165–178.

36. HJ Hecht, HM Kalisz, J Hendle, RD Schmid, D Schomburg. *J Mol Biol* 229: 153–172, 1993.

37. H Qiu, M Caffrey. *Biomaterials* 21: 223–234, 2000.

38. J-R Li, Y-K Du, P Boullanger, L Jiang. *Thin Solid Films* 352: 213–217, 1999.

39. J Heberle, G Büldt, E Koglin, JP Rosenbusch, EM Landau. *J Mol Biol* 281: 587–592, 1998.

40. D Voet, JG Voet, CW Pratt. *Fundamentals of Biochemistry*. New York: Wiley, 1999, p. 106.

41. K Larsson, RP Rand. *Biochim Biophys Acta* 326: 245–255, 1973.

42. A Nilsson, A Holmgren, G Lindblom. *Biochemistry* 30: 2126– 2133, 1991.

43. A Nilsson, A Holmgren, G Lindblom. *Chem Phys Lipids* 69: 219–227, 1994.

44. A Nilsson, A Holmgren, G Lindblom. *Chem Phys Lipids* 71: 119–131, 1994.

45. R Wallin, S Engström, CF Mandenius. *Biocatalysis* 8: 73–80, 1993.

46. J Barauskas, Z Talaikyte, V Razumas, T Nylander. *Chemija* 4: 46–51, 1997.

47. J Koolman, K-H Röhm. *Color Atlas of Biochemistry*. Stuttgart: Thieme, 1996, pp 250–251.

48. PA Adams. In: J Everse, KE Everse, MB Grisham, Eds. *Peroxidases in Chemistry and Biology*. Vol. 2. Boca Raton, FL: CRC Press, 1991, pp. 171–199.

49. V Razumas, T Arnebrant. *J Electroanal Chem* 427: 1–5, 1997.

50. AJ Bard, LR Faulkner. *Electrochemical Methods. Fundamentals and Applications.* New York: Wiley, 1980, pp. 142–146.

51. N Oyama, T Ohsaka. In: RW Murray, Ed. *Molecular Design of Electrode Surfaces.* New York: Wiley, 1992, pp. 333–402.

52. J Koolman, K-H Röhm. *Color Atlas of Biochemistry.* Stuttgart: Thieme, 1996, pp. 298–299.

53. J Barauskas, V Razumas, Z Talaikyte, A Bulovas, T Nylander, D Tauraite, E Butkus. *Chem Phys Lipids* 123: 87–97, 2003.

54. MJ Eddowes. In: AEG Cass, Ed. *Biosensors. A Practical Approach.* Oxford: IRL Press, 1990, pp. 211–263.

55. F Scheller, F Schubert. *Biosensors.* Amsterdam: Elsevier, 1992, pp. 66–84.

56. D Voet, JG Voet, CW Pratt. *Fundamentals of Biochemistry.* New York: Wiley, 1999, pp. 326–331.

8

NMR Characterization of Cubic and Sponge Phases

OLLE SÖDERMAN AND BJÖRN LINDMAN

CONTENTS

ABSTRACT

We present an overview of applications of NMR to bicontinuous cubic and sponge phases. Our point of departure is a historical account of how the concept of bicontinuity evolved in self-assembly systems. We introduce the NMR method and discuss the NMR approaches that are suitable for obtaining information about bicontinuous phases. In particular, we stress the use of NMR relaxation and NMR diffusion data as suitable in this context. Examples of applications of these approaches are then given, and we outline how relaxation data as a function of phase composition and also Larmor frequency at one particular composition may be interpreted. Finally, we present diffusion data for both sponge phases and cubic phases and point out that such data imply a close similarity in the microstructure of the ordered cubic and disordered sponge phase.

8.1 INTRODUCTION AND SOME HISTORICAL REMARKS

Surfactant self-assembly systems, which include liquid-crystalline phases and isotropic solutions, can fruitfully be divided into those that have discrete self-assembly aggregates and those where the aggregates are connected in one, two, or three dimensions. Regarding lamellar phases, the two-dimensional connectivity was already appreciated at a very early stage. The same holds true for the ("normal" and "reverse") hexagonal phases, although erroneous models of linearly associated spherical micelles, "pearls-on-a-string," can be found in the literature; such a linear association was also, again incorrectly, advanced to explain droplet growth in microemulsions.

The general acceptance of connectivity for these anisotropic phases has stood in sharp contrast to a great difficulty in obtaining an acceptance of bicontinuity for other phases. This is partly related to the fact that contrary to these anisotropic phases, it has been much more difficult to structurally characterize the different isotropic phases found in simple and complex surfactant systems: cubic liquid crystals, solutions in binary surfactant water systems, and microemulsions. Indeed, in particular for microemulsions, various interpretations can be found in the literature of investigations by different techniques. As can easily be determined, this is related to difficulties of interpretation of experimental findings, as the same results have sometimes been interpreted in completely opposite ways. In fact, very few experimental observations make possible a distinction between discrete and connected structures. The first real verification was due to observations of molecular self-diffusion over macroscopic distances. Electrical conductivity offers a partial in-sight in providing information on the extension of aqueous domains. Fluorescence quenching can provide information on the growth of nonpolar domains, but a probe has to be introduced. More recently, cryogenic transmission electron microscopy has developed into a very important tool for imaging different surfactant phases.

The senior author (BL) came into contact with the problem of bicontinuity at a very early stage of his career, as an undergraduate student at the Royal Institute of Technology in Stockholm at a study visit at the Laboratory of Surface Chemistry; this laboratory (later renamed the Institute for Surface Chemistry) had just been founded by Professor Per Ekwall. During the visit, the students were shown different samples of surfactant liquid-crystalline phases, and I still remember noting the large differences in flowing and optical properties between lamellar, hexagonal, and cubic phases.

As a rather new Ph.D. student, I started some years later NMR studies of different surfactant phases in collaboration with Ekwall and his coworker Krister Fontell. These studies focused initially on the ion binding by quadrupole relaxation (1).

Apparently, Ekwall was pleased with the information provided by the then-rather-novel tool of NMR so, when he was approached by Gray and Winsor about a book to be titled *Liquid Crystals and Plastic Crystals* (2), he suggested to them that I should write a chapter on the applications of NMR (3). Therefore, together with a fellow Ph.D. student, with great ambition I started to penetrate the literature. I became confused when I came to the cubic phases. As we know, cubic phases can be located in different concentration ranges in a phase diagram, inter alia between the micellar solutions and the normal hexagonal phase, and between the hexagonal and the lamellar phases. Regarding the structure, two of the leaders of the field, Winsor and Luzzati, had advanced very different and conflicting views. According to Winsor, all cubic phases must be built up of discrete spherical aggregates (4); a principal piece of evidence was the narrow NMR signals (long spin–spin relaxation times), which would exclude any extended structures (rod micelles give broad signals). Luzzati, on the other hand, from x-ray studies, favored structures with infinitely connected surfactant aggregates, thus bicontinuous structures (5); for the dilute cubic phase he had suggested a "mixture" model with both discrete and infinite aggregates (6).

Both Winsor's and Luzzati's ideas were in direct conflict with a monotonic change in aggregate structure with surfactant concentration; we would now discuss changes in the "critical packing parameter" or spontaneous curvature of the surfactant film. Winsor's NMR arguments came from plastic crystals, where reorientation of globular aggregates produces narrow NMR signals; he correctly noted that reorienting globular micelles would give an analogous NMR effect. However, my discussions with Håkan Wennerström revealed that this is not the full picture. Thus also lateral diffusion over an extended surfactant aggregate can for certain cases lead to averaging of the interactions of the nuclear spins and thus to long relaxation times; this would be predicted to occur for bicontinuous cubic phases. For this reason, NMR relaxation is not directly able to distinguish between surfactant phase structures based on discrete aggregates and connected "bicontinuous" structures.

Having gained some experience from self-diffusion, using NMR in Karlsruhe, Germany (7) and using the capillary tube technique with radioactive tracers in Montpellier, France (8), I soon realized that the surfactant self-diffusion would be very different for discrete aggregates and for connected structures. This would thus be an interesting possibility for solving the problem of the structure of cubic liquid-crystalline phases.

These plans advanced considerably when Krister Fontell in 1972 visited Lund and gave a seminar on the structure of cubic phases (9). In his overview, he presented a phase diagram of dode-cyltrimethylammonium chloride by Balmbra and Clunie with two cubic phases (10). Such a system immediately struck me as ideal for testing the structure problem by diffusion.

The radioactive tracer approach, which proved to be so success-ful for studies of micellar solutions, did not work out: it could not be adapted directly to cubic phases because of their extremely high viscosities. Instead, attempts with spin-echo NMR showed this to be an ideal experimental approach, and in one afternoon the problem was solved. The cubic phase, which is more dilute in surfactant, was found to be characterized by very slow surfactant diffusion and thus must consist of (more or less stationary) discrete aggregates. In the more concentrated cubic phase, surfactant diffusion was found to be more than an order of magnitude faster (see Figure 8.8). This finding, surprising from other starting points, could only be understood if the surfactant molecules could diffuse freely over macroscopic distances; thus surfactant aggregates are connected over large distances (11). Slightly later, Lindblom and coworkers elaborated the approach with the important monoolein–water system and made a quantitative comparison of the diffusion of lipid in the cubic phase with the lateral diffusion in the lamellar phase (12).

Regarding the cubic phases built up of discrete aggregates, Krister Fontell found the spherical shape inconsistent with his small angle x-ray diffraction data (13). The junior author (OS) addressed a confirmation of this using NMR relaxation (14,15). Indeed, a more detailed analysis of NMR bandshapes and relaxation data confirmed that the discrete micelles can be described as short prolates with an axial ratio of around 1.5:1 (15).

Already in his first work with Ekwall, the senior author had come into contact with "microemulsions" and performed NMR studies on such systems (16). (However, Ekwall considered this term superfluous and a misnomer: "these solutions are nothing but isotropic solutions of surfactant self-assemblies.") The field appeared then rather confused. Many authors were confused about the stability, thermodynamic or kinetic, of microemulsions and did not relate their occurrence to phase diagrams. Regarding microstructure, the situation was even more controversial; most authors lacked insight into other surfactant self-assemblies and, therefore, failed to base their studies on well-established features. Notable exceptions were, in addition to Ekwall, Friberg and Shinoda. They both performed pioneering phase diagram work and presented convincing arguments that the generally accepted droplet picture of microemulsions was only part of the story.

Using a similar approach as for cubic phases, it was quite straightforward to address the problem of microemulsion structure. Thus, by measuring oil and water self-diffusion, it should be quite easy to establish whether oil or water or neither of them is confined to discrete domains, droplets. In the first work on microemulsion structure by self-diffusion, using both tracer techniques and NMR spin-echo measurements, it was clearly shown that, in addition to droplet microemulsions, over wide ranges of composition they can be bicontinuous (17): this is manifested by both oil and water diffusion being rapid, not much less than the self-diffusion of the neat liquids.

Microemulsions are multicomponent systems, with typically at least three to five components. In the first study, using radiotracers and classical NMR methodology, each component must be studied in a separate experiment on a separate sample with suitable component labeling. Both the labeling and the huge experimental series required considerably slowed down progress.

At about the time when the senior author was finishing the first paper on microemulsion structure, he received from The Swedish

Royal Academy of Sciences a manuscript to review about an improved methodology for measuring self-diffusion by NMR by using a Fourier transformation in the NMR spin-echo experiment, Peter Stilbs and his student Michael Moseley (18) showed it to be possible in a single fast experiment to measure the self-diffusion of all components even for a complex multicomponent solution [for a review, see (19)]. This was immediately seen as the remedy in the microemulsion structure project, and on the same day, contact was made with Stilbs and the first experiments planned. The first publication followed quickly (20,21). Stilbs generously installed his technique in Lund, where it was the basis of intense work on the structure of microemulsions and other systems for a decade (22–28). The foundations of an understanding of microemulsion structure were laid by Olsson, Söderman, Guéring, Ceglie and many others, with Shinoda as a main contributor of interesting surfactant systems. The approach is still one of the cornerstones of research in Lund, although now with improved experimental facilities.

Another problem soon to be attacked was that of the microstructure of the sponge phases (sometimes called L_3 or anomalous phase). Isotropic solutions in simple surfactant–water mixtures were for a long time considered synonymous with solutions of discrete surfactant micelles. Indications of a more complex situation were given by the clouding and phase separation into two solutions of nonionic surfactants at elevated temperature and the observation of an "anomalous" phase by Lang and Morgan (29). Others to make similar observations were Mitchell et al. (30) and Fontell (31), the latter showing that addition of electrolyte to ionic surfactant systems can create an additional isotropic solution phase.

In his Ph.D. thesis project, Per-Gunnar Nilsson investigated micellar sizes in nonionic surfactant systems. This was then a matter of very important scientific dispute. The much-increased light and other scattering with increasing temperature was attributed to critical effects alone; both critical effects and micellar growth would give an increased scattering, but light scattering can not distinguish between the two phenomena. However, here NMR was expected to provide the solution. Since NMR monitors molecular events it is insensitive to critical effects, and both NMR relaxation and self-diffusion were successful in resolving this issue. It was found for nonionic surfactants with shorter oxyethylene chains that the self-diffusion becomes much retarded as temperature is increased. This could in principle be due to either micellar growth or association between small micelles; however, complementing with NMR relaxation it was straightforward to demonstrate that real growth takes place (32).

It was natural to make a corresponding investigation addressing micellar size in the sponge phase; this was performed by combining surfactant and water self-diffusion (33). Water diffusion was much reduced compared to classical micellar solutions. In fact it was close to two-thirds of the value for neat water. This is what is to be expected from the obstruction effect for a solution of large disclike aggregates. The surfactant diffusion was, on the other hand, found to be much more rapid than what was observed for previously studied micellar solutions. This rapid diffusion was taken to indicate that solutions contain disc micelles that are strongly attracting and therefore allow fast exchange between micelles. This interpretation was wrong, as was soon pointed out by our Ph.D. students.

This is an interesting example of "scientific blindness"; being focused on one problem, you do not use your knowledge and cannot see the larger picture. In this project, we were focused on the problem of micellar size and demonstrating the power of the NMR techniques in competition with scattering techniques to resolve central issues on surfactant self-assembly. It is true that the aggregates are disclike, but they are not discrete. Had we approached the problem in terms of microstructure and connectivity, as we did for cubic phases and microemulsion, we would have directly realized that the L_3 solutions, in contrast to the conventional micellar solutions, are connected and not discrete. In retrospect, these systems are extremely nice illustrations of bicontinuity. Both the water and surfactant self-diffusion coefficients are close to two-thirds of the values for the neat liquids, corresponding to an ideal zero-mean-curvature bicontinuous structure.

Our personal experience, as outlined above, of the development of the field of bicontinuous surfactant systems may make a suitable background for our description of principles and findings of NMR studies of bicontinuous cubic phases and related systems, illustrative for comparison, i.e., sponge phases and microemulsions.

We will start our treatise by providing a brief overview of NMR and how it can be applied to resolve problems of cubic phases. We will show how so-called static NMR parameters may tell directly about whether the phase is isotropic or anisotropic. This has to be discussed with reference to the molecular motions. Although all liquid and liquid crystalline phases in surfactant systems are characterized by almost as fast molecular reorientations and conformational changes as in simple liquids and solutions (on the order of

10 psec), all directions are not sampled with the same probability for the anisotropic phases. This (very slight, order parameters 0.01 to 0.1) preferential orientation accounts for the nonzero static parameters, such as dipolar and quadrupolar couplings.

We will then proceed to describe how dynamic NMR parameters, NMR relaxation, give information on the averaging of the interactions of the nuclear spins as a function of time. If averaging results solely from the rotation of a spherical aggregate, the (exponential) averaging is described by a single correlation time. However, the situation is more complex, with partial averaging being due to much faster local motions; this results in a two-step averaging and the necessity to introduce two correlation times. The "slow motion" (nanosecond time scale) is due not only to aggregation reorientation but also to translational motions over a curved surface.

Of particular relevance for our problems is that for nonspherical aggregates, the relaxation pattern will be more complex and will contain information on the degree of anisometry of the aggregate, such as axial ratio for prolate and oblate micelles.

Molecular self-diffusion also reports on aggregate size, but in particular it is of interest for us because of the information it conveys about confinement: a molecule that is located in domains that extend over macroscopic distances diffuses fast while one that is confined to closed domains will be very inhibited in its diffusion. While self-diffusion coefficients may be measured in several different ways, the Fourier transform pulsed field gradient spin-echo NMR technique (PFG NMR) is by far the fastest, most versatile, and most precise.

After describing how NMR can provide key information on bicontinuous surfactant phases, we present a number of examples of systems studied. It will be found fruitful to compare cubic liquid-crystalline phases with other isotropic surfactant systems, in particular sponge phases. To facilitate the presentation we will use the abbreviation BCP(s) for bicontinuous cubic phase(s) and SP(s) for sponge phase(s)

8.2 THE NUCLEAR MAGNETIC RESONANCE TECHNIQUE

8.2.1 Introduction

NMR is an extremely versatile technique, and a large number of NMR parameters exist that convey information on many different aspects of a wide variety of different systems. In NMR, one studies the physical properties of the nuclear spin system, which in turn are dependent on molecular properties. It is through the coupling between the spin system and the molecular degrees of freedom that relevant molecular information is obtained. It is fruitful to divide the experimentally obtainable NMR parameters into static and dynamic parameters. Below we will describe each of these two classes separately, with special emphasis on what is relevant for cubic and sponge phases.

8.2.2 Static NMR Parameters

The static parameters are obtained from the observed resonance frequencies, which are determined by the energy levels in the spin system. Due to the rotational symmetries of the cubit unit cell and the liquid-like properties of the sponge phase, these phases exhibit high-resolution spectra that contain little information about the microstructure of the phases. This is in contrast to spectra of phases of lower symmetries (such as hexagonal and lamellar phases), which are dominated by static effects; for these the NMR band shape directly reflects the underlying microstructure. The fact that the NMR spectra are "liquidlike" from cubic and sponge phases is useful, however, since NMR diffusometry can easily be applied (see below).

There is, however, one instance where the NMR band shape can be used to obtain information about the underlying microstructure of BCPs, and this is for the case of oriented and polymerized BCPs (34). Thus, if we can suppress the surfactant lateral diffusion by polymerizing the surfactant and obtain a single crystal of the cubic phase, then the band shape will report on the structure of the underlying dividing surface.

8.2.3 Dynamic NMR Parameters

Nuclear spin relaxation is typically induced by molecular motions, which render the spin Hamiltonian to become time dependent. Specific for surfactant systems is the fact that the relevant interactions (in most cases dipole–dipole or quadrupole) are first partially averaged by local motions (35,36). Then, global motion further reduces the interaction. In surfactant systems it is in many cases sufficient to consider only two types of global motion, viz. aggregate tumbling and lateral diffusion along curved surfaces. Specific for BCPs and SPs is the fact that aggregate reorientational motion is a very slow process on the NMR time scale (which is essentially given by the inverse of the residual interaction, expressed in sec^{-1}) and as a consequence, the only global motion needed to be considered is the surfactant diffusion along the curved dividing surface. As noted above, due to the symmetry of BCPs and SPs, the interactions are averaged to zero and a liquidlike spectrum is obtained. The NMR relaxation times, on the other hand, can distinguish between cubic and spherical symmetry (37). Therefore, NMR relaxation data from BCPs may convey information about the geometry of the dividing surface and also about the rate of molecular diffusion along the dividing surface. Early attempts used crude models, in which the NMR relaxation was described in terms of surfactant diffusion over the curved surface of a cube inscribed in the cubic unit cell. Later, Halle et al. developed the theory for spin relaxation in BCPs (38).

8.2.4 NMR Diffusometry

One of the most useful NMR techniques in the study of surfactant systems is the determination of self-diffusion coefficients by means of the Fourier-transform pulsed field-gradient NMR technique (39). The technique is commonly referred to as PGSE NMR or NMR diffusometry (henceforth referred to as NMRD). As in any measurement of particle migration in an equilibrium system, the molecules have to be labeled in some fashion. In NMRD, this is achieved through the application of a spatially varying magnetic field in the form of pulsed (magnetic) field gradients. With reference to the pulse sequence in Figure 8.1, the important thing to note is that the mean square displacement (MSD) during the diffusion time Δ of the molecules is determined. The second half of the echo is Fourier-transformed, and thus the MSD of all molecules that have separate NMR signals may be determined in one experiment. Thus, the approach is component resolved.

The most general analysis of NMRD experiments uses the fact that the echo attenuation, E, for a particular molecular species is given by the Fourier transform of the diffusion propagator $P(\mathbf{r}_0, \mathbf{r}_1, t)$ (40), with respect to the reciprocal space vector (which can be regarded as the scaled area of the gradient pulse) $\mathbf{q}$, defined as:

$$\mathbf{q} = (2\pi)^{-1}\gamma G\delta \tag{8.1}$$

where γ is the gyromagnetic ratio and the rest of the quantities are defined in Figure 8.1. E is then given by

$$E = \int\int \rho(\mathbf{r}_0) P(\mathbf{r}_0, \mathbf{r}_1, \Delta) e^{i2\pi\mathbf{q}\cdot(\mathbf{r}_1-\mathbf{r}_0)} d\mathbf{r}_0 \mathbf{r}_1 \tag{8.2}$$

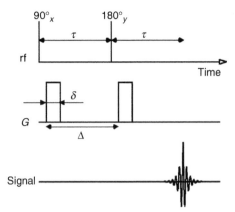

Figure 8.1   The basic PFG NMR pulse sequence. Two gradient pulses, each of length δ and amplitude G, are inserted in a spin echo sequence. The time between the leading edges of the gradient pulses is Δ, which corresponds to the diffusion time.

where $\rho(\mathbf{r}_0)$ is the equilibrium spin-density, and the integral is taken over all starting $\mathbf{r}_0$ and finishing $\mathbf{r}_1$ positions. $P(\mathbf{r}_0, \mathbf{r}_1, \Delta)$ is the solution to the diffusion equation with a delta function initial condition and appropriate boundary conditions. Thus, Fourier-transformation of the echo decay yields directly $P(\mathbf{r}_0, \mathbf{r}_1, \Delta)$. For systems in which $P(\mathbf{r}_0, \mathbf{r}_1, \Delta)$ is a Gaussian function, solution of Equation (8.2) yields the following equation for the (normalized) echo attenuation

$$E(\Delta, \delta, g) = \exp\left(-\gamma^2 G^2 \delta^2 \Delta D\right) \tag{8.3}$$

and the diffusion coefficient is obtained by regressing echo intensities for a given component onto Equation (8.3). The use of Equation (8.3) assumes that no diffusion occurs during the gradient pulse. For finite gradient pulses, the diffusion time Δ should be replaced by $(\Delta - \delta/3)$.

The range of diffusion coefficients that can be measured with NMRD is very wide, ranging from rapid diffusion of small molecules in low-viscosity solvents (typical values of $D = 1 \times 10^{-9}$ m^2 sec^{-1}) to polymers in the semidilute concentration regime or in the melt (down to values of $D \approx 1 \times 10^{-16}$ m^2 sec^{-1}).

Measurements of D of order 10^{-16} m^2 sec^{-1} require gradients of large magnitudes and place severe demands on the experimental setup (39). What often limits the lowest value of D that can be measured is the value of the tranverse relaxation time, T$_2$. As a general rule, slow diffusion is often found in systems that also show rapid transverse relaxation. As a consequence, the echo intensity gets severely damped by T_2-relaxation in such systems. One solution to this problem is to use stimulated echo experiments, which are sensitive to the spin-lattice relaxation rate T_1. The latter is often considerably longer in aggregated surfactant systems. However, it should be remarked that rapid relaxation is very seldom a problem in cubic and sponge phases, on account of the rather narrow NMR lines (implying long T_2-values) most often observed in such systems.

Finally, we note that the time scale of the experiment is such that the root MSD of the typical constituents of BCPs and SPs as determined by NMRD is in the μm-regime. This means that a molecule samples many structural units (the unit cell in the case of a BCP). The microstructure is then reflected in the reduction of the diffusion coefficients from the reference value (lateral diffusion along a bilayer) due to the fact that the surfactant must diffuse along the dividing surface. This is contrary to NMR relaxation (discussed above) since here the measured relaxation parameters reflect the diffusion rate along the dividing surface of one unit cell and consequently should be comparable to the reference value.

8.3 RESULTS

8.3.1 Introduction

For some 20 years it has been recognized that microstructures in bicontinuous phases are appropriately described by results from differential geometry. Thus the microstructure is described in terms of a multiply connected bilayer (for the case of most cubic and sponge phases, while a monolayer description is used in bicontinuous microemulsions) (41), and the features of the bilayer are captured by using results pertaining to triply periodic minimal surfaces (TMPS). One such surface is shown in Figure 8.2. The surface divides space into two volumes, and the bilayer is draped around the surface. Since the following discussion will be based on this structural model, we will say a few words about it. First, we note that for a normal structure (i.e. one where the curvature of the hydrophilic/hydrophobic interface is towards oil), the bilayer is composed of water, which drapes the film. For a reversed structure, the bilayer consists of surfactant (and other nonpolar material, if present), which drapes the film. The description in terms of a TMPS is reasonable at moderate volume fractions of bilayer, but the association to a bilayer

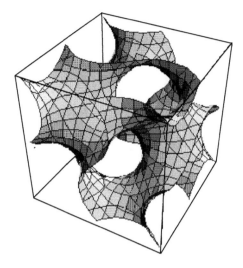

Figure 8.2 A patch of the gyroid surface. For the case of a normal structure, water drapes the surface, while for a reversed structure, the surface is the midpoint of the surfactant bilayer. (Figure courtesy of Andrew Fogden.)

becomes less clear at high bilayer volume fractions. Here, the interconnected rod (ICR) model originally introduced by Luzatti and coworkers gives a better description. In this model, the structure is described in terms of a network of connected rods, built up of surfactant (plus nonpolar material) in the normal case and water in the reversed case (42). The ICR structure is depicted in Figure 8.3.

The presence of a second solution phase in surfactant–water systems, distinct from the normal micellar phase and with different properties, came up in work by Ekwall, Fontell, Shinoda, Tiddy, Lang, and others. It received different notations, "L_3 phase", "surfactant phase", "anomalous phase," etc., and was thoroughly examined with respect to its occurrence. For example, it occurs at temperatures above the cloud point for nonionic surfactant systems and at high salinities for ionic surfactant systems. Subsequently, an analogous behavior (L_4 phase) has been identified for oil-rich systems.

The aggregate structure for these phases remained obscure for a long period, but here NMR techniques proved to be particularly powerful. The observations of narrow NMR signals, i.e., long transverse relaxation times, showed the absence of phase anisotropy.

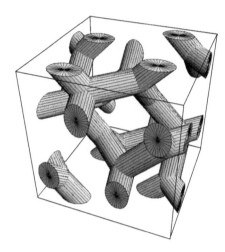

Figure 8.3 A picture of the interconnected rod (ICR) structure. For a normal structure, the rods are made up of surfactants, while for a reversed structure, the rods are filled with water. (Figure courtesy of Andrew Fogden.)

Surfactant and water self-diffusion data were striking and gave the clue to the microstructure. Both components diffuse very fast, in fact only moderately slower than the neat liquid components. As shown in Figure 8.4, over very wide concentration ranges the self-diffusion coefficients are very close to two-thirds of the values of the neat components (33,43,44). In the early work, the relative diffusion coefficients were compared to the obstruction effects of particles of different shapes (45). It was found that the water self-diffusion is inconsistent with spherical or prolate shapes of the surfactant self-assembly structures but in agreement with predictions for large oblates. Later theoretical work by Anderson and Wennerström (46) has shown the observations to be according to predictions for minimal surface structures. Several subsequent self-diffusion investigations of various systems have convincingly shown that the diffusion coefficients provide direct support for a connected bilayer structure, and quantitative analyses have been made in terms of precisely defined bicontinuous structures. The self-diffusion data and theoretical calculations suggest a close structural and topological analogy between L_3 phases, bicontinuous microemulsions, and the class of bicontinuous cubic liquid-crystalline phases. In particular, the analogy

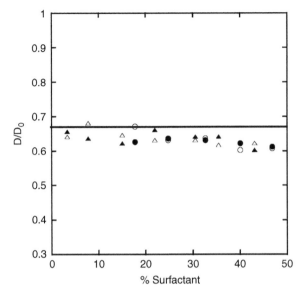

Figure 8.4 Relative surfactant (open symbols) and water self-diffusion coefficients (filled symbols) for the sponge phase in C12E3 (circles) and C12E4 (triangles). D_0 values refer to neat liquids of water and surfactant, respectively. (Adapted from B Lindman, U Olsson, H Wennerström, P Stilbs. *Langmuir* 9:625–626, 1993.)

between L_3 and bilayer continuous cubic phases is revealed in self-diffusion studies (47).

8.3.2 Bicontinuous Cubic Phases

8.3.2.1 Bandshape Studies

As pointed out above, the NMR spectra from a BCP do not in general contain any line-splittings (dipolar- or quadrupole-splittings) but consist of rather narrow NMR lines resembling those obtained from simple liquids. This is due to the symmetry of the phase and the liquidlike properties of the surfactant environment, which ensure rapid surfactant diffusion. The fact that a system with very high viscosity gives rise to a liquidlike NMR spectrum is sometimes confusing but is as mentioned a consequence of the cubic symmetry and rapid surfactant diffusion. Thus the NMR spectrum as such conveys little information about the phase structure. However, there is one instance where the bandshape conveys direct information about the microstructure. If the surfactant lateral diffusion is stopped through polymerization of the surfactant in the cubic structure and the cubic phase is aligned (i.e., in the form of a single crystal) then the bandshape is a direct signature of the phase microstructure. This fact was first pointed out by Anderson (34) who also presented calculations for bandshapes of a number of cubic phases. The starting point is the expressions for the Gauss map and Gaussian curvature of the underlying minimal surface, and the bandshape is calculated for a number of different minimal surfaces. An example is shown in Figure 8.5.

As far as we are aware, no experimental results along these lines have been presented, perhaps due to the conditions of polymerization and alignment. The polymerization of cubic phases has been achieved. A more difficult condition is the necessity to work with aligned samples. There are methods available to align other surfactant self-assembly structures. Application of magnetic or electric fields or the cooling (or heating) of aligned lamellar or hexagonal

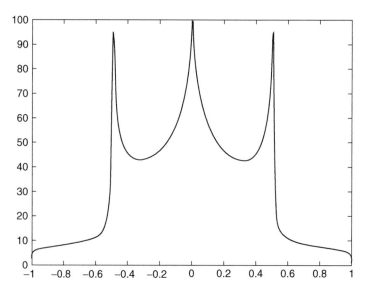

Figure 8.5 Calculated bandshape for a polymerized Schwarz P or D cubic phase with the crystallographic axis parallel to the static field of the NMR magnet. The NMR spectrum for a nonpolymerized corresponding cubic phase is a single narrow band. (Adapted from D Anderson. *Colloq Phys (Paris)* C 7:1, 1990.)

phases such that a cubic phase is induced are among the possibilities (48). One difficulty lies in the very high viscosity of cubic phases. Since the determination of cubic phase structures is in general a difficult problem, Anderson's suggestion offers a potentially useful approach to the investigation of microstructure in cubic phases.

8.3.2.2 NMR Relaxation

In this section we will limit ourselves to discussing ^{2}H NMR relaxation studies, where we assume that the ^{2}H nucleus has been incorporated into the surfactant (or cosurfactant) by isotopic labeling. This nucleus is relaxed by the quadrupolar interaction, which leads to reasonably simple expressions for the longitudinal (R_1) and transverse (R_2) relaxation rates (49)

$$R_1 = \frac{3\pi^2}{40} \chi^2 \left(2\, J(\omega_0) + 8\, J(2\omega_0)\right) \tag{8.4}$$

$$R_2 = \frac{3\pi^2}{40} \chi^2 \left(3\, J(0) + 5\, J(\omega_0) + 2\, J(2\omega_0)\right) \tag{8.5}$$

where ω_0 is the Larmor frequency, χ is the quadrupolar coupling constant and $J(\omega)$ is a reduced spectral density function.

On account of the fact that the rapid internal molecular motions in the surfactant film are anisotropic, the quadrupolar interaction is averaged in two steps. This forms the basis for the so-called two-step model used to describe NMR relaxation in aggregated surfactant systems. Within this model, the spectral density $J(\omega)$ function takes the form (35,36)

$$J(\omega) = A + S^2 J_s(\omega_0) \tag{8.6}$$

where the first term on the right-hand side refers to the constant contribution from the rapid internal motion (assumed to be within the extreme narrowing regime) and the second term on the right-hand side refers to the relaxation induced by the surfactant diffusion along the dividing surface. S is the order parameter, which determines the proportion of the interaction averaged by the latter process. In earlier relaxation work on BCPs, $J_s(\omega)$ was taken from the lateral diffusion over a sphere inscribed in the cubic unit cell. Later, Halle et al. derived the time correlation functions (TCF) (we recall that the $J(\omega)$ is the Fourier transform of the appropriate TCF) for diffusion along the dividing surface of a TMPS (38). For a powder sample, with a random distribution of unit cell orientations, the isotropically averaged TCF is biexponential. However, Halle et al. propose a single exponential approximation, which yields the following form for $J_s(\omega)$

$$J_s(\omega) = \frac{2\tau_{iso}}{1+\omega^2\tau_{iso}^2}$$ (8.7)

with the correlation time τ_{iso} given by

$$\tau_{iso} = \frac{a^2\alpha}{D_s}$$ (8.8)

D_s is the lateral diffusion coefficient, a is the size of the cubic unit cell, and α is a constant, the value of which depends on the chosen TMPS (38).

We will start by showing data that illustrate the large relaxation effects due to the lateral diffusion over the dividing surface. In the system didodecyltrimethyl ammonium bromide (DDAB)/decanol/water, there is a BCP that extends from 30 wt% to 75 wt% water. The molar ratio of decanol to DDAB in the region of existence of the cubic phase is around one. The phase is reversed. In Figure 8.6, the R_2 values of specifically deuterium-labeled decanol (with the CH_2 group adjacent to the CH_2OH group replaced with CD_2) are plotted vs. the volume fraction of water. Clearly, there is a large effect on the value of R_2. We note first that if $\omega^2\tau_{iso}^2 \gg 1$, then the $J_s(0)$ term dominates R_2 and we obtain

$$R_2 = \frac{3\pi^2}{20}\chi^2\left(5A + 3S^2\tau_{iso}\right)$$ (8.9)

As the volume fraction of water increases, the cubic unit cell size and hence the value of τ_{iso} increases. If D_s, S, and A remain constant (which is reasonable since these properties should depend only weakly on the unit cell size), the increase in R_2 directly reflects the increasing unit cell size. Further analysis, for instance the calculation of D_s, requires that the space group of the BCP and the unit cell size as a function of water volume fraction are known, for instance through x-ray experiments.

We now turn to frequency (ω_0) dependence of the relaxation. As noted above, the model presented in (38) for relaxation in a BCP assumes single exponential irreducible TCFs. This leads to a biexponential TCF for a nonoriented cubic phase. However, the deviation

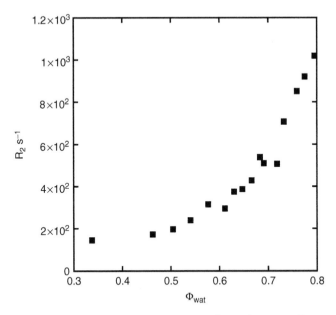

Figure 8.6 The deuterium transverse relaxation rate R_2 as a function of water content for specifically deuterated decanol in the BCP formed in the DDAB/decanol/water system.

from a single exponential TCF is small, and if one assumes a single-exponential TCF, Equation (8.7) is obtained for the BCP. To investigate the underlying assumptions leading to Equation (8.7), one needs to perform NMR relaxation experiments at varying frequencies. Such data for the BCP found in the binary system potassium octanoate (KC8)/water is presented in Figure 8.7. The solid line is the result of a fit to a spectral density function based on Equation (8.6), Equation (8.7), and Equation (8.8). It is clear that a single exponential TCF is in agreement with the data. The obtained value for τ_{iso} is 6.6 ± 0.2 n sec. From independent x-ray studies the space group for the BCP in the KC8 system has been determined to Ia3d (55). The TMPS that is compatible with this space group is the gyroid, often denoted G. For the G surface, the constant α in Equation (8.8) takes the value $\alpha = 0.00272$, and for the composition used in obtaining the data presented in Figure 8.7, the unit cell size, determined from x-ray studies, is 60 Å. Using Equation (8.8), the value of the lateral diffusion coefficient of the surfactant over the dividing surface can be evaluated, and the result is $D_s = 5.6 \times 10^{-11}$ m^2 sec^{-1}. We will

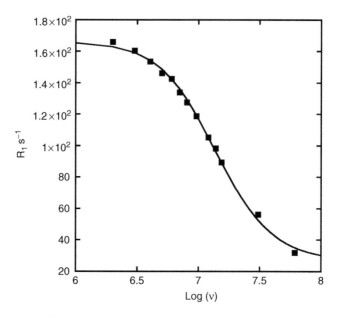

Figure 8.7 The deuterium longitudinal relaxation rate as a function of the Larmor frequency for the BCP formed in the potassium octanoate/water system (sample composition: potassium octanoate/water 68/32 by weight).

return to this result below, when we discuss NMR diffusion studies of BCP.

Before leaving this section we note that also other NMR relaxation studies of BCPs have been reported. Kuhner et al. present extensive frequency-dependent ^{1}H NMR data for the potassium dodecanoate/water system (50), and Burnell et al. have presented ^{1}H NMR relaxation data for a nonionic surfactant system over a large frequency range and interpreted the data in terms of the ICR model discussed above (51).

8.3.3 Diffusion Studies of Bicontinuous Cubic Phases and Sponge Phases

In this section we will discuss diffusion studies of both bicontinuous and sponge phases, for reasons that will become apparent. We start by discussing perhaps the most obvious application of NMR diffusometry in the study of cubic phases, namely to decide whether they

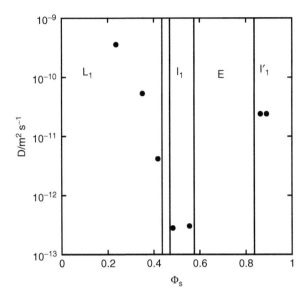

Figure 8.8 Diffusion coefficients for the surfactant in the DTAC/water system as a function of volume fraction of surfactant. Also included are the phase boundaries between the various phases. L_1 denotes micellar solution, I_1 a micellar cubic phase, I_1' a BCP, and *E* a hexagonal phase. (Adapted from T Bull, B Lindman. *Mol Cryst Liq Cryst* 28:155–160, 1974.)

are discrete or bicontinuous. In Figure 8.8 we present some classical data due to Bull and Lindman (11). In the binary system dodecyltrimethyl ammonium chloride (DTAC)/water, two cubic phases exist (10). The data in Figure 8.8 refer to surfactant diffusion and are presented as a function of the volume fraction of surfactant. Clearly, the two cubic phases show drastically different diffusion coefficients, differing by some two orders of magnitude. This shows without doubt that the first cubic phase (I_1 in Figure 8.8) is discrete and composed of (slightly elongated) micelles (14). The second one (denoted I_1') is bicontinuous. Thus, NMR diffusometry provides a rapid and unequivocal way of deciding whether a particular cubic phase is bicontinuous or not.

To interpret the value of the diffusion coefficient, we turn to the BCP found in the KC8/water system (discussed above). A PFG NMR study of a sample of the same composition and at the same temperature as used in the NMR relaxation study above gave a value of the surfactant diffusion coefficient of 4.8×10^{-11} $m^1 sec^{-1}$. This is then a quantity that can be directly compared to the value of D_s obtained from the relaxation studies. There are, however, two differences. First, in the NMR diffusometry approach, the length scale is in the μm regime, and thus the diffusion over many unit cells is obtained. In the relaxation measurements, the diffusion is measured over one unit cell. Thus, any defects or dislocations in the crystal structure affect the two approaches differently. Secondly, the value from NMRD is obtained from the mean square displacement, and thus it is influenced by the fact that the surfactant has to follow the dividing surface. This obstruction effect has been discussed by Anderson and Wennerström. They present data for a number of different structures, based both on TMPS and ICR. In general, for a system such as that of KC8/water, the obstruction effect is close to two thirds, and the value from the NMRD study should be multiplied by the inverse of this number to be directly comparable to those derived from NMR relaxation results. Thus we obtain from PFG NMR for the lateral surfactant lateral diffusion over the dividing surface 7.2×10^{-11} m^1 sec^{-1}. This is slightly higher than the value for D_s obtained from the relaxation study above ($D_s = 5.6 \times 10^{-11}$ $m^2 sec^{-1}$), but given the approximations inherent in the models, this is perhaps to be expected.

It is of interest to compare the value of D_s with what would be expected from predictions based on continuum fluid mechanics (52). For a cylinder of length ℓ and radius R embedded in a planar fluid layer of thickness ℓ and viscosity η, the lateral diffusion is given by:

$$D_{LAT} = \frac{kT}{4\pi\eta} \left[\ln \varepsilon - \gamma + \frac{8}{\pi\varepsilon} - \frac{2}{\varepsilon^2} \ln \varepsilon + O(\varepsilon^{-2}) \right] \qquad (8.10)$$

where $\gamma = 0.5722$ is Eulers constant and the parameter ε is given by:

$$\varepsilon = \frac{\eta}{R\bar{\eta}} \qquad (8.11)$$

where $\bar{\eta}$ is the average viscosity of the two media on either side of the surfactant monolayer.

Using the reasonable values ℓ = 11 Å, R = 3.6 Å, $\bar{\eta}$ = 0.75 × 10^{-3} $m^2 s^{-1}$ and $\eta = 1 \times 10^{-3}$ $m^2 sec^{-1}$, we obtain $D_{LAT} = 3.8 \times 10^{-10}$ $m^2 sec^{-1}$, which is roughly five times larger than that determined from NMR. This result indicates that the lateral diffusion of the surfactant in the monolayer is not given by the viscosity of the film, but rather dependent on the interaction in the head-group region. After all, the system is very highly concentrated in surfactant, and interactions between two adjacent films are probably important (the surfactant concentration corresponds approximately to 4 M).

Our final example of NMR diffusion studies of BCPs and SPs pertains to a comparison between the two phases. It has proven to be fruitful to relate the microstructure in the disordered SP to that in the ordered BCP. In the AOT/brine system, presented in Figure 8.9, there is a cubic phase at high surfactant volume fractions that is transformed into a SP phase upon dilution with brine (31). The cubic phase is of the reversed type. Given in Figure 8.10 and Figure 8.11 are the water and surfactant diffusion as measured by the PFG NMR approach, presented as a function of the surfactant volume fraction [data from (47)]. As is customary, the reduced water diffusion coefficient, (i.e., D_W/D_W^0 where D_W^0 is the bulk diffusion coefficient of pure water at identical conditions) is presented. A characteristic and important feature of the data is the smooth variation of the water and surfactant self-diffusion coefficients as one crosses the SP-BCP transition. D_W/D_W^0 decreases roughly linearly with Φ_S and approaches a value ot about two thirds at infinite dilution. The surfactant diffusion shows a minor decrease with Φ_S. The decrease is less than a factor of six over a change in Φ_S by a decade.

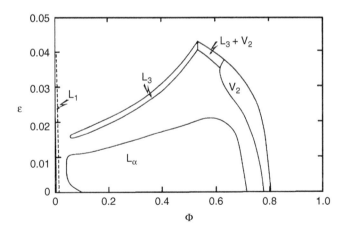

Figure 8.9 A schematic partial phase diagram of the AOT–water–NaCl system. Φ denotes the volume fraction of surfactant and ε denotes the weight fraction of NaCl in the water. V_2 is a BCP of the reversed type, while L_α is a lamellar phase. L_3 denotes the SP. (Adapted from B Balinov, U Olsson, O Söderman. *J Phys Chem* 95:5931–5936, 1991.)

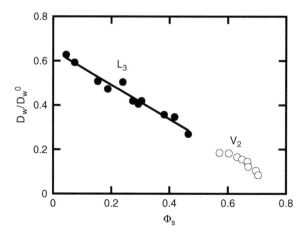

Figure 8.10 Variation of the reduced water self-diffusion coefficient with the volume fraction of surfactant, Φ_S. Data correspond to the SP (filled symbols) and the BCP (open symbols). The solid line is a linear fit to the data in the SP (see text for details). The obtained relation is $D_w/D_W^0 = 0.65 - 0.77\,\Phi_S$. (Adapted from B Balinov, U Olsson, O Söderman. *J Phys Chem* 95:5931–5936, 1991.)

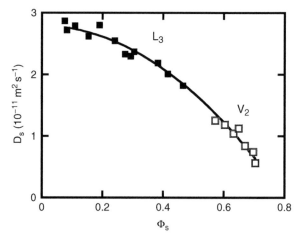

Figure 8.11 Variation of the surfactant self-diffusion coefficient with the volume fraction of surfactant, Φ_S. Data correspond to the SP (filled symbols) and the BCP (open symbols). The solid line is the prediction for diffusion over a TMPS (the Schwartz's minimal surface) assuming a value of $D_S^0 = 4 \times 10^{-11}$ m^2 sec^{-1} (see text for details). (Adapted from B Balinov, U Olsson, O Söderman. *J Phys Chem* 95:5931–5936, 1991.)

Since the diffusion of both surfactant and water depends on the microstructure, the smooth variation implies that the microstructure in the two phases is similar, and since we know that the cubic phase is bicontinous, it follows that the sponge phase is also bicontinuous. Moreover, the fact that the reduced water diffusion at infinite dilution is two thirds is only compatible with a bilayer structure, either in the form of large disklike micelles (45) or a multiply connected bilayer structure (46). However, the former can be excluded since large discrete aggregates would lead to very slow surfactant diffusion.

A more detailed analysis of the data in Figure 8.10 and Figure 8.11 can be made using the results presented in (46). At lower surfactant concentrations ($\Phi_S \leq 0.4$), the water and surfactant diffusion coefficients may be written

$$D_W = D_W^0(a - b\Phi_S)$$
$$D_s = D_S^0\left(a' - b'\Phi_S^2\right)$$

(8.12)

where D_S^0 is the surfactant lateral diffusion coefficient at infinite dilution (i.e., the lateral diffusion in a bilayer at otherwise equal conditions) and a, a', b, and b' are constants. While a is approximately two thirds, a' is analytically proven to be two thirds in (46). The values of b and b' depend on the topology, i.e., which family of TMPS one chooses to describe the structure. Two complications arise when using the relations in Equation (8.12). First, for the water, a quantitative analysis requires that hydration of the surfactant headgroups be taken into account. As is clear from Figure 8.10, the extrapolated value is in agreement with Equation (8.12), but the slope is steeper than the one predicted on the basis of results in (46). This is due to hydration effects. Second, for the surfactant the quantity D_S^0 is concentration dependent. The predictions of Equation (8.12) using data for b' for the P-surface and a value of $D_S^0 = 4.2 \times 10^{-11}$ m^2 sec^{-1} are included in Figure 8.11. The value of D_S^0 corresponds to the surfactant diffusion in a bilayer at infinite dilution. In the binary AOT/water lamellar phase at roughly $\Phi_S = 0.6$, a value of the lateral diffusion of 2.7×10^{-11} m^2 sec^{-1} of AOT has been reported (53), showing that there is a slight concentration dependence of D_S^0 at moderate surfactant concentrations (also apparent from the solid line in Figure 8.11). As Φ_S increases, the concentration dependence of D_S becomes stronger (cf. Figure 8.11). Here Equation (8.12) can no longer be used. Instead results based on the ICR model apply. Such results are presented in (47), and applied to the data in

Figure 8.11 they indicate that the rather strong variation of D_S with Φ_S in the cubic phase is due mainly to the concentration dependence of D_S^0.

Finally, we note that in some cases a BCP region may actually consist of several different structures. The water diffusion in different structures is slightly different owing to different values of the constant b in Equation (8.12), and this fact can be used to detect the transformation between different cubic structures (54).

ACKNOWLEDGMENTS

We acknowledge the fruitful and pleasant cooperation over the years with scientists in Lund and elsewhere. The work described here has received financial support from a number of sources, of which the most important are the Swedish Science Foundation, the Swedish Foundation for Strategic Research, and the Swedish Agency for Innovation Systems.

REFERENCES

1. B Lindman, P Ekwall. *Mol Cryst* 5:79–93, 1968.

2. GW Gray, PA Winsor, Eds. *Liquid Crystals and Plastic Crystals.* Chichester: Ellis Horwood Publishers, 1974.

3. Å Johansson, B Lindman. In GW Gray, PA Winsor, Eds. *Liquid Crystals and Plastic Crystals.* Chichester: Ellis Horwood Publishers, 1974, 192–230.

4. PA Winsor. *Chem Rev* 68:1, 1968.

5. V Luzzati, PA Spegt. *Nature* 215:701, 1967.

6. A Tardieu, V Luzzati. *Biochim Biophys Acta* 219:11, 1970.

7. HG Hertz, B Lindman, V Siepe. *Ber Bunsenges Phys Chem* 73:542–549, 1969.

8. B Lindman, B Brun. *J Colloid Interface Sci* 42:388–399, 1973.

9. K Fontell. In GW Gray, PA Winsor, Eds. *Liquid Crystals and Plastic Crystals.* Chichester: Ellis Horwood Publishers, 1974, 80–109.

10. R Balmbra, J Clunie. *Nature* 222:1159, 1969.

11. T Bull, B Lindman. *Mol Cryst Liq Cryst* 28:155–160, 1974.

12. G Lindblom, K Larsson, L Johansson, K Fontell, S Forsen. *J Am Chem Soc* 101:5465–5470, 1979.

13. K Fontell, KK Fox, E Hansson. *Mol Cryst Liq Cryst* 1:9, 1985.

14. O Söderman, H Walderhaug, U Henriksson, P Stilbs. *J Phys Chem* 89:3693–3701, 1985.

15. O Söderman, U Henriksson. *J Chem Soc Faraday Trans* 1 83:1515, 1987.

16. B Lindman, P Ekwall. *Kolloid-Z Z Polym* 234:1115–1123, 1969.

17. B Lindman, N Kamenka, TM Kathopoulis, B Brun, PG Nilsson. *J Phys Chem* 84:2485–2490, 1980.

18. P Stilbs, ME Moseley. *Chem Scripta* 13:26–28, 1979.

19. P Stilbs. *Prog NMR Spectroscopy* 19:1–45, 1987.

20. P Stilbs, ME Moseley, B Lindman. *J Magn Reson* 40:401–404, 1980.

21. B Lindman, P Stilbs, ME Moseley. *J Colloid Interface Sci* 83:569–582, 1981.

22. D Chatenay, P Guéring, W Urbach, AM Cazabat, D Langevin. In KL Mittal, P Bothorel, Eds. *Surfactants in Solution.* New York: Plenum, 1987, 1373–1381.

23. P Guering, B Lindman. *Langmuir* 1:464–468, 1985.

24. JO Carnali, A Ceglie, B Lindman, K Shinoda. *Langmuir* 2:417– 423, 1986.

25. U Olsson, K Shinoda, B Lindman. *J Phys Chem* 90:4083–4088, 1986.

26. B Lindman, K Shinoda, M Jonströmer, A Shinohara. *J Phys Chem* 92:4702–4706, 1988.

27. B Lindman, K Shinoda, U Olsson, D Anderson, G Karlström, H Wennerström. *Coll Surfaces* 38:205–224, 1989.

28. K Shinoda, M Araki, A Sadaghiani, A Khan, B Lindman. *J Phys Chem* 95:989–993, 1991.

29. JC Lang, RD Morgan. *J Chem Phys* 73:5849, 1980.

30. DJ Mitchell, GJT Tiddy, L Waring, T Bostock, MP McDonald. *J Chem Soc Faraday Trans* 1 79:975, 1983.

31. K Fontell. In: *Colloidal Dispersions and Micellar Behavior.* Washington, D.C.: American Chemical Society, 1975, 270.

32. PG Nilsson, H Wennerström, B Lindman. *J Phys Chem* 87:1377– 1385, 1983.

33. PG Nilsson, B Lindman. *J Phys Chem* 88:4764–4769, 1984.

34. D Anderson. *Colloq Phys (Paris)* C 7:1, 1990.

35. H Wennerström, B Lindman, O Söderman, T Drakenberg, JB Rosenholm. *J Am Chem Soc* 101:6860–6864, 1979.

36. B Halle, H Wennerstrom. *J Chem Phys* 75:1928–1943, 1981.

37. B Halle. *Liq Cryst* 12:625–639, 1992.

38. B Halle, S Ljunggren, S Lidin. *J Chem Phys* 97:1401–1415, 1992.

39. O Söderman, P Stilbs. *Prog NMR Spectroscopy* 26:445, 1994.

40. PT Callaghan. *Principles of Nuclear Magnetic Resonance Microscopy.* Oxford: Clarendon Press, 1991.

41. LE Scriven. *Nature* 263:123, 1976.

42. V Luzzati, A Tardieu, T Gulik-Krzywicki, E Rivas, F Reiss-Husson. *Nature* 220:485, 1968.

43. P-G Nilsson, H Wennerström, B Lindman. *Chem Scripta* 25:67–72, 1985.

44. B Lindman, U Olsson, H Wennerström, P Stilbs. *Langmuir* 9:625–626, 1993.

45. B Jönsson, H Wennerström, PG Nilsson, P Linse. *Colloid Polym Sci* 264:77–88, 1986.

46. D Anderson, H Wennerström. *J Phys Chem* 94:8683, 1990.

47. B Balinov, U Olsson, O Söderman. *J Phys Chem* 95:5931–5936, 1991.

48. PO Quist, B Halle, I Furo. *J Chem Phys* 95:6945–6961, 1991.

49. A Abragam. *The Principles of Nuclear Magnetism.* Oxford: Clarendon Press, 1961.

50. W Kuhner, E Rommel, F Noack, P Meier. *Z Naturforsch* 42:127–135, 1987.

51. EE Burnell, D Capitani, C Casieri, AL Sefre. *J Phys Chem B* 104:8782–8791, 2000.

52. BD Hughes, BA Pailthorpe, LR White. *J Fluid Mech* 110:349, 1981.

53. G Lindblom, H Wennerström. *Biophys Chem* 6:167, 1977.

54. P Ström, D Anderson. *Langmuir* 8:691, 1992.

55. K Fontell, L Mandell, P Ekwall. *Acta Chem Scand* 22:3209–3223, 1968.

Section 3

Applications

9

Synthesis of Controlled-Porosity Ceramics Using Bicontinuous Liquid Crystals

STEPHEN E. RANKIN

CONTENTS

9.1 INTRODUCTION

Interest in lyotropic surfactant mesophases has expanded into the field of materials science over the last ten years with the development of ordered mesoporous ceramic (OMC) materials that resemble "fossilized" lyotropic liquid crystals (1,2). The ability to control nanometer-level pore ordering in these ceramics has created exciting opportunities for the application of liquid crystal science in new materials such as low dielectric constant electrical interlayers (3), tailored adsorbents (4), chromatographic packing materials (5), molecular sieve membranes (6), catalysts and catalyst supports (7), enzyme supports (8), sensors, and optical components (9). Bicontinuous pore morphologies hold many advantages for these applications but have been studied rarely due to the relative difficulty of their synthesis (10).

OMC materials are synthesized by the liquid-phase reactions of inorganic precursors (e.g., dissolved silicate ions or alkoxysilanes) in the presence of surfactants such as alkyltrimethylammonium halides $(C_nH_{2n+1}N(CH_3)_{3+}X^-)$, illustrated schematically in Figure 9.1.

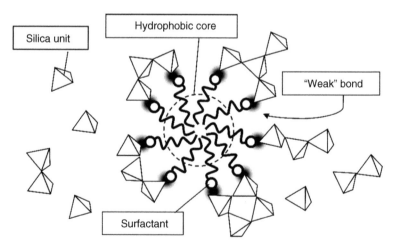

Figure 9.1 Schematic illustration of the coassembly of silica building blocks with structure directors such as surfactants. In a polar solvent, the apolar parts of the surfactant assemble to form one microphase, while silicates grow and interact with the polar head groups of the surfactant.

Researchers first formed these materials by hydrothermal precipitation-driven assembly (PDA) crystallization processes (1,2,11), but more recently, evaporation-driven assembly (EDA) methods have been developed for making films (12,13), lines (14), and aerosol particles (15). Weak interactions (van der Waals, screened ionic, or hydrogen bonding) between the polar parts of the surfactants and the silicates (tetrahedra) provide the driving force for coassembly.

If an alkoxysilane is used as the silica source, the alkoxide groups must first be hydrolyzed (Equation 9.1). Alternatively, a distribution of silanol (SiOH) functional siloxane species is prepared by dissolving silica at high pH. Silica grows from either of these molecular precursors by condensation reactions between silanol groups (Equation 9.2). Since the silica precursors are tetrafunctional, a highly crosslinked silica network forms. The type of precursor, presence of alcohols, and pH of the synthesis solution can strongly influence the structures of the silica oligomers present during the synthesis process (16).

$$Si(OR) + H_2O \leftrightarrow Si(OH) + ROH \qquad (9.1)$$

$$Si(OH) + Si(OH) \leftrightarrow Si - O - Si + H_2O \qquad (9.2)$$

In the presence of surfactant templates, the products of these reactions strongly resemble lyotropic liquid crystal mesophases (17), where the ceramic partially or fully occupies one microphase (15). The most widely studied OMC is MCM-41, a class of materials consisting of two-dimensional hexagonal close-packed (HCP) cylindrical channels running through silica [analogous to the H_1 surfactant phase (18)]. Synthesis conditions have been adjusted to produce materials with channel sizes between 2.7 and 30 nm, with wall thicknesses between 0.8 and 4.5 nm (19,20). However, the one-dimensional channels of the two-dimensional HCP structure can be easily blocked, and pore orientation by interfaces limits the accessibility of the channels.

When a surface is present that interacts preferentially with either component (polar or apolar) of a lyotropic two-dimensional HCP liquid crystal, the micelles align parallel to that interface (21). Figure 9.2 illustrates this effect with the results of lattice Monte Carlo simulations of a nonionic surfactant in water. The simulation technique follows that of Larson (22,23), and is described elsewhere (21). The composition of surfactant in Figure 9.2 forms a two-dimensional HCP phase in bulk solution, and confining this solution between parallel hydrophilic walls orients the cylindrical micelles parallel to those walls. The same thing happens with hydrophobic walls (21).

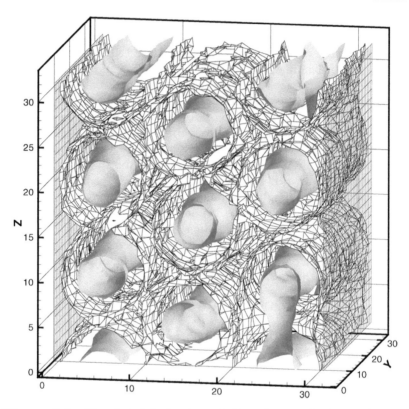

Figure 9.2 Simulated behavior of 60 vol% nonionic surfactant solution in water, confined between parallel hydrophilic walls. In these lattice Monte Carlo simulations, H_4T_4 surfactants consist of four hydrophilic (head) beads and four hydrophobic (tail) beads. Solvent molecules occupy one lattice site. Solid gray surfaces indicate the positions of surfactant tails, and black mesh the positions of the surfactant heads. The grey mesh to the left and right of the image shows the position of the walls. (From SE Rankin, AP Malanoski, FB van Swol. *Mater Res Soc Symp Proc* 636:D121–D126, 2001. Copyright 2001 Materials Research Society.)

Similar effects of surface chemistry on mesophase orientation have been observed in simulations of block copolymer thin films (24–26). Experiments confirm that both hydrophilic (mica) (27) and hydrophobic (graphite) (28) surfaces align MCM-41 pores parallel to the substrate in a thin film.

The alignment of cylindrical pores parallel to surfaces makes them unavailable for separation, sensing, and catalytic applications, especially in thin films. Some perpendicular alignment of cylindrical channels in thin films prepared at low pH has been achieved by promoting the formation of curved nuclei or defects (27,29), but complete control over pore orientation has not decisively been achieved. The cylindrical channels also bend easily, creating curved fingerprint-like disclination defects (30) that may compartmentalize the pore space.

Bicontinuous pore morphologies solve these problems. Silica or another metal oxide polymerizes in one of the continuous microphases, creating a material with sufficient structural stability. The other microphase is initially occupied by surfactant, but after calcination or extraction, a pore network is left behind that is better connected both internally and at the surface of the material than the two-dimensional HCP structure. For instance, bicontinuous surfactant-templated silica membranes formed by hydrothermal growth (31) or EDA (6) on porous alumina supports have been shown to display selective permeability for small gases.

With this technological advantage in mind, I will review here the formation of controlled-porosity ceramics with bicontinuous solid/pore morphologies. Surfactant templating of ceramics has been reviewed before (10,32–34), but this chapter will focus on recent developments in understanding the synthesis of *bicontinuous* pore structures. For more information about applications of the materials, the references in the first paragraph are good starting points for this rapidly expanding field. This chapter will address only silica as the solid material. Transition metal oxides (7,35) and organic functionalization (36,37) are also widely used in applications and are reviewed elsewhere.

9.2 BICONTINUOUS OMC STRUCTURES

9.2.1 Gyroid Structure (Ia3d Symmetry)

Simultaneous with the discovery of MCM-41, materials were isolated that resemble other lyotropic surfactant mesophases, including a lamellar (L_α) phase and a cubic phase (2). The first cubic phase made (MCM-48) has the structure of a bicontinuous isotropic (V_1) liquid crystal mesophase (18), with an Ia3d space group (38). The V_1 phase occurs at surfactant concentrations between those giving L_α and two-dimensional HCP cylindrical phases (39). The surface dividing microphases in the V_1 phase is ideally a triply periodic minimal "gyroid" surface with zero mean curvature everywhere because it is made up of saddle-like sections (40).

In MCM-48, silica occupies the triply periodic minimal surface between two continuous pore networks. The available pore structure is illustrated in Figure 9.3, which shows where the surfactant tails are found in a lattice Monte Carlo simulation (21) of 72 vol% symmetric surfactant in water. The structure consists of a regular arrangement of Y-shaped triple branches connected by pore channels.

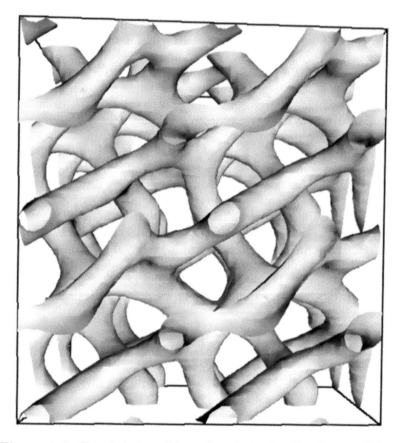

Figure 9.3 Simulated position of surfactant tails after equilibration for a solution of 72 vol% H_4T_4 in water, found using the lattice Monte Carlo simulation method described in Reference 21. This is a gyroid mesophase, emphasizing the bicontinuous nature of the phase.

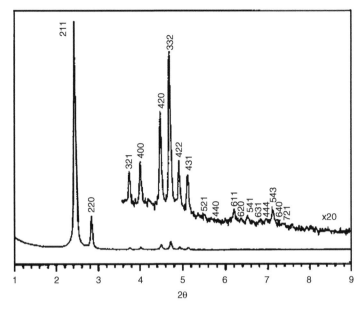

Figure 9.4 Synchrotron x-ray diffraction for as-made MCM-48 powder synthesized using $CH_3(CH_2)_{15}N(CH_3)_3Cl$ as the structure director at 100°C. (Reproduced with permission from Q. Huo et al., *Chem Mater* 8:1147–1160, 1996. Copyright 1996 Am. Chem. Soc.)

Experimentally, the Ia3d structure in MCM-48 and related materials is characterized by x-ray diffraction (XRD) (41) and by transmission electron microscopy (TEM) projections. Figure 9.4 illustrates well-resolved XRD results for large single MCM-48 crystals, where 18 different reflections could be indexed (42). Figure 9.5 shows the projections of the gyroid structure as seen by TEM in four directions, along with the theoretical projections of the structure (43). Nitrogen adsorption has been used to confirm that the surface area/volume relationship in MCM-48 materials follows that expected for the Ia3d space group (44), and the pore structure has been confirmed by three-dimensional reconstruction of the electron density by TEM (45).

Since MCM-48 was first discovered, a number of refinements have been made. The conditions where cubic particles form tend to be more limited than for MCM-41, but reliable synthesis methods have been found (46). Under the right conditions, MCM-48 grows

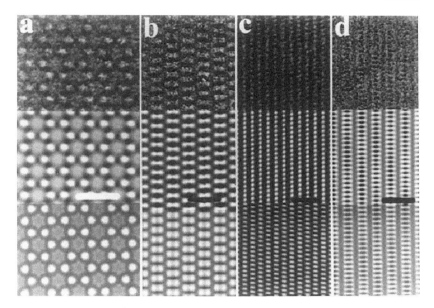

Figure 9.5 TEM images of MCM-48 in various projections. At the top of each column is the raw TEM image, below it a processed image, and below that the predicted image, based on the projection of the gyroid surface along (a) [111], (b) [311], (c) [432], and (d) [335] directions. (Reproduced with permission from V Alfredsson and MW Anderson, *Chem Mater* 8: 1141–1146, 1996. Copyright 1996 Am. Chem. Soc.)

as well-defined, faceted cubosomes (47) and particles (48). I will discuss the conditions and mechanism leading to the formation of this structure below.

9.2.2 Three-Dimensional Hexagonal Structure (P6₃/mmc Symmetry)

The mesophase structures of surfactant-templated silicates need not be those typically observed in binary surfactant/solvent phases. For charged surfactants, the growing silicate network provides a high-density charged environment. For all surfactants, the silicates can bind strongly and anisotropically with the surfactant headgroups, and the network may develop stress as it cures. These factors suggest that new phases, not observed for the surfactant solutions alone, may be observed in the coassembled silica product (12).

In 1995, Huo et al. (49) reported the first example of a surfactant-templated ceramic without a binary solution analogue: SBA-2, a three-dimensional hexagonal close-packed (space group P6$_3$/mmc) pore silica material formed by using double ammonium-headed cationic surfactants. Shortly thereafter, the same type of symmetry was found in aqueous nonionic polyethylene oxide surfactant mixtures (50). This structure is made by close packing of globular micellar aggregates (18). Ordinarily, the micelles in these types of phases remain discrete rather than forming a continuous microphase. However, in most ceramic materials with this structure, the spherical (or elliptical) cages formed by the surfactants are interconnected. The accessibility of these pores can be determined by small-angle x-ray scattering (SAXS) (51) and nitrogen adsorption (13). Interconnected pores may form either because the micelles were packed together closely enough to touch during synthesis, or because the walls between micelles were thin enough that they collapse or sinter away during surfactant removal.

The three-dimensional hexagonal phase was prepared in a thin film geometry by Tolbert et al. using the same double-headed quaternary amine surfactant (52). They prepared the films by growing them on immersed mica substrates under hydrothermal conditions. This is a significant breakthrough in the synthesis of templated ceramic thin films because the c axis in these films is aligned perpendicular to the substrate. This orientation provides a short-pathlength molecular-sieve membrane. Tolbert et al. speculate that this alignment is aided by [1] an ordered array of admicelles on mica and [2] acidic conditions, which allow the silica-surfactant aggregates to grow epitaxially from the wall layer (52). A film with the same symmetry could be formed from this surfactant by dip coating from an ethanol solution as well (53).

The formation of films with this symmetry is not limited by the choice of surfactant, however. Lu et al. (12) generated a three-dimensional hexagonal film from an initially ill-ordered dip-coated film during the removal of the surfactant, cetyltrimethylammonium bromide (CTAB). They used an acid-catalyzed ethanol solution for coating. The three-dimensional hexagonal structure formed when the films were left attached to the substrate, but the primitive cubic structure formed when they were detached. They propose that stress development during calcination contributes to three-dimensional hexagonal structure formation (12).

Grosso et al. also formed oriented three-dimensional hexagonal porous ceramic films by dip coating with CTAB as a template under similar conditions and identified the phase before surfactant

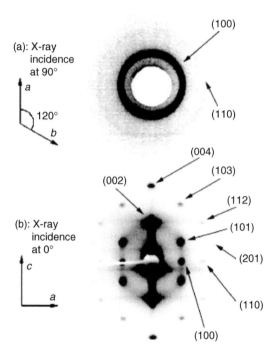

Figure 9.6 Two-dimensional x-ray scattering patterns for oriented, three-dimensional hexagonal close-packed pore films. The x-ray beam (a) is transmitted normal to the film or (b) meets the film edge-on. In (a), the presence of only two diffuse rings indicates that the structure consists of randomly oriented grains lateral to the film. The spots in (b) show that the three-dimensional hexagonal mesophase is oriented with the c axis normal to the substrate. (D Grosso, AR Balkenende, P-A Albouy, M Lavergne, L Mazerolles, F Babonneau. *J Mater Chem* 10:2085–2089, 2000. Reproduced by permission of The Royal Society of Chemistry.)

removal (54). The excellent orientation in their films was confirmed by contrasting x-ray diffraction of the films oriented perpendicular or parallel to the x-ray beam, as illustrated in Figure 9.6. When the x-ray beam is perpendicular to the film (Figure 9.6a), rings are observed that indicate randomly rotated hexagonal crystallites. For the parallel orientation (Figure 9.6b), sharp spots are seen, corresponding to the three-dimensional hexagonal structure oriented normal to the substrate. A study of the effects of surfactant content and silicate aging by Besson et al. (55) suggests that the three-dimensional hexagonal phase forms as a thermodynamically controlled phase in CTAB/silica mixtures, as discussed more below.

9.2.3 Other Cubic Structures

In addition to the widely synthesized MCM-48 (gyroid) structure, Huo et al. synthesized cubic close-packed cage structures with Pm3n symmetry by using large-headgroup cationic surfactants under acidic conditions [SBA-1 and SBA-6] (56). This structure is analogous to the I_1 phase of cetyltrimethylammonium chloride in formamide (57). Direct three-dimensional imaging of the pore structure by TEM (58), illustrated in Figure 9.7, shows that the cages in this material are interconnected by narrow throats. The dimensions of the cages of SBA-1 are on the order of 15 to 22 Å, while the throats are about 2 Å, and the coordination of the cages is 12. Presumably, the cages are formed by close packing of globular arrays of spherical or ellipsoidal micelles, around which the silica grows.

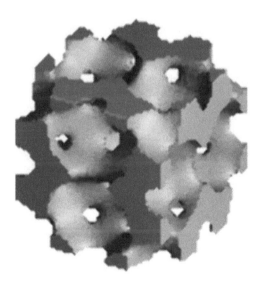

Figure 9.7 Three-dimensional electron density reconstruction from transmission electron micrograph of SBA-6, which has a close-packed globular structure with space group Pn3m. (From Y Sakamoto, M Kaneda, O Terasaki, D Zhao, JM Kim, GD Stucky, HJ Shin, R Ryoo. *Nature* 408:449–453, 2000. Copyright 1999 Nature Publishing Group.)

The cubic Pn3m structure has also been found to form in thin films prepared by acid catalysis. Lu et al. suggested that sometimes a primitive cubic film appears as an intermediate structure before a three-dimensional hexagonal structure forms (12). Grosso et al. (59) produced a Pn3m cubic product in thin films from CTAB. However, they saw evidence that the three-dimensional hexagonal film forms first, before the Pn3m structure forms. More recently, Grosso et al. showed that either the three-dimensional hexagonal or the Pn3m structure can be the product, depending on the precise conditions under which a film is formed under acidic conditions from CTAB and TEOS in ethanol (60).

Two other types of cubic symmetry have been found in packed globular micelle templated ceramics: first, a body-centered cubic ($Im\bar{3}m$ space group) array of cages connected by narrow throats (58). It is formed in powder form under acidic conditions from triblock copolymer Pluronic F127 ($EO_{106}PO_{70}EO_{106}$ where EO = ethylene oxide and PO = propylene oxide). The same surfactant can be used with solvent evaporation to form thin films with the same structure (61). Second, face-centered cubic ($Fm\bar{3}m$ space group) interconnected arrays of pores have recently been shown to be synthesized from nonionic templates such as $C_{18}EO_{10}$ (C = methylene) (62) and $EO_{39}(BO)_{47}EO_{39}$ (BO = butylene oxide) (63).

9.2.4 Disordered Bicontinuous Networks

In addition to ordered bicontinuous ceramics, disordered bicontinuous "sponge" structures can be formed. Ryoo et al. (64) synthesized and confirmed the bicontinuous nature of disordered pore networks from CTAC and sodium silicate under basic conditions. McGrath et al. (65) synthesized materials mimicking the disordered bicontinuous L_3 phase, using cetylpyridinium chloride as surfactant, with hexanol as cosurfactant and a high salt (NaCl) concentration. In this case, the L_3 phases were allowed to form and equilibrate first for several days before the silica precursor (tetramethoxysilane) was added. This is perhaps one of the clearest cases of mesophase templating, since the phase was present before adding the silica, and the product structure almost exactly matches the surfactant dimensions. While there is no long-range order, the characteristic pore sizes between 12.5 and 35 nm can be controlled by adjusting the surfactant concentration. Similar direct templating of a preexisting phase has been demonstrated for frozen bicontinuous oil/water microemulsions (66).

Thermodynamically stable L_3 morphologies are very similar to the interface structure formed during spinodal decomposition (67). This structure is a metastable morphology and is beyond the scope of this review. However, it is worth noting that other routes to disordered bicontinuous morphologies are available that are based on spinodal decomposition. Nakanishi's group has found that the balance of rates between phase separation and silica polycondensation can be controlled to form bicontinuous porous silica morphologies with controlled micron-scale size, without using any surfactant as a template (68).

9.3 SYNTHESIS MECHANISM

There are significant differences in the usual conditions for PDA and EDA. PDA is inherently heterogeneous and therefore can be controlled by manipulating the nucleation, growth, and transformation of silica-surfactant aggregates. PDA most often is done in basic conditions, where negatively charged silicate oligomers are present in solution.

EDA can be a homogeneous process, although interfacial influences may play a significant role in the formation and stabilization of the product phase. EDA is often used to make thin films (by coating), particles, and patterned structures. EDA is usually done under acidic conditions, to slow the rate of silica formation and prevent the formation of particulate defects. Because of the differences in the conditions of formation under PDA and EDA, we will discuss the synthesis mechanisms separately. In both cases, however, CTAB has been most widely employed and studied as a template, so this section will focus mainly on this surfactant.

9.3.1 Precipitation-Driven Assembly (PDA)

9.3.1.1 Liquid Crystal Templating?

By analogy with zeolites (69), the MCM family of materials was first synthesized under hydrothermal, alkaline conditions (1,2). Under these conditions, the silicates should be anionic, soluble species, and coulombic interactions with the cationic headgroups of the surfactants (such as CTAB) drives the formation and stabilization of an ordered phase. Initially, the mechanism was described as liquid crystal templating (LCT) (1,2), based on the analogy between the solid MCM phases and the phases observed in aqueous CTAB solutions.

The LCT mechanism implies that silica polymerizes in a pre-formed polar microphase. However, several observations contradict this hypothesis. In the dilute aqueous surfactant solutions used for most PDA methods, small angle x-ray scattering [SAXS] (70,71), small angle neutron scattering (72), and ^{2}H NMR of α-deuterated CTAB (72) show that two-dimensional HCP mesoporous silica structures form from solutions that contain no two-dimensional HCP phase prior to silicate addition. Phase investigations of aqueous CTAB support this finding: isotropic micellar structures are present below 20 wt% CTAB (73) [typical synthesis conditions use less than 5% CTAB].

Attard and coworkers have further explored the liquid crystal templating mechanism by first forming a concentrated surfactant mesophase and adding a silica precursor (74,75). Even though the structure was the same before adding silica and in the ceramic product, methanol formed from tetramethoxysilane destroyed the HCP order, and only after removing the methanol did the structure regrow (74,75). This suggests that LCT may be an adequate explanation for EDA (see below) but not PDA.

9.3.1.2 Cooperative Assembly

A more apt model of PDA is cooperative assembly of silicates and surfactants. Within 3 min after the addition of a silicon alkoxide at room temperature under some conditions, SAXS has shown that an ordered phase appears (70,71). More conclusively, both SAXS and ^{2}H NMR have demonstrated the formation of reversible, thermodynamically stable silicate–surfactant mesophases at a pH where the silicate oligomers do not polymerize (76). Macroscopic phase separation can occur for these systems, forming a surfactant/silica–rich phase and a water-rich phase (76). Therefore, partitioning of solutes between the two phases plays a role in the evolution of the structure. This mechanism is also consistent with the observation that different solid pore structures can be synthesized from solutions containing the same surfactant concentration by varying the silica:surfactant ratio (77). The mechanism is shown schematically in Figure 9.8, where silicates and surfactants assemble in dynamic equilibrium with a dilute aqueous phase.

The silicates present in alkaline solutions are distributed with respect to size, structure, and charge. As pH increases, the magnitude of the negative charge on each silicon increases (78), but the size of the silicate oligomers decreases because large ones become unstable (79). At moderately high pH, silicates act as highly charged

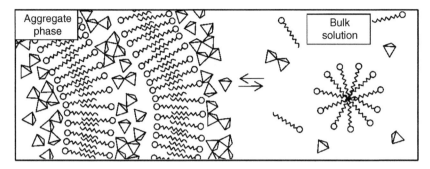

Figure 9.8 Schematic illustration of precipitation-driven assembly (PDA), in which nucleation, growth, and rearrangement of silicate oligomers and surfactants into ordered aggregates takes place from, and in contact with, a solution of the component ingredients. This process is inherently heterogeneous.

but reactive counterions for cationic surfactants. Stucky and coworkers have pointed out that strong binding (possibly multidentate) is likely between highly charged silicates and cationic surfactants (11). Binding of multivalent counterions provides a strong driving force for micellization, micelle elongation, and phase separation of surfactants (80).

Under some conditions, silica-encapsulated cylindrical micelle "nanotubes" have been captured by cryo-TEM (70) and observed in products of low-temperature MCM-41 synthesis with cetylpyridinium chloride (81). The aggregation of these nanotubes explains the formation of faceted MCM-41 particles (82). Ia3d cubic pore structures, however, are always isolated during a phase transformation process rather than forming by direct micelle assembly.

We should at least partially be able to understand and predict surfactant-silicate coassembly by aggregate geometry. One valuable concept is the packing parameter of the surfactant, g (83):

$$g \equiv \frac{V}{a_0 L}$$

where V is the volume occupied by a surfactant tail, a is the cross-sectional area of the head group, and L is the length of the surfactant tail. As g increases, the preferred curvature of the surfactant interface decreases, and based on geometric arguments, one can predict the shape of micelles in dilute solution: spherical for $g < 1/3$, elongated (cylinders) for $1/3 < g < 1/2$, bicontinuous layers for $1/2 < g < 1$, and lamellae for $g = 1$. Values of $g > 1$ should give rise to "inverted" structures in organic solvents.

The packing parameter can be used to rationalize how changing conditions and surfactants affect product structure. For instance, Huo et al. (42) showed that as the size of the ammonium headgroup decreases in hexadecylammonium surfactants, the product changes from one made of globular close-packed micelles (SBA-1) to a two-dimensional HCP (cylindrical) structure (SBA-3). Stucky and coworkers have proposed that charge density matching between the silicate and surfactant determines the effective area available per headgroup (11,84). The expectation that silica charge increases with increasing pH explains why the transition from HCP to lamellar products occurs as synthesis pH increases: surfactants pack more closely to match the increasing charge density, and a_0 decreases, causing g to increase (11).

The cooperative crystallization process is dynamic and may involve transformations between silica–surfactant mesophases. The time scale for silica polycondensation usually exceeds the time for self-assembly, so structures rearrange under changing conditions at a rate limited by silica network rearrangement and growth (85–87). For example, at room temperature a lamellar phase has been observed by *in situ* XRD that later evolves (11) or recrystallizes (88) into a two-dimensional HCP pore structure (a L2H transition). Silica–surfactant coassembly is actually a labile process, and hexagonal-to-lamellar (H2L) (85,86), gyroid-to-lamellar (G2L) (89), and hexagonal-to-gyroid (H2G) (87,90) transformations have been observed.

The L2H transition is consistent with decreasing charge density as silica polymerizes. A continuous structure change occurs, which is consistent with "puckering" of existing layers (11,86) to form a perforated lamellar intermediate (91). The H2L transition, on the other hand (induced by thermal or solvent swelling of the surfactant tails) occurs through a discontinuous transition. For binary surfactant solutions, the H2L transition proceeds through the Ia3d phase (91). The activation energy for both transitions is consistent with rearrangement of the silica network being the rate limiting step (85). If silica rearrangement or dissolution is not possible, certain structures may become "frozen" even though they are metastable. We will discuss in the next subsection how the Ia3d structure is captured during phase transformations.

The cooperative assembly concept has been quite successful at explaining trends in synthesis variables and suggesting new directions for materials development, but two conceptual problems prevent g

from being a truly predictive tool. First, the effective value depends on complex, temperature-dependent partitioning of solvents and counterions between headgroups, outer surfactant layers, and the surfactant core (92). Second, even with constant g, global volumetric constraints are also important (18). For a constant g value between 0.5 and 1, mesh, bicontinuous monolayer, or bicontinuous bilayer structures are possible, depending on the volume fraction of apolar microphase (18).

In addition to these thermodynamic issues common to all self-assembling systems, new challenges arise beyond shifting charge density because of the growing silicate network. As noted above, the network charge density decreases as silanols condense together. The network also shrinks during polycondensation. Internal stress within the silica as it solidifies may play a role in mesophase transformations and the shape of crystallites. Ozin's group has pursued this hypothesis and showed similarities between the shapes of HCP silica synthesized under acidic conditions and shapes that would be expected for crystallites with anisotropic stress (93).

In spite of these limitations, the cooperative assembly mechanism helps to explain how to control product structure by adjusting the [1] surfactant structure, [2] pH, [3] surfactant:silica ratio, [4] temperature, and [5] counterions. In addition to modeling the formation of ordered materials from anionic silicate (I^-) and cationic surfactant (S^+), cooperative assembly explains the success of all combinations of charged (56) and neutral (94) surfactants and inorganic species (7). When the inorganic ions and headgroups are opposite in sign, the inorganics act as competitive (95) counterions for the headgroup. When the inorganic ions and headgroups agree in sign, the inorganics act as a concentrated salt, and counterions of opposite charge mediate to allow cooperative assembly (56). Nonionic surfactants behave more like they do in aqueous solution (96), because they are less susceptible to salt effects than their ionic counterparts are. However, for good ordering with nonionic surfactants, it is important to promote hydrogen bonding with the silica (94) by choosing conditions (acidic usually) where alkoxysilanes are hydrolyzed to a high extent (97).

9.3.1.3 Bicontinuous Structures by PDA

Because quantitative models of silicate–surfactant–solvent phase behavior are not yet available, the next best approach to designing bicontinuous pore structures uses geometric parameters of the

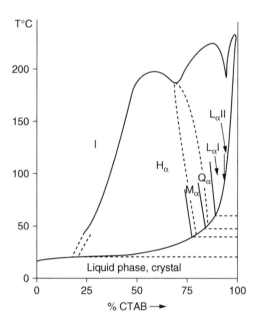

Figure 9.9 Measured phase diagram for hexadecyl (cetyl) trimeth-ylammonium bromide (CTAB) in water. I, H_α, M_α, Q_α, and L_α repre-sent isotropic lamellar, hexagonal close-packed cylinders and p2 monoclinic (deformed hexagonal), cubic Ia3d, and lamellar phases, respectively. (Reproduced with permission from X Auvray et al., *J Phys Chem* 93:7458–7464, 1989 Copyright 1989 Am. Chem. Soc.)

surfactant and its phase diagram under conditions as similar as possible to those of the synthesis. CTAB is a common templating surfactant, and its phase diagram in water is illustrated in Figure 9.9 (73). Only isotropic micelles and a two-dimensional HCP microemul-sions form at room temperature. Therefore, it is not surprising that an Ia3d porous ceramic synthesis at room temperature has only been reported once (98). At higher temperatures, a two-dimensional HCP phase, Ia3d cubic phase, and lamellar phase form as CTAB content increases. This is the typical sequence of ordered phases observed for binary surfactant solutions (23).

Although a region of Ia3d bicontinuous phase exists in the CTAB–water phase diagram (Figure 9.9), forming silica with the Ia3d pore structure (MCM-48) is challenging. In all documented cases, the bicontinuous cubic pore structure is observed as an inter-mediate between other ordered pore structures, so the synthesis time and temperature are extremely important variables, in addition to the surfactant:silica ratio, the pH, and the silica source.

Many successful syntheses increase the packing parameter (g) by using tetraethoxysilane (TEOS) as a silica source (11,46,47,72,77, 99–102). For each mole of hydrolyzed alkoxysilane (Equation 9.1), up to four moles of ethanol are produced, which can move preferentially into the outer layers of the surfactant microphase, increasing the effective volume of the surfactant tails. The role of ethanol has been confirmed by using dissolved colloidal silica as the ceramic source and adding ethanol just before hydrothermal treatment, leading to MCM-48 (100,101). Conversely, allowing ethanol to evaporate when TEOS is used as the precursor prevents MCM-48 from forming (46,101). However, the amount of ethanol must not be too large or the Ia3d symmetry is lost (100,101). Use of an alkoxysilane is preferable because mass transfer limitations can reduce the level of ordering when ethanol is added externally (100). In addition, because hydrolysis is rate limiting at high pH (16), residual alkoxy groups on the silicates should remain even at the point of network formation. They probably reduce the charge density of the silicate over what would be expected for a fully hydrolyzed surface. Other lower alcohols (101) or alkoxyalkylamines (49) can also be used to generate MCM-48.

MCM-48 is always produced by a transition (direct transformation or competitive crystallization) from another ordered phase. Some researchers isolated or specifically prepared a HCP structure as the precursor phase (100,102) while others found that a lamellar phase appears between the hexagonal and Ia3d structure (46,90). Tolbert et al. studied the process in more detail *in situ* by allowing the hexagonal phase to form at room temperature for a certain time (t_{HCP}) and then raising the temperature to induce a transformation (85).

There is an optimal length of time for the room temperature HCP synthesis step. If t_{HCP} is too long, the hexagonal structure transforms to a lamellar phase, presumably because of a high activation energy barrier for transformation of a highly condensed silica aggregate to the Ia3d phase. Tolbert et al. (85) propose that the topologically simpler H2L transition is favored in this case. This transition may occur by merging between cylindrical channels, which may lead to an Ia3d phase eventually but has a good chance of getting "frozen" on the way (87). A higher transformation temperature can overcome this obstacle. With shorter t_{HCP}, the H2G transformation occurs with a lamellar intermediate phase, consistent with

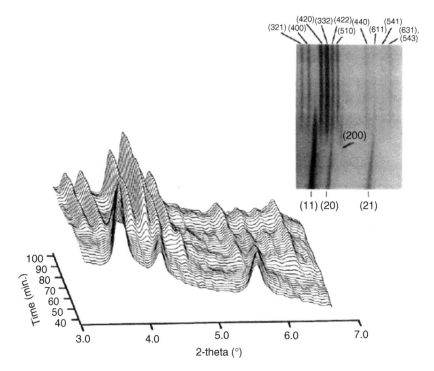

Figure 9.10 *In situ* x-ray diffraction results during transformation from MCM-41 to MCM-48 at 150°C. Only a limited set of higher-order peaks is shown. The inset shows a contour map of intensity versus time with the peaks indexed. The peaks at the bottoms are close-packed hexagonal (p6m) and at the top are cubic (Ia3d). The (200) peak from a lamellar intermediate appears briefly. (Reproduced with permission from CC Landry et al., *Chem Mater* 13:1600–1608, 2001. Copyright 2001 Am. Chem. Soc.)

a combination of cylinder merging and branching mechanisms (87). Figure 9.10 shows an example of these *in situ* XRD studies starting from the two-dimensional HCP phase, with an intermediate lamellar phase, and ending with the Ia3d phase. At very short t_{HCP}, the lamellar phase is again obtained. This probably happens because while the activation energy barriers for transformation are low (the silica has not condensed much) the negative charge density is high in the silica, favoring lamellae. At longer times or higher pH, further condensation may lead to another L2H2G transformation.

Research has also shown that the Ia3d geometry can be produced without an alcohol additive. Some reports involved the use of tetramethylammonium cations (48,103), whose role is unknown and may be similar to that of other organic additives. MCM-48 can be formed without *any* organic additives, either using partially neutralized sodium silicate as the SiO_2 source and sulfate counterions (104), or by using colloidal silica and cetyltrimetylammonium hydroxide as the base rather than sodium hydroxide (105). In the former case, the SiO_2/OH⁻ratio needs to be between 2.3 and 2.9 (104). In the latter synthesis with only CTAOH, SiO_2 and water, at 150°C after 24 h, a region was identified between hexagonal and lamellar products where MCM-48 was produced (105).

Alkylammonium surfactants have relatively little propensity to form bicontinuous cubic mesophases, as illustrated by the CTAB phase diagram in Figure 9.9. Surfactants that are more able to form Ia3d phases in water have been shown to be more effective templates for MCM-48. The first example is cetylbenzyldimethylammonium chloride, which produces MCM-48 at a wide range of surfactant:silica ratios (106). Probably, the benzyl group is able to "fold" and enter into the outer layers of the surfactant microphase, increasing the g value (10). Huo et al. (42) introduced the concept of using Gemini surfactants of structure $[\{H(CH_2)_nN\}_2(CH_2)_m{}^+]$. They showed that as m increases for n = 16 under hydrothermal alkaline conditions, the product is first lamellar (m = 3,4) and becomes two-dimensional hexagonal (m = 6,7,8,10). This makes sense since a_0 increases and g decreases in this series. With the highest spacer length (m = 12), the product becomes MCM-48.

One can rationalize that the longer spacer must be solubilized in the tails of the surfactant aggregates, leading to an increase in the V/a_0 ratio, which promotes the formation of a cubic phase (42). A subsequent study of synthesis variables (107) showed that cubic pores are formed for n = 16 or 18 and m = 10 or 12. The apparent degree of ordering of the products depends on synthesis treatment time and temperature but not the actual pore geometry (107). A bicontinuous cubic structure can be formed even at room temperature (42). Independent studies of Gemini surfactants show that the surfactant headgroup area passes through a maximum at m = 10 (when n = 12) (108). For m < 10, repulsive interactions between the two ammonium groups keep the spacer extended and sitting within the headgroup area in an aggregate. For m > 10, the headgroup spacing is determined by charge repulsion and the spacer moves into the tail aggregate, increasing g and favoring the Ia3d structure.

9.3.2 Evaporation-Driven Assembly (EDA)

In evaporation-driven assembly (15), a solution is initially prepared containing an excess of volatile solvent. One prepares a material by casting or forming the solution into a desired product shape and evaporating the solvent. While there may be interfacial nucleation or concentration gradients, evaporation drives the formation of and transitions between ordered silica–surfactant phases, rather than partitioning between a precipitated phase and a solution.

EDA presents significant advantages for the development of new applications for these materials. With PDA, one can hope to control only the shape, size, and crystallinity of discrete particles. On the other hand, EDA allows an unlimited variety of product shapes. The first EDA materials produced were continuous thin films formed by spin coating of surfactant–silica solutions in ethanol (109). Soon thereafter, dip coating was also used to prepare solid-supported thin films (12,13,61). Freestanding films can be produced by allowing silica–surfactant coassembly at the air–water interface (110,111). Because they are thin but have very uniform pores, these films have great potential for membrane separations, sensors, catalytic layers, and low dielectric constant electrical insulation (112).

One still can use EDA to prepare particles by drying aerosol droplets (113). Hexagonal, bicontinuous cubic and lamellar particles have been obtained but with the advantage (over PDA) that the synthesis and drying process only takes several minutes, instead of several days (113).

In addition to these cast or preformed structures, EDA synthesis solutions can be used as an "ink" to create detailed patterns by selective wetting during coating (15,114), dip pen lithography (14), ink jet printing (14), or rubber stamp printing (115). Hierarchical three-dimensional self-assembled structures (115) can be formed by using colloid crystal templating (116) to generate ordered macropores, with another level of mesoporous ordering in the walls that form between colloidal particles.

The EDA mechanism more closely follows the liquid crystal template mechanism originally proposed by Kresge et al. (1) for PDA. If one dries the material slowly enough that concentration gradients and precipitation are avoided, the composition of the material is uniform, and transitions occur sequentially from a molecular solution to micellar aggregates, and finally into a coassembled liquid-crystalline phase (15). After one ordered phase forms, there may be continued changes in the liquid-crystalline phase as continued drying and condensation occur.

There is a major difference in the type of bicontinuous structures that have been isolated during PDA and EDA. PDA structures have mainly been Ia3d cubic structures, while EDA structures have been formed by packing of globular micelles (with three-dimensional hexagonal or one of several three-dimensional cubic symmetries—see above). Several possible reasons for this difference exist. The first is obviously the pH of the synthesis—this affects the distribution of silicates, their rate of polymerization and their charge. The second is the presence of large amounts of solvent (ethanol, usually) which can affect the coassembly. Finally, there are a group of factors associated with the coating process itself: the presence of anisotropic interfaces, rapid drying, concentration gradients, and shear. We discuss the first two factors first, before summarizing what is known about evaporation-driven processes.

9.3.2.1 Acidic Sol–Gel Silica

Avoiding precipitation requires that EDA be carried out under acidic rather than basic conditions. Brinker et al. have found that operating near the isoelectric point of silica (around pH 2) works well because the silica condensation rate is minimized (16). Other researchers had success at lower pH values, where cationic silicates should be present (52,54,61). Acid-catalyzed synthesis can be made to work with cationic surfactants, so the positive charges on surfactants and silicates may be mediated by a negative layer of counterions.

Under acidic conditions, cyclization plays an important role in structure development of silica. Researchers have found vibrational (117) and NMR (118–120) spectroscopic evidence for the presence of six- to ten-atom siloxane rings in acid-catalyzed silica sols. The kinetic trends observed for the evolution of various silicon sites and the unusually high conversion at gelation for acidic sol–gel solutions can only be explained by extensive cyclization (121–123).

This cyclization is so severe that even 100% formation of single-ring species cannot model the high gel conversion of silica (about 82%; random branching theory (124) would predict 33%). There must be polycyclic and cagelike species present to explain the gel conversion of tetraethoxysilane (125–127). Equilibrium and kinetic studies have shown that the four-site (eight-atom) ring is favored over other siloxane rings (128), and XRD from acid-catalyzed sol–gel silica suggests a high number of these four-membered rings (129). Therefore, the double four-ring silicate is likely to be a prominent species under acidic sol–gel conditions, just as it and other polycyclic silicates are present at high pH (78).

Since silicate structure differences may be minor, one real difference between acidic and basic conditions may be that under acidic conditions, hydrolysis is fast and condensation slow, while under basic conditions condensation and silica dissolution are more rapid, and hydrolysis is rate limiting (16). This might explain why acidic conditions are favored for film formation. Due to fast condensation, alkaline silica–surfactant solutions often produce particles. The charge carried by the silicates (neutral or positive under acidic conditions) is the other significant difference from most PDA processes.

9.3.2.2 Solvent Effects

Polar organics such as alcohols act as cosolvents in aqueous surfactant solutions. If the solvent is somewhat apolar, it will penetrate into and swell the core of the micelle (130,131). As discussed above, swelling surfactant tails allow ethanol to promote Ia3d cubic structure formation in PDA. Brinker et al. suggest that structure development during EDA can be understood based on surfactant–solvent–water ternary phase diagrams at the temperature of synthesis (12). This is certainly a useful tool for predicting the compositions giving hexagonal close-packed structures (132), and for tracing the composition evolution within a thin film. However, the dried silica–CTAB thin films are prone to forming close-packed globular micelle phases (55) that are not observed in the CTAB–ethanol–water phase diagram at room temperature (133). *In situ* investigations (see below) have shown that the final ordered phase usually forms after all of the ethanol has left the film, so the main effect of the solvent is to control the drying and polymerization rate of the product.

9.3.2.3 Coating Flow Effects

The flow geometry experienced while processing an EDA material may produce significant effects in coatings, as discussed by Hurd and Steinberg (134). The first new features are large interfaces in contact with the coating solution — both the vapor–liquid interface and the substrate–liquid interface. Usually, the substrate interface is silicon dioxide and is polar, while the vapor phase is apolar. Both types of interfaces align anisotropic mesophases (hexagonal cylinders and lamellae) parallel to themselves (21) (see Section 9.1). This may cause the ordered mesophase to grow by propagation from both the substrate and the evaporating film surface. Lamellar structures in aerosol droplets form as concentric spheres, supporting this hypothesis of vapor–liquid interfacial alignment (113).

Evaporative synthesis under flow also introduces gradients of concentration and velocity (shearing). A parabolic film height profile is observed near the point of 100% dryness during dip coating (135). This gives extremely rapid drying near the drying line and therefore a large concentration gradient (136) and shear stress (134), with a maximum stress appearing just before the drying line. Shearing may align mesostructures in the direction of the coating (137), not only for lamellar structures but also for cubic phases (134). Concentration gradients may also exist normal to the vapor–liquid interface if the rate of evaporation exceeds that of diffusion within the coating (138). This is the basis for growth of freestanding films at vapor–liquid interfaces (34). These gradients may induce nucleation of the ordered film at the vapor interface. The specific effects of these coating flows are still not completely understood, but the EDA mechanism is beginning to be better understood due to *in situ* experiments.

9.3.2.4 *In Situ* Studies of EDA Mechanism

The EDA process is much more photogenic than the PDA process. Dip coating, in particular, affords spatial resolution of the process of film formation because the distance above the coating reservoir is proportional to the drying time. This allows one to directly study the EDA process by a variety of techniques. Imaging ellipsometry (139) gives the thickness profile, and fluorescent probes may be added to determine the ethanol/water ratio (140) or local molecular mobility (141) as a function of coating height. Long-range structural order can be determined as a function of coating time *in situ* by synchrotron XRD experiments (142).

Figure 9.11a shows an example of *in situ* imaging ellipsometry during dip coating of an acidic solution of prereacted TEOS and CTAB in ethanol, with a fluorescent probe molecule added to indicate the onset of micelle formation (12). The black interference fringes indicate increments of film thickness as a function of height above the coating reservoir. In Figure 9.11b, the actual thickness is plotted versus the height, which is proportional to time, above the reservoir. The amount of fluorescence depolarization (P), which indicates the onset of micellization, is also shown. The fast pace of this synthesis method is illustrated by the surfactant concentration scale, which indicates a tenfold increase of surfactant concentration in under 3 sec near the drying line. Using the same technique for sodium dodecyl sulfate templating, M. Huang et al. (141) found (1) preferential evaporation of ethanol leading to water enrichment in the final stages of coating, (2) transformation to a lamellar phase well after

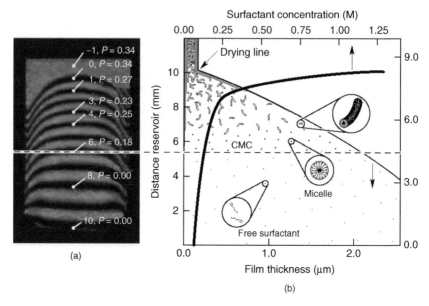

Figure 9.11 (a) Ellipsometry image taken *in situ* during dip coating to form a HCP film structure. Each black fringe corresponds to an increment of film thickness, and P is the fluorescence depolarization, indicating that a fluorescent probe enters into a micellar aggregate. (b) Schematic illustration of evaporation-driven assembly (EDA), showing film thickness and corresponding surfactant concentration as a function of distance/time above reservoir. (From Y Lu, R Ganguli, CA Drewien, MT Anderson, CJ Brinker, W Gong, Y Guo, H Soyez, B Dunn, MH Huang, JI Zink. *Nature* 389:364–368, 1997. Copyright 1997 Nature Publishing Group.)

the onset of micellization, and (3) evidence for molecular mobility persisting for many hours after preparing the films.

Lu et al. (12) used *ex situ* TEM observation of gradations in structure to conclude that the ordered structures grow from both the vapor–liquid and substrate–liquid interfaces during acid-catalyzed templating of CTAB in ethanol. Also as drying occurs with higher surfactant contents, they propose that lamellae form first near the interfaces and transform first to (primitive) cubic and then to the three-dimensional hexagonal structure, similar to the L2H transformation observed for PDA particles (11). However, this order of phase appearance has not been observed by other researchers and may have been influenced by stress development during calcination (11).

Grosso et al. recently studied the details of the dip coating process during surfactant templating by *in situ* synchrotron XRD (59,60). They observed a preliminary diffraction peak attributed to micelle formation near the interface. Only after several seconds of drying, bicontinuous ordered phases begin to form throughout the film (as verified by two-dimensional XRD and TEM of the product films). They studied the effect of CTAB/TEOS ratio in the coating solution for sols that were aged for a week at room temperature under acidic conditions. Figure 9.12 presents the dynamic *in situ* SAXS results. In all cases, the initial scattering from the micelle layer is seen, followed by the formation of ordered phases.

For the lowest CTAB/TEOS ratio (0.08) (Figure 9.12a), the micelles transform epitaxially into a two-dimensional hexagonal cylindrical structure. However, the HCP ordering is very weak and confined to thin layers just near the air and substrate interfaces. There-fore, this does not seem to be an equilibrated bulk structure, and another investigation (55) of a wide array of conditions has suggested that the two-dimensional HCP structure is actually the bulk product at much higher surfactant ratios.

At CTAB/TEOS = 0.10 (Figure 9.12b), the three-dimensional hexa-gonal phase is the first and only phase observed. At CTAB/ TEOS = 0.12 (Figure 9.12c), the three-dimensional hexagonal phase forms briefly, but the cubic Pm3n phase is the ultimate product. The ellip-sometry of Grosso et al. (60) also shows that the drying occurs in three stages consistent with other investigations of dip coating: (1) preferential, fast evaporation of ethanol, (2) slower loss of water, and (3) shrinkage due to condensation of the completely dry film. Their studies suggest that micelles form near the interfaces during the first stage but that the organized films emerge after the end of the second drying stage (60).

A study of humidity effects on EDA films (143) has shown that the length of the second drying step (loss of water) can be controlled by the relative humidity, and that the final water content can be manipulated by the humidity to change the product phase structure. Higher water favors a cubic (Pm3n) phase over the two-dimensional or three-dimensional hexagonal structure (143). For a certain time period after the film is completely dry, the mesostructure can be modulated. This suggests that the silica network is still labile at this point, and that self-assembly thermodynamics may dictate the struc-ture. Similarly, Alonso et al. found that more ethanol in the vapor phase during drying can promote the three-dimensional hexagonal-to-cubic (H2C) phase transformation (144). The ethanol in the vapor increases

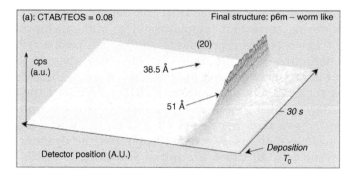

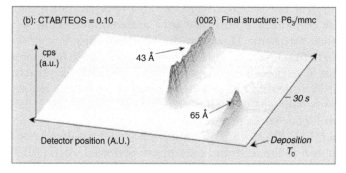

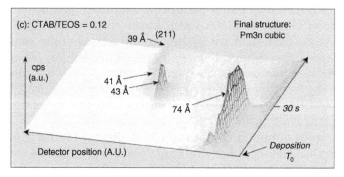

Figure 9.12 *In situ* x-ray diffraction results during dip coating of TEOS/CTAB/ethanol/water/acid solutions resulting in three types of ordered phases. CTAB/TEOS ratios are indicated. From full two-dimensional scattering images, the authors assign the peak with d-spacing of 38.5 Å to the (20) plane of a two-dimensional HCP structure; 43 Å to the (002) plane of a three-dimensional HCP structure; and 39 Å to the (211) plane of a primitive cubic structure. (Reproduced with permission from D. Grosso et al., *Chem Mater* 14:931-939, 2002. Copyright 2002 Am. Chem. Soc.)

the ethanol content in the final "dry" film and may promote the H2C transition because of either slower silica polycondensation or easier transformation due to reduced viscosity in the film.

The study of the phases produced for different amounts of aging of the silica sol and different CTAB/Si ratios by Besson et al. (55) is consistent with the idea that thermodynamics of micelle assembly controls the mesostructure while silica polymerization controls the rate of phase formation. Figure 9.13 is the reported synthesis field diagram, showing that as the CTAB/Si ratio increases, the order of phases is three-dimensional hexagonal, then Pn3m, then two-dimensional HCP. This is consistent with the behavior expected for micelle phases as their concentration is increased (18). However, there is an optimal range of aging times for the silica sol. If it is too short, the silica oligomers are either dense (55) or volatile, giving a high volume

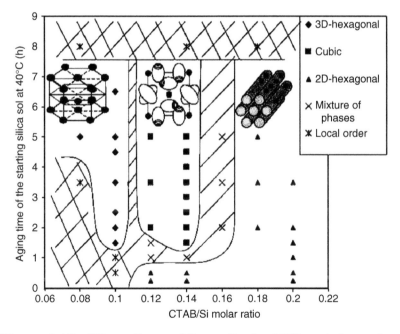

Figure 9.13 "Phase diagram" for synthesis of different phases from CTAB and TEOS by dip coating of acidic ethanol solutions. S Besson, T Gacoin, C Ricolleau, C Jacquiod, J-P Boilot. *J Mater Chem* 13:404–409, 2003. Reproduced by permission of The Royal Society of Chemistry.

fraction of surfactant. If it is too long, the oligomers gel too quickly, leading to a porous product without long-range order.

The reason that close-packed micelle mesophases can be observed in these films is probably the acid catalysis of the reaction. Positively charged silanols, even in the presence of an intermediate layer of counterions, will exert a driving force to minimize the charge density at the cationic surface of micellar aggregates. This increases the area per surfactant, thus decreasing g over what it would be in a dilute solution. This is the opposite effect from the charge density matching in alkaline-produced materials. Before the ethanol is gone, there is evidence that elongated cylindrical micelles form (60). As the ethanol leaves at a high enough surfactant:TEOS ratio, globular micelles form and pack together. The high concentration of silicate cations must cause large headgroup repulsion, producing a highly curved micellar aggregate. Similarly, finding close-packed globular phases of cationic surfactants in formamide has been attributed to efficient counterion dissociation and large headgroup charge (57,145).

For nonionic surfactants, these charge-matching considerations disappear (75), and for instance the mesostructures of porous titania thin films have been controlled by using the phase diagram of nonionic surfactant Pluronic P123 in water to predict the volume fraction of surfactant required to form a desired phase (146). Because the relationship between surfactant thermodynamics and porous product is straightforward, globular cubic phase formation can be promoted by selecting a nonionic template surfactant with a large polar headgroup (61,63,96).

9.4 CONCLUSIONS AND SYNTHESIS CONDITIONS

The synthesis of bicontinuous porous ceramics by surfactant templating is a complex process, but in this chapter I have tried to summarize the state of the art of understanding the process of their formation. The process can primarily be understood to be cooperative assembly of silicate species and surfactants in the presence of solvents and counterions.

Under alkaline conditions, where the PDA mechanism operates, polydentate silicate anions compete with other counterions to coassemble with surfactants and form ordered meso-phases. The concept of surfactant packing parameter and its response to surfactant concentration, surfactant structure, charge density matching, and cosolvent addition can qualitatively be used to explain the appearance of the bicontinuous cubic Ia3d structure between two-dimensional HCP and lamellar phases. Product phases can be related to phase diagrams of the template surfactants, especially for nonionic surfactants, where charge density and counterion binding do not interfere. However, the forces allowing the Ia3d phase to exist must be delicately balanced, and usually the only way to capture the phase is to prevent transformation into the next (usually lamellar) mesophase.

Because of the complex set of conditions under which they form, I summarize here some of the conditions that have been used to produce Ia3d particles. The products should show clear evidence for the Ia3d geometry by XRD, with many narrow peaks indicating product quality. Also, the products should have a high specific surface area because impure phases (collapsed lamellar structures) increase mass but not surface area. Finally, fast or energy-efficient methods are of particular interest for the long-term development of these materials.

Table 9.1 summarizes selected Ia3d bicontinuous particles formed by PDA. Included in the table are synthesis techniques giving MCM-48 quickly (100), at low temperature (with Gemini surfactants) (42,107), or with inexpensive silica precursors (101,104). All give products of high quality at the indicated conditions. Also, studies have been made of the range of compositions yielding MCM-48. Sayari mapped the conditions giving high-quality MCM-48 crystals from Cab-O-Sil fumed silica at 140°C after 40 h, with 0.317 $(CH_3)_4NOH$ (TMAOH) per Si added in all cases (48). Peña et al. found the compositions giving MCM-48 from the simple ternary system of Aerosil-200, CTAOH, and water after 24 h at 150°C (105). Shaded compositions yielding MCM-48 in Figure 9.14 for both studies are fairly close and occupy narrow bands bordered by HCP and, in the study of Peña et al. (105), lamellar phases.

EDA can also be understood in terms of cooperative assembly. Because homogeneous products are usually desired by EDA, synthesis is conducted at low pH, where silica polycondensation is slowest. After ethanol removal, the silica oligomers are positively charged. The effect of positively charged surfactants is analogous to that of added salts, rather than competitive counterions. One can still correlate the product phases to the phase diagram of the templating surfactant, especially for nonionic surfactants where salt effects are weak.

Typical conditions for the formation of EDA materials are similar to those originally reported by Brinker's group (12). First, a silica sol is prepared by mixing together tetraethoxysilane with limited water (3 water/Si) in an ethanol solution with a pH somewhat less than 4. The solution is aged at elevated temperature for some

Table 9.1 Synthesis Methods for Bicontinuous Silicates with Ia3d Symmetry by PDA

Surfactant	SiO$_2$ Source	Surf:Si	OH:Si	H$_2$O:Si	Synthesis Conditions	Product	Ref.
CTAB	TEOS	0.12	0.5	1109	3 h 25°C, 3 h 150°C	"Well-ordered"	100
Gemini C18/C12 spacer	TEOS	0.06	0.6	150	4 days in basic solution + 2 days in water, both at 100°C	Eight indexed XRD peaks, SBET = 1600 m²/g	107
CTAB	Colloidal silica Ludox HS40	0.71	0.5	100	3.57 Ethanol/Si added, 4 days at 100°C, product removed at 67°C	12 XRD peaks indexed, 0.3 μm single crystals	101
Gemini C22/C12 spacer	TEOS	0.06	0.69	150	1 day room temperature	~11 XRD peaks visible	42
CTAB	Sodium silicate	0.20	0.38	120	36 h at 150°C	Eight XRD peaks, SBET = 1116 m²/g	104

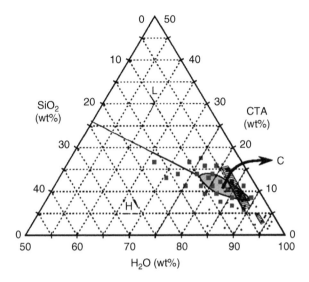

Figure 9.14 Composition space giving rise to cubic Ia3d porous particles. The gray squares are original points for CTAOH/silica/water from Pena et al. (From ML Peña, Q Kan, A Corma, F Rey. *Microporous Mesoporous Mater* 44–45:9–16, 2001) and the small black circles are superimposed points for CTAB/silica + TMAOH/water from Sayari (From A Sayari. *J Am Chem Soc* 122:6504–6505, 2000). The shaded region gives MCM-48 products after 24 h at 150°C. The lined region gives MCM-48 products after 40 h at 140°C.

time and more ethanol, acid, water, and the surfactant are added. This solution is again aged for some time before spin or dip coating. At this pH, the initial aging steps should give rise to a distribution of hydrolyzed silica oligomers containing many ring structures (127). The product structure is a function of both the age of the silica sol and the surfactant:Si ratio, and Figure 9.13 gives a typical set of ordered phases observed.

For cationic surfactants, ordered close-packed micellar phases are observed that may be absent from the binary water–surfactant phase diagram. The formation of these phases is driven by forcing the surfactants to coexist with a high concenration of charged material. The charge density of the cationic headgroups is minimized by increasing the area, which drives the formation of spherical micelles. The micelles formed early during the synthesis seem to be cylindrical, but a spherical micellar phase forms during the drying. If conditions are orchestrated where films remain flexible after initial deposition, transformation from one micellar phase (three-dimensional hexagonal) to another (cubic) is possible. Still, the order of product phases observed as the surfactant concentration in the final film increases is consistent with the order seen in other surfactants able to form close-packed spherical micelle ordered phases (Figure 9.13).

As this chapter has shown, great progress has been made in the last decade at finding ways to synthesized controlled-porosity ceramics using bicontinuous liquid crystals. It remains challenging to simultaneously control both the coassembly process and the rate of silica polycondensation. The most promising insights into the synthesis of these structures have come from *in situ* studies of long-range order and coating, systematic studies of composition, and creative design of surfactants. These bicontinuous structures hold great promise for continued research and development.

ACKNOWLEDGMENT

This material is partially based on work supported by the National Science Foundation under Grant No. DMR-0210517.

REFERENCES

1. CT Kresge, ME Leonowicz, WJ Roth, JC Vartuli, JS Beck. *Nature* 359:710–712, 1992.

2. JS Beck, JC Vartuli, WJ Roth, ME Leonowicz, CT Kresge, KD Schmitt, CTW Chu, DH Olson, EW Sheppard, SB McCullen, JB Higgins, JL Schlenker. *J Am Chem Soc* 114:10,834–10,843, 1992.

3. G Wirnsberger, P Yang, BJ Scott, BF Chmelka, GD Stucky. *Spectrochim Acta A* 57:2049–2060, 2001.

4. X Feng, GE Fryxell, LQ Wang, AY Kim, J Liu, KM Kemner. *Science* 276:923–926, 1997.

5. KW Gallis, JT Araujo, KJ Duff, JG Moore, CC Landry. *Adv Mater* 11:1452–1455, 1999.

6. C-Y Tsai, S-Y Tam, Y Lu, CJ Brinker. *J Membr Sci* 169:255–268, 2000.

7. JY Ying, CP Mehnert, MS Wong. *Angew Chem Int Ed* 38:56–77, 1999.

8. C Lei, Y Shin, J Liu, EJ Ackerman. *J Am Chem Soc* 124:11,242–11,243, 2002.

9. RC Hayward, P Alberius-Henning, BF Chmelka, GD Stucky. *Microporous Mesoporous Mater* 44:619–624, 2001.

10. MS Morey, A Davidson, GD Stucky. *J Por Mater* 5:195–204, 1998.

11. A Monnier, F Schüth, Q Huo, D Kumar, D Margolese, RS Maxwell, GD Stucky, M Krishnamurty, P Petroff, A Firouzi, M Janicke, BF Chmelka. *Science* 261:1299–1303, 1993.

12. Y Lu, R Ganguli, CA Drewien, MT Anderson, CJ Brinker, W Gong, Y Guo, H Soyez, B Dunn, MH Huang, JI Zink. *Nature* 389:364–368, 1997.

13. D Zhao, P Yang, DI Margolese, BF Chmelka, GD Stucky. *Chem Comm* 2499–2500, 1998.

14. HY Fan, YF Lu, A Stump, ST Reed, T Baer, R Schunk, V Perez-Luna, GP Lopez, CJ Brinker. *Nature* 405:56–60, 2000.

15. CJ Brinker, YF Lu, A Sellinger, HY Fan. *Adv Mater* 11:579–585, 1999.

16. CJ Brinker, GW Scherer. *Sol-Gel Science: The Physics and Chemistry of Sol-Gel Processing*. New York: Academic Press, 1990.

17. HT Davis, JF Bodet, LE Scriven, WG Miller. In N Boccara, Ed. *Physics of Amphiphile Layers*. New York: Springer-Verlag, 1987.

18. ST Hyde. In K Holmberg, Ed. *Handbook of Applied Surface and Colloid Chemistry*. New York: John Wiley & Sons, 2001, pp. 299–332.

19. P Selvam, SK Bhatia, CG Sonwane. *Ind Eng Chem Res* 40:3237–3261, 2001.

20. DY Zhao, JL Feng, QS Huo, N Melosh, GH Fredrickson, BF Chmelka, GD Stucky. *Science* 279:548–552, 1998.

21. SE Rankin, AP Malanoski, FB van Swol. *Mater Res Soc Symp Proc* 636:D121–D126, 2001.

22. RG Larson, LE Scriven, HT Davis. *J Chem Phys* 83:2411–2420, 1985.

23. RG Larson. *J De Physique II* 6:1441–1463, 1996.

24. M Kikuchi, K Binder. *J Chem Phys* 101:3367–3377, 1994.

25. GT Pickett, AC Balazs. *Macromolecules* 30:3097–3103, 1997.

26. Q Wang, QL Yan, PF Nealey, JJ de Pablo. *J Chem Phys* 112: 450–464, 2000.

27. H Yang, N Coombs GA Ozin. *J Mater Chem* 8:1205–1211, 1998.

28. H Yang, N Coombs, I Sokolov, GA Ozin. *J Mater Chem* 7:1285–1290, 1997.

29. HW Hillhouse, JW van Egmond, M Tsapatsis. *Chem Mater* 12:2888–2893, 2000.

30. H Yang, GA Ozin, CT Kresge. *Adv Mater* 10:883–887, 1998.

31. N Nishiyama, DH Park, A Koide, Y Egashira, K Ueyama. *J Membr Sci* 182:235–244, 2001.

32. KJ Edler, SJ Roser. *Int Rev Phys Chem* 20:387–466, 2001.

33. U Ciesla, F Schüth. *Microporous Mesoporous Mater* 27:131–149, 1999.

34. DM Dabbs and IA Aksay. *Annu Rev Phys Chem* 51:601–622, 2000.

35. A Sayari, P Liu. *Microporous Mater* 12:149–177, 1997.

36. A Stein, BJ Melde, RC Schroden. *Adv Mater* 12:1403–1419, 2000.

37. A Sayari, S Hamoudi. *Chem Mater* 13:3151–3168, 2001.

38. MW Anderson. *Zeolites* 19:220–227, 1997.

39. MC Holmes. *Curr Opin Colloid Interface Sci* 3:485–492, 1998.

40. ST Hyde. *Curr Opin Colloid Interface Sci* 1:653–662, 1996.

41. R Schmidt, M Stocker, D Akporiaye, EH Torstad, A Olsen. *Microporous Mesoporous Mater* 5:1–7, 1995.

42. Q Huo, DI Margolese, GD Stucky. *Chem Mater* 8:1147–1160, 1996.

43. V Alfredsson, MW Anderson. *Chem Mater* 8:1141–1146, 1996.

44. PI Ravikovich, AV Neimark. *Langmuir* 16:2419–2423, 2000.

45. M Kaneda, T Tsubakiyama, A Carlsson, Y Sakamoto, T Ohsuna, O Terasaki, S H Joo, R Ryoo, *J Phys Chem B* 106:1256–1266, 2002.

46. J Xu, Z Luan, H He, W Zhou, L Kevan. *Chem Mater* 10:3690–3698, 1998.

47. V Alfredsson, MW Anderson, T Ohsuna, O Terasaki, M Jacob, M Bojrup. *Chem Mater* 9:2066–2070, 1997.

48. A Sayari. *J Am Chem Soc* 122:6504–6505, 2000.

49. Q Huo, R Leon, PM Petroff, GD Stucky. *Science* 268:1324–1327, 1995.

50. M Clerc. *J Phys II Fr* 6:653–662, 1996.

51. K Yu, B Smarsly, CJ Brinker. *Adv Funct Mater* 13:47–52, 2003.

52. SH Tolbert, TE Schaeffer, J Feng, PK Hansma, GD Stucky. *Chem Mater* 9:1962–1967, 1997.

53. DY Zhao, PD Yang, DI Margolese, BF Chmelka, GD Stucky. *Chem Commun* 2499–2500, 1998.

54. D Grosso, AR Balkenende, P-A Albouy, M Lavergne, L Mazerolles, F Babonneau. *J Mater Chem* 10:2085–2089, 2000.

55. S Besson, T Gacoin, C Ricolleau, C Jacquiod, J-P Boilot. *J Mater Chem* 13:404–409, 2003.

56. QS Huo, DI Margolese, U Ciesla, PY Feng, TE Gier, P Sieger, R Leon, PM Petroff, F Schüth, GD Stucky. *Nature* 368:317–321, 1994.

57. X Auvray, M Abiyaala, P Duval, C Petipas, I Rico, A Lattes. *Langmuir* 9:444–448, 1993.

58. Y Sakamoto, M Kaneda, O Terasaki, D Zhao, JM Kim, GD Stucky, HJ Shin, R Ryoo. *Nature* 408:449–453, 2000.

59. D Grosso, F Babonneau, GJdAA Soler-Illia, P-A Albouy, H Amenitsch. *Chem Commun* 748–749, 2002.

60. D Grosso, F Babonneau, P-A Albouy, H Amenitsch, AR Balkenende, A Brunet-Bruneau, J Rivory. *Chem Mater* 14:931–939, 2002.

61. D Zhao, P Yang, N Melosh, J Feng, BF Chmelka, GD Stucky. *Adv Mater* 10:1380–1385, 1998.

62. Y Sakamoto, I Diaz, O Terasaki, D Zhao, J Perez-Pariente, JM Kim, G D Stucky. *J Phys Chem B* 106:3118–3123, 2002.

63. J Matos, M Kruk, LP Mercuri, M Jaroniec, L Zhao, T Kamiyama, O Terasaki, TJ Pinnavaia, Y Liu. *J Am Chem Soc* 125:821–829, 2003.

64. R Ryoo, JM Kim, CH Ko, CH Shin. *J Phys Chem* 100:17,718–17,721, 1996.

65. KM McGrath, DM Dabbs, N Yao, IA Aksay, SM Gruner. *Science* 277:552–556, 1997.

66. SD Sims, D Walsh, S Mann. *Adv Mater* 10:151–154, 1998.

67. H Jinnai, Y Nishikawa, M Ito, SD Smith, DA Agard, RJ Spontak. *Adv Mater* 14:1615–1618, 2002.

68. H Kaji, K Nakanishi, N Soga. *J Non-Cryst Solids* 185:18–30, 1995.

69. RM Barrer. *The Hydrothermal Chemistry of Zeolites.* London: Academic Press, 1983.

70. O Regev. *Langmuir* 12:4940–4944, 1996.

71. M Linden, SA Schunk, F Schuth. *Angew Chem Int Ed* 37:821–823, 1998.

72. A Firouzi, D Kumar, LM Bull, T Besier, P Sieger, Q Huo, SA Walker, JA Zasadzinski, C Glinka, J Nicol, D Margolese, GD Stucky, BF Chmelka. *Science* 267:1138–1143, 1995.

73. X Auvray, C Petipas, R Anthore, I Rico, A Lattes. *J Phys Chem* 93:7458–7464, 1989.

74. GS Attard, JC Glyde, CG Göltner. *Nature* 378:366–368, 1995.

75. NRB Coleman, GS Attard. *Microporous Mesoporous Mater* 44–45:73–80, 2001.

76. A Firouzi, F Atef, AG Oertli, GD Stucky, BF Chmelka. *J Am Chem Soc* 119:3596–3610, 1997.

77. JC Vartuli, KD Schmitt, CT Kresge, WJ Roth, ME Leonowicz, SB McCullen, SD Hellring, JS Beck, JL Schlenker, DH Olson, EW Sheppard. *Chem Mater* 6:2317–2326, 1994.

78. AV McCormick, AT Bell. *Catal Rev Sci Eng* 31:97–127, 1989.

79. J Šefčík, AV McCormick. *AIChE J* 43:2773–2784, 1997.

80. G Gunnarsson, B Jönsson, H Wennerström. *J Phys Chem* 84:3114–3121, 1980.

81. Z Yuan, W Zhou. *Chem Phys Lett* 333:427–431, 2001.

82. C-F Cheng, H He, W Zhou, J Klinowski. *Chem Phys Lett* 244:117–120, 1995.

83. JN Israelachvili. *Intermolecular and Surface Forces*. New York: Academic Press, 1992.

84. QS Huo, DI Margolese, U Ciesla, DG Demuth, PY Feng, TE Gier, P Sieger, A Firouzi, BF Chmelka, F Schüth, GD Stucky. *Chem Mater* 6:1176–1191, 1994.

85. SH Tolbert, CC Landry, GD Stucky, BF Chmelka, P Norby, JC Hanson, A Monnier. *Chem Mater* 13:2247–2256, 2001.

86. AF Gross, EJ Ruiz, VH Le, SH Tolbert. *Microporous Mesoporous Mater* 44–45:785–791, 2001.

87. CC Landry, SH Tolbert, KW Gallis, A Monnier, GD Stucky, P Norby, JC Hanson. *Chem Mater* 13:1600–1608, 2001.

88. A Matijasic, A-C Voegtlin, J Patarin, JL Guth, L Huve. *Chem Commun* 1123–1125, 1996.

89. YS Lee, D Surjadi, JF Rathman. *Langmuir* 16:195–202, 2000.

90. S Pevzner, O Regev. *Microporous Mesoporous Mater* 38:413–421, 2000.

91. SS Funari, G Rapp. *Proc Natl Acad Sci USA* 96:7756–7759, 1999.

92. R Aveyard, BP Binks, PDI Fletcher. In E Wyn-Jones, Ed. *The Structure, Dynamics, and Equilibrium Properties of Colloidal Systems*. Dordrecht, The Netherlands: Kluwer, 1990, pp. 557–581.

93. I Sokolov, H Yang, GA Ozin, CT Kresge. *Adv Mater* 11:636–642, 1999.

94. PT Tanev, TJ Pinnavaia. *Science* 267:865–867, 1995.

95. B Echchahed, M Morin, S Blais, A-R Badiei, G Berhault, L Bonneviot. *Microporous Mesoporous Mater* 44–45:53–63, 2001.

96. DY Zhao, QS Huo, JL Feng, BF Chmelka, GD Stucky. *J Am Chem Soc* 120:6024–6036, 1998.

97. JM Kim, Y-J Han, BF Chmelka, GD Stucky. *Chem Commun* 2437–2438, 2000.

98. K Schumacher, M Grün, KK Unger. *Microporous Mesoporous Mater* 27:201–206, 1999.

99. P Behrens, A Glaue, C Haggenmuller, G Schechner. *Solid State Ionics* 101–103:255–260, 1997.

100. KW Gallis, CC Landry. *Chem Mater* 9:2035–2038, 1997.

101. JM Kim, SK Kim, R Ryoo. *Chem Commun* 259–260, 1998.

102. AA Romero, MD Alba, W Zhou, J Klinowski. *J Phys Chem B* 101:5294–5300, 1997.

103. CA Fyfe, G Fu. *J Am Chem Soc* 117:9709–9714, 1995.

104. Y Liu, A Karkamkar, TJ Pinnavaia. *Chem Commun* 1822–1823, 2001.

105. ML Peña, Q Kan, A Corma, F Rey. *Microporous Mesoporous Mater* 44–45:9–16, 2001.

106. M Morey, A Davidson, H Eckert, G Stucky. *Chem Mater* 8:486–492, 1996.

107. P Van Der Voort, M Matthieu, F Mees, EF Vansant. *J Phys Chem B* 102:8847–8851, 1998.

108. E Alami, G Beinert, P Marie, R Zana. *Langmuir* 9:1465, 1993.

109. M Ogawa. *J Chem Soc Chem Commun* 1149–1150, 1996.

110. H Yang, N Coombs, I Sokolov, GA Ozin. *Nature* 381:589–592, 1996.

111. H Yang, N Coombs, O Dag, I Sokolov, GA Ozin. *J Mater Chem* 7:1755–1761, 1997.

112. S Pevzner, O Regev and R Yerushalmi-Rozen. *Curr Opin Colloid Interface Sci* 4:420–427, 2000.

113. YF Lu, HY Fan, A Stump, TL Ward, T Rieker, CJ Brinker. *Nature* 398:223–226, 1999.

114. AA Darhuber, SM Troian, JM Davis, SM Miller, S Wagner. *J Appl Phys* 88:5119–5126, 2000.

115. PD Yang, T Deng, DY Zhao, PY Feng, D Pine, BF Chmelka, GM Whitesides, GD Stucky. *Science* 282:2244–2246, 1998.

116. OD Velev, AM Lenhoff. *Curr Opin Colloid Interface Sci* 5:56–63, 2000.

117. JD Barrie, KA Aitchison. *Mater Res Soc Symp Proc* 271:225–230, 1992.

118. F Brunet, B Cabane. *J Non-Cryst Solids* 163:211–225, 1991.

119. LW Kelts, NJ Armstrong. *J Mater Res* 4:423–433, 1989.

120. JC Pouxviel, JP Boilot, JC Beloeil, J Lallemand. *J Non-Cryst Solids* 89:345–360, 1987.

121. F Devreux, JP Boilot, F Chaput. *Phys Rev A* 41:6901–6909, 1990.

122. A Tang, R Xu, S Li, Y An. *J Mater Chem* 3:893–896, 1993.

123. SE Rankin, CW Macosko, AV McCormick. Chem Mater 10:2037, 1998.

124. PJ Flory. *Principles of Polymer Chemistry*. Ithaca, NY: Cornell University Press, 1953.

125. LV Ng, P Thompson, J Sanchez, CW Macosko, AV McCormick. *Macromolecules* 28:6471–6476, 1995.

126. SE Rankin, LJ Kasehagen, AV McCormick, CW Macosko. *Macromolecules* 33:7639–7648, 2000.

127. J Šefčík, SE Rankin. *J Phys Chem B* 107:52–60, 2002.

128. SJ Clarson, JA Semlyen. *Siloxane Polymers*. Englewood Cliffs, NJ: Prentice Hall, 1993.

129. K Kamiya, T Kohkai, M Wada, T Hashimoto, J Matsuoka, H Nasu. *J Non-Cryst Solids* 240:202–211, 1998.

130. H-P Lin, Y-R Cheng, S-B Liu, C-Y Mou. *J Mater Chem* 9:1197–1201, 1999.

131. MT Anderson, JE Martin, JG Odinek, PP Newcomer. *Chem Mater* 10:311–321, 1998.

132. M Klotz, A Ayral, C Guizard, L Cot. *J Mater Chem* 10:663–669, 2000.

133. K Fontell, A Khan, B Lindstrom, D Maciejewska, S Puang-Ngern. *Colloid Polym Sci* 269:727–742, 1991.

134. AJ Hurd, L Steinberg. *Granular Mater* 3:19–21, 2001.

135. AJ Hurd, CJ Brinker. *Mater Res Soc Symp Proc* 180:575–581, 1990.

136. AJ Hurd. *Adv Chem Ser* 234:433–350, 1994.

137. NA Melosh, P Davidson, P Feng, DJ Pine, BF Chmelka. *J Am Chem Soc* 123:1240–1241, 2001.

138. RA Cairncross, LF Francis, LE Scriven. *AIChE J* 42:55–67, 1994.

139. AJ Hurd, CJ Brinker. *J Phys (Paris)* 49:1017–1025, 1988.

140. F Nishida, JM McKiernan, B Dunn, JI Zink, CJ Brinker, AJ Hurd. *J Am Ceram Soc* 78:1640–1648, 1995.

141. MH Huang, BS Dunn, JI Zink. *J Am Chem Soc* 122:3739–3745, 2000.

142. D Grosso, P-A Albouy, H Amenitsch, AR Balkenende, F Babonneau. *Mater Res Soc Symp Proc* 628:CC6.17.1–6, 2000.

143. F Cagnol, D Grosso, GJdAA Soler-Illia, EL Crepaldi, F Babonneau, H Amenitsch, C Sanchez. *J Mater Chem* 13:61–66, 2003.

144. B Alonso, AR Balkenende, P-A Albouy, D Durand, F Babonneau. *New J Chem* 26:1270–1272, 2002.

145. TA Bleasdale, GJT Tiddy, E Wyn-Jones. *J Phys Chem* 95:5385–5386, 1991.

146. PCA Albertius, KL Frindell, RC Hayward, EJ Kramer, GD Stucky, BF Chmelka. *Chem Mater* 14:3284–3294, 2002

10

Controlled Release from Cubic Liquid-Crystalline Particles (Cubosomes)

BEN J. BOYD

CONTENTS

10.1 INTRODUCTION

A drug delivery platform based on submicron particles, which can provide controlled release of drug after intravenous (IV) infusion, has been the subject of much research in recent decades. Despite some success with long-circulating liposomes providing a means of controlling the release of some hydrophilic drugs (1), the ideal controlled-release injectable delivery system that operates independently of drug properties has not been achieved to date.

The bicontinuous cubic phase has the potential to fulfill a number of the requirements of the ideal delivery system. This phase comprises both hydrophilic and hydrophobic domains; drugs distribute between the lipidic bilayer of the cubic phase and the internal and external aqueous solution depending on the partition coefficient of the drug between the different domains (2). The bicontinuous cubic phase formed by glyceryl monooleate (GMO), in particular, provides a slow-release matrix for compounds of widely varying properties (3). For example, bicontinuous cubic phase permits sustained release of both charged and uncharged hydrophilic and lipophilic molecules (4–10), peptides and proteins (6,11,12), and even large molecules such as hemoglobin (13). Its ability to protect labile compounds has been demonstrated by protecting glucose from glucose oxidase (14). A more extensive compilation of works describing the use of bicontinuous cubic phase, as well as other lyotropic liquid-crystalline phases in drug delivery, has been provided by Drummond and Fong (15). The phase behavior can also be manipulated to a limited extent to provide low viscosity precursors, which facilitates ease of handling and administration from a syringe (8). These attributes are all desirable in an injectable drug delivery platform technology; however, the nondispersed bicontinuous cubic phase is not suitable for intravenous injection.

The bicontinuous cubic phase of GMO can be dispersed into stable, submicron-sized particles, the well-known "cubosomes" (16–19). These particles consist of fragments of cubic phase, most often sterically stabilized using a polyethylene oxide–polypropylene oxide block copolymer, with a particle size distribution centered around 200 to 300 nm. A number of methods have been proposed for their preparation, including the phase equilibration approach (20), the high energy dispersion approach (21), and the hydrotrope precursor method (22).

All of the aforementioned advantages of bulk (nondispersed) cubic phase, combined with the ability to prepare submicron dispersions of a size suitable for IV injection, suggest that GMO may provide a submicron drug delivery platform useful as a controlled release, intravenous drug delivery system. In the case of lipophilic drugs, there is currently no drug delivery system on the market that can provide a controlled release, intravenously injected delivery system for these compounds. The potential for cubosomes to provide such a delivery system was hypothesized as early as 1990 (6).

For intravenous administration, drug administration often takes the form of a slow infusion due to solubility limitations or toxicity of the free drug in solution. Perhaps the most significant benefit cubosomes could provide in intravenous administration regimens would be to replace a slow infusion with a rapid, convenient, bolus injection, following which the cubosomes slowly release their drug payload, while circulating in a "stealthlike" way in the bloodstream in a manner similar to that of stealth liposomes. The release rate would be required to match that of the infusion, maintaining the free drug at the level required to provide a therapeutically useful drug concentration in the plasma over the required time period.

Cubosomes also offer the opportunity to formulate hydrophilic and amphiphilic compounds such as peptides in a submicron form for oral delivery. The promise of controlled release, together with the mucoadhesive nature of cubic phase (23), and the ability to protect labile compounds from the acidic and enzymatic environments in the gastrointestinal tract (14), provide a compelling motivation for research into this application of cubosomes. For oral delivery of drugs, generally a once-a-day administration is desired, so for a controlled release application the cubosomes would ideally be required to protect the loaded drug from degradation and release their payload over a 24-h period.

In this discussion, the existing literature on drug release from cubosomes will be reviewed and considered in light of known diffusion behavior of drugs in bicontinuous cubic phases. Data obtained in this laboratory on drug release from a range of lipophilic and hydrophilic compounds are also presented to augment the limited data that have been reported in the literature. Finally, the possible future developments in the use of cubosomes for controlled drug delivery are discussed.

10.2 PREVIOUSLY PUBLISHED STUDIES ON CONTROLLED RELEASE FROM CUBOSOMES

Despite the apparent potential for cubosomes to provide a superior drug delivery system and the high level of interest in the drug loading and release behavior of bulk cubic phase in the scientific literature, there is surprisingly very little published data on the drug loading and drug release characteristics of cubosomes. The existing literature is summarized in Table 10.1.

10.2.1 *In Vitro* Studies

To our knowledge only one report to date claims slow release of a lipophilic compound from cubosomes (24). In this report, the *in vitro* release of rifampicin was measured from cubosomes prepared by the hydrotrope precursor method, taking advantage of dilution trajectories described in detail by Spicer et al. (22). The rate of release of rifampicin was measured using the equilibrium dialysis method. In this method, the cubosome formulation is contained within a sealed section of dialysis tubing, and following immersion in a large volume of release medium, the rate of appearance of drug in the release medium is measured. The data obtained showed zero-order release of rifampicin from the cubosome dispersion over a period of 10 to 12 days. A control solution of rifampicin in phosphate-buffered

Table 10.1 Reported *In Vitro* and *In Vivo* Studies Involving Drug Release from Cubosomes

Drug	Experimental Detail	Duration of Release/Effect	Ref.
Rifampicin	*In vitro* release using dialysis	>10 days	24
Bromocresol green	*In vitro* release using dialysis	>10 days	24
Somatostatin	*In vivo* plasma levels in rabbit after IV administration	Sustained somatostatin plasma levels for >6 h vs. control	25
Insulin	*In vivo* blood glucose and insulin levels in rat after oral administration	Blood glucose maintained below basal levels for >6 h vs. IV control	28

saline (PBS) exhibited a more rapid release rate over only a few days, leading to the conclusion that there was evidence in support of slow release from the cubosome sample.

In the same report, a similar release study with bromocresol green, a model hydrophilic compound was also reported. This study concluded the release of bromocresol green was occurring over approximately the same timescale as that of rifampicin. In the case of bromocresol green, the release profile was more reminiscent of first-order release.

10.2.2 *In Vivo* Studies

The analytical challenges associated with measuring drug levels in plasma *in vivo* are numerous. Even the simplest *in vivo* studies are often compromised by the ability to measure drug in the nanogram per milliliter range in the presence of many endogenous impurities. In intravenous applications, the measurement of drug release from particles such as cubosomes, liposomes, or nanoparticles is further confounded, first by the need to differentiate between colloidally bound and free drug levels, and second by the anticipated low levels of free drug released over time compared to total drug dose administered. Perhaps as a consequence of these difficulties, *in vivo* studies reported thus far have instead measured either the total drug in circulation or the pharmacodynamic endpoint or effect, as an indicator of the release of drug from the formulation. This makes it difficult to draw definitive conclusions regarding actual release rates from the particles themselves. Nevertheless, *in vivo* studies with cubosomes have been conducted.

The peptide somatostatin has been studied *in vivo* in a rabbit model (25). In this study somatostatin was administered intravenously either as a bolus dose of free peptide or as an injection of peptide-loaded cubosome dispersion, with a concentration of somatostatin in the injected dispersion of 7 μg/mL.

While free somatostatin given as a bolus dose in solution was shown to be rapidly eliminated from the circulation with a half-life of less than a minute, the corresponding cubosome dispersion provided for sustained somatostatin plasma levels for up to 6 h. An early "burst release" component was observed, attributed to free somatostatin in the dispersion, while the remaining drug was apparently retained inside the cubosomes and cleared more slowly from the circulation.

It is well known that particulate delivery systems, such as nonpegylated liposomes, are rapidly removed from plasma by the reticuloendothelial system (RES) (26). The authors surmised that the sustained plasma levels of somatostatin may be attributed to long circulation times of the cubosomes, made possible by the poly-oxyethylene chains of the surface stabilizer (Poloxamer 407) providing a "stealthlike" coating on the cubosome surface, thereby avoiding RES uptake (27).

The efficacy of insulin-loaded "nanocubicles" given orally to diabetic rats has also been studied (28). The so-called nanocubicles are essentially GMO-based cubosomes prepared by the hydrotrope precursor method similar to that used in the *in vitro* studies described earlier. Oral administration of insulin in the form of nanocubicles resulted in a sustained hypoglycaemic effect for over 6 h, while the same dose of insulin as an oral solution had little effect on blood glucose levels. The mucoadhesive nature of GMO and protection of insulin from proteolytic enzymes were highlighted as factors providing the positive result.

In light of promising studies reporting delayed drug release from cubosomes, the question as to why so little work has been published in peer-reviewed literature demands closer scrutiny.

10.3 DRUG RELEASE FROM BULK NONDISPERSED CUBIC PHASE IN RELATION TO CUBOSOMES

As mentioned above, drug release profiles have been determined for a range of drugs from the bulk cubic phase of GMO (3). It is significant that in almost every case the drug release was found to obey Higuchi kinetics for drug diffusion through a matrix (29). Most studies reported were carried out with a sample of preformed cubic phase contained within a holder of some description, presenting a well-defined surface to the release medium. The equation describing drug release from these type of systems is derived from a solution to Fick's Law, expressed in the form:

$$Q = [D_m C_d (2A - C_d)t]^{1/2}$$

where Q = amount of drug released per unit area of matrix, D_m = diffusion coefficient of the drug in the matrix, A = initial amount of drug in unit volume of matrix, C_d = solubility of drug in matrix, and t = time. Where true Fickian diffusion applies, a plot of mass released through a defined surface area against square root of time results in a straight line, the slope of which is related to the diffusion coefficient through the matrix.

It is important to note that Q is dependent on the surface area that the matrix presents to the release medium. Accordingly, a sample of bulk cubic phase of fixed volume, containing drug, when reduced to an equivalent dispersion of small particles results in a significant increase in surface area. If drug loading, solubility, and diffusion through the cubic phase matrix are assumed to be constant, flux through the surface of the matrix into the release medium should increase dramatically compared to the nondispersed bulk cubic phase of equivalent volume.

Data are available on release of somatostatin from both cubosomes (25) and nondispersed cubic phase (12), and the above concept can be applied to attempt to understand the likely mechanism of release operating in the cubosome case. The release of somatostatin from the bulk nondispersed cubic phase has been shown to obey the Higuchi kinetics described above and provided a release rate of around 1 $\mu g/(mL*min^{1/2})$. Although experimental conditions were not detailed, the assumption that a modified USP paddle method was used with 500 mL release medium (11), results in a release rate of 500 $\mu g/min^{1/2}$. Figure 10.1 illustrates the estimated slope of

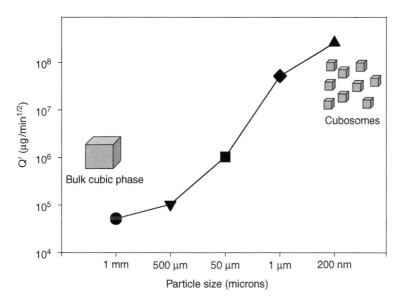

Figure 10.1 Anticipated effect of particle size on slope of release versus square root of time plot for somatostatin from cubic phase when broken down to particles of decreasing size.

the Q versus square root of time curve (denoted Q') when 1 mL of the bulk phase is broken down into cubosomes of decreasing particle size. For the purpose of the calculation, it was assumed that the holder in which the cubic phase is loaded presents a surface area of 1 cm² to the release medium. Thus orders of magnitude change in the release rate are expected on production of the cubosome dispersion from the bulk phase.

In the case of a cubosome dispersion, Table 10.2 outlines the steps in the calculation used to estimate the time required for release of somatostatin. For the *in vivo* study, only 0.15 mL of cubic phase was injected in the form of cubosomes. The calculation outlined above therefore indicates that the release of somatostatin from cubosomes should be complete in fractions of a minute.

These concepts are supported by findings of NMR investigations into the penetration of europium ion into cubosomes from the surrounding solution (30). In this case the NMR spectrum was characterized by a single shifted peak, indicating that the entire structure was permeable almost instantaneously to the europium ion, so that diffusion into or out of the matrix was extremely rapid.

Based on the simple diffusion considerations outlined above, drug release from cubosomes would be expected to be very rapid, even for a moderately large peptide such as somatostatin. However, while the data at hand still support the notion that cubosomes somehow manage to retain drug molecules within the structure for an extended period of time, there is still a need for the cubosome-based

Table 10.2 Calculations of Time for Release of Somato-statin from Cubosomes Based on Release Data from Bulk Cubic Phase

Property	Bulk Cubic Phase	Cubosomes
Volume of cubic phase (cm³)	1[a]	0.15[b]
Total surface area (cm²)	1[a]	77,942[c]
Q' (μg/min$^{1/2}$)	500	3.9×10^7
Mass somatostatin loaded (μg)	—	10.5[b]
Estimated time for 100% release (min)	—	1.0×10^{-11}

[a] Assumed volume and configuration of bulk phase in Ericsson, B., Loander, S., and Ohlin, M. *Proc Int Symp Control Rel Bioact Mat* 15: 382–383, 1988.
[b] From Engström, S., Ericsson, B., and Landh, T. *Proc Int Symp Control Rel Bioact Mat* 23: 89–90, 1996.
[c] Based on monodisperse cubosome particles, diagonal corner-to-corner diameter 200 nm.

controlled drug delivery concept to be investigated *in vitro* by appropriate experimental means.

The absence of data disproving the controlled release from cubosomes is as lacking as data in support of controlled release from cubosomes. This may be in part due to the lack of accepted methodologies to study drug release from colloidal particles in general (31). The following section of this chapter is a discussion of the merits of using dialysis and pressure ultrafiltration to study drug release from cubosomes.

10.4 STUDIES OF DRUG RELEASE FROM CUBOSOMES IN THIS LABORATORY

10.4.1 Dialysis as a Method for Measuring Drug Release from Cubosomes

In this laboratory the potential of cubosomes to provide controlled release of a drug has been given some experimental consideration, particularly in the case of lipophilic compounds for which no sustained-release delivery systems exist. The notion that rifampicin might be slowly released from cubosomes was sufficient motivation to investigate the existing reports. Some of the data from experiments described in the next two sections have been recently published (32) and have also been presented in poster form (33).

Cubosomes were prepared by the method of Gustafsson et al. (21) using a drug, Myverol 18-99 (Swift Chemicals, Australia), and Poloxamer 407 (BASF) molten precursor. Cryo-TEM revealed the presence of cubic faceted particles, with a small number of co-existing vesicular structures, and SAXS confirmed the presence of cubic phase with the *Im3m* space group in the dispersion, both consistent with previous reports (18,19,22).

Initial attempts to repeat the release studies with rifampicin from cubosomes by previous workers (24) were unsuccessful due to instability of the rifampicin in solution over extended time periods during the equilibrium dialysis experiment.

As an alternative, the release of diazepam as a model lipophilic compound was investigated in a side-by-side diffusion cell, using the same membrane as in the work reported for rifampicin (3500 MWCO, Spectrapor 3). A volume of cubosome dispersion was diluted into water in the donor compartment such that the dilution factor was 50-fold. A high level of dilution was used because highly lipophilic drugs partition primarily into the lipidic domains of cubosomes, and some form of dilution is required to stimulate release. This process is also representative of the probable handling of a cubosome dispersion during an intravenous administration, where dilution occurs on injection directly into the bloodstream or on dilution in an infusion bag. Appearance of the drug in the receptor compartment was found to be slow, in this case occurring over 24 h (Figure 10.2A). Complete release was not attained due to the low aqueous solubility and high partition coefficient of diazepam (32), which dictated non-sink conditions during the experiment despite the dilution in the donor compartment.

In order to separate the processes of drug release from the cubosomes and free drug transfer through the membrane, a simple drug solution was placed in the donor compartment at twice the concentration of drug found in the receptor compartment from the cubosome experiment. The appearance of the drug in the receptor compartment followed the same time course, whether a simple solution or cubosome dispersion was added to the donor compartment (Figure 10.2A). This indicates that the transfer of the free drug across the membrane, and not release of drug from the cubosomes, was governing the appearance of drug on the receptor side of the membrane. This data also revealed that the drug concentration in the receptor compartment reflected the free solution in the cubosome dispersion only at long times, in this case 24 h, when a resolution of the order of minutes was required for the purposes of these studies. As a consequence, equilibrium dialysis was concluded to be unsuitable as a method for measuring lipophilic drug release from cubosomes.

10.4.2   Pressure Ultrafiltration—A Proposed Method for Measuring Drug Release from Colloidal Particles Such As Cubosomes

A number of reports have been published highlighting the advantages of pressure ultrafiltration and disadvantages of dialysis-based methods for measuring drug release from colloidal systems such as submicron emulsions (35–37). The pressure ultrafiltration technique has been used in this laboratory to investigate the release of lipophilic drugs from a number of lipid-based delivery systems, including cubosomes (32).

Pressure ultrafiltration cells consist of a reservoir and a membrane. The cubosome dispersion is diluted directly into the release medium in the reservoir, which allows release of drug into the

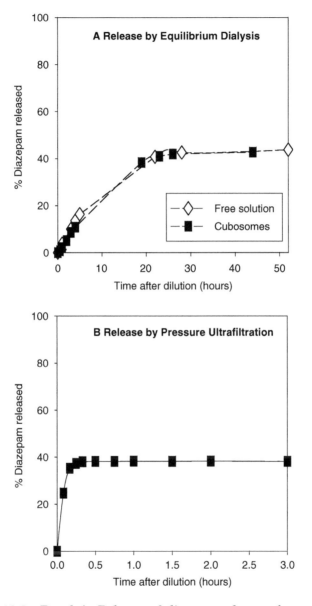

Figure 10.2 Panel A: Release of diazepam from cubosomes and from free solution determined using the equilibrium dialysis method; Panel B: Release of diazepam from cubosomes determined using the pressure ultrafiltration method.

surrounding solution without the impediment of a membrane. Lipophilic drugs partition out of the cubosomes into the surrounding solution. With poorly water-soluble drugs under partition control, it is very difficult to obtain sink conditions without the use of added solubilizers such as surfactants. In the case of cubosome release studies, the use of surfactants is not recommended since they are likely to influence the structure of the meso-phase. Without added solubilizers, complete release of lipophilic drugs from lipid-based delivery systems requires large dilution volumes, which can result in reduced analytical sensitivity. As a consequence, a 50-fold dilution of the dispersion in the release medium was used as standard, which while limiting the equilibrium extent of drug release to below 100% in all cases, still provided sufficient analytical sensitivity to illustrate the release characteristics from the cubosomes. At required time points, the cell was capped and air pressure was applied to force a small fraction of the free drug solution through the 10,000 MWCO membrane (YM 10, Millipore, Australia), leaving the colloid-bound drug and the remaining drug in solution in the reservoir. The first 1000 μL of filtrate was discarded to ensure saturation of the membrane and to account for the dead volume of the cell before collecting a small 100 μL sample for HPLC determination of free drug content. This provides a sample representative of the free drug in solution over time and can be conducted with a resolution of minutes, allowing better investigation of release kinetics than the dialysis method. Fresh water was added to replace the volume removed during sampling.

10.4.2.1 Release of Lipophilic Compounds

10.4.2.1.1 Diazepam

In order to compare the pressure ultrafiltration method to the equilibrium dialysis method described previously, the same cubosome dispersion containing diazepam was used to investigate drug release by the pressure ultrafiltration method. Figure 10.2B shows that the release of diazepam rapidly reaches around 40% and then stops when pressure ultrafiltration is used. The plateau of the release profile is ascribed to the fact that the dilution is not sufficient to provide an infinite sink, as was the case for the equilibrium dialysis experiments. Nevertheless, for the purpose of elucidating whether release of drug from the particles is likely to be therapeutically useful, it is not necessary to obtain 100% release. In the case of diazepam, if 40% of the drug is released almost immediately, then the likelihood of cubosomes providing any sort of sustained effect for the additional 60% of the drug is unlikely. Hence it is justified to claim that, at least in the case of diazepam under these conditions, cubosomes must be classified as a burst release delivery system in the same manner as submicron emulsions (35).

10.4.2.1.2 Rifampicin, Propofol, and Griseofulvin

Figure 10.3 illustrates the release of rifampicin, propofol, and griseofulvin (all diluted 50-fold in water). The extent of release varied among the drugs, even though the dilution factor was the same, due to the differences in drug lipophilicity. Although the partition coefficients of the drugs between the cubic phase and water were not measured directly, the rank order of extent of release for the nonionizable drugs was consistent with their octanol–water partition coefficients listed next to the relevant curve in Figure 10.3, with rifampicin as the exception. The release data for diazepam are

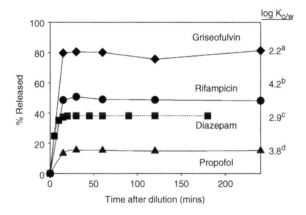

Figure 10.3 Release of a range of lipophilic drugs from cubosomes using the pressure ultrafiltration method. The octanol/water partition coefficients for each drug are indicated to the right of the appropriate curve. [a]Leo, A., Hansch, C., and Elkins, D. *Chem Rev* 71: 525–616, 1971; [b]Taillardat-Bertschinger, A., Marca Martinet, C.A., Carrupt, P.-A., Reist, M., Caron, G., Fruttero, R., and Testa, B. *Pharm Res* 19: 729–737, 2002; [c]calculated using the KowWin software program located at http://esc.syrres.com; [d] Hansch, C., Leo, A., and Hoekman, D.H. Exploring QSAR, Washington, D.C.: American Chemical Society, 1995.

again plotted for comparison purposes. It can be seen that the lower the partition coefficient, the lower the affinity of the drug for the cubosome, and hence a greater extent of release. Rifampicin is expected to be partly ionized at neutral pH, which may explain the greater extent of release than the octanol–water partition coefficient would predict.

In all cases the release process was complete within minutes. As was the case for diazepam, this behavior implies that complete release would be even more rapid in a sink situation and that the lipophilic drugs will not be retained inside the cubosomes on exposure to dilution or sink conditions. Thus it is apparent from this study that burst release from cubosomes occurs for lipophilic drugs that are under partition control.

The finding that rifampicin is released in a burst fashion from cubosomes when measured using pressure ultrafiltration does not agree with the previous report (24). However, it would appear that the shortcomings of the equilibrium dialysis measurements may have influenced the conclusions drawn from the results in the aforementioned study.

10.4.2.2 Release of Hydrophilic Compounds

10.4.2.2.1 Glucose

The release behavior of hydrophilic compounds from cubosomes following dilution is also of great interest considering the numerous therapeutic peptides and monoclonal antibodies that might benefit from such a platform. The ability of cubosomes to retain somatostatin, a large peptide, within the particle for extended time periods appears to be limited based on conclusions reached above. The release of glucose as a model hydrophilic compound from cubosomes was therefore also investigated using pressure ultrafiltration as part of these studies.

Loading a hydrophilic drug into the cubosomes is not possible using the method described for lipophilic drugs earlier, which requires the drug to be soluble in the molten lipid and polymer mixture. However, consideration of the GMO phase diagram allows formulation of a mixture containing a small proportion of water (or in this case glucose solution). If only 5% water is added to the lipid mixture, the system is in the L_2 phase region rather than the cubic region (38), which provides a low-viscosity liquid precursor that can be injected into water to form a cubosome dispersion. In this study, the glucose is incorporated into the cubic phase formed on exposure to the dispersing medium, and subsequent release properties can be investigated.

For this experiment the dispersion was subjected only to the Ultraturrax dispersion step, with the dispersion formed directly in the ultrafiltration reservoir via immersion of the dispersing tool in the release medium. This allowed the determination of free glucose immediately after formation of the cubosome dispersion. Not surprisingly, virtually all of the glucose was released rapidly on formation of the cubosomes, as illustrated in Figure 10.4. Note that the percentage glucose in the release medium is plotted rather than the percentage released, as it is not known if the glucose actually resides inside the structure at any time after forming the cubosomes, as it may be excluded during manufacture. However, because the glucose is included in the L_2 precursor, it is likely that most of the glucose would be initially retained in the cubic phase internal aqueous

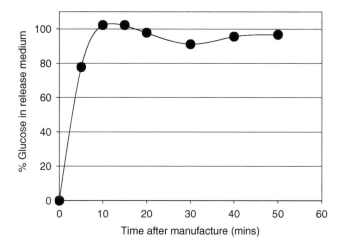

Figure 10.4 Release of glucose from cubosome dispersion by pressure ultrafiltration. Precursor containing glucose (0.5% w/w), Poloxamer 407 (10% w/w), water (5% w/w), and Myverol 18-99K (remainder) was heated to 60°C and injected into water at room temperature in pressure ultrafiltration cell. Myverol content in resulting dispersion was 10% w/w. Release samples were obtained by pressurising the ultrafiltration cell and were analyzed for glucose content by RI-HPLC.

domains, and subsequent release may be rapid, as in the case of lipophilic compounds, due to the large surface area of the particles.

Unpublished data obtained in this laboratory for the release of glucose from the bulk nondispersed GMO–water cubic phase reveal that the release rate is similar in magnitude to that found for somatostatin (12). Consequently, release from the cubosomes was expected to be extremely rapid, for the reasons discussed in Section 10.3.

10.5 THE FUTURE FOR CUBOSOMES AS A CONTROLLED-RELEASE DELIVERY SYSTEM

It would appear that the promise of controlled release of drug molecules from cubosomes, originally hypothesized on the basis of the solid gellike characteristics and tortuous structure of the bulk cubic phase, may not be attainable without further modification of the cubosome structure.

Strategies such as ionic and covalent bonding of drugs to functionalized lipids within the structure may well prevent the burst release of drug from the structure. A patent application has recently been published that claims the use of functionalized "anchor" lipids incorporated into the cubosomes for the purpose of interacting with the active to increase loading and facilitating controlled release (39). The incorporation of ketoprofen is exemplified, but no *in vitro* release data are provided. Such modifications certainly influence the partition coefficient of the drug between the cubic phase and the aqueous solution, which would be evident in the extent of drug release in a nonsink situation; whether the diffusion coefficient of the drug would be altered to the degree that the release rate is dramatically reduced is not yet known. While this type of approach may eventually lead to a controlled-release cubosome formulation, such approaches somewhat diminish the elegance and simplicity of the original cubosome concept and start to impact on the perceived but hitherto unrealized benefits of cubosomes over other delivery systems such as liposomes, solid lipid nanoparticles, and simple emulsions.

In developing a new functional lipid for an intravenous application, one must bear in mind that the excipient will be treated as a new drug, so the benefits need to be well established, both *in vitro* and *in vivo*, before embarking on a formal development effort using the new excipient.

Of the existing excipient database, there are no materials approved for intravenous use that form a cubic phase that is stable to dilution in excess water (40,41). The two best-known materials for forming reverse bicontinuous cubic phase, GMO and phytanetriol (42), are not approved for use in injectable products. Although the toxicology of these compounds has not been fully assessed for intravenous use, the oleic acid formed by hydrolysis of GMO has been implicated in pulmonary toxicity (43) and has also been shown to form calcium oleate crystals *in vivo* (44). Phytanetriol has only been used in cosmetic products, and its systemic toxicity has not been reported. There is a definite need for new materials with suitable toxicological profiles with which to formulate these types of systems.

10.6 CONCLUSIONS

Current *in vivo* data for drug release indicate that cubosomes may have some application for sustained-release intravenous drug delivery. The limited *in vitro* data published on drug release from cubosomes also support this finding. However, consideration of the large surface area of cubosomes, together with *in vitro* data presented in this chapter from this laboratory indicate that the sustained drug levels *in vivo* may be due not to retarded release from the particles but rather to some other physiological effect. New approaches such as the reported use of functionalized lipid components doped into the cubic phase matrix may provide a mechanism for retarding the release of drug molecules, which may facilitate the development of an intravenous cubosome-based, sustained-release drug delivery product.

ACKNOWLEDGMENTS

Darryl Whittaker is thanked for valuable suggestions regarding this manuscript. Some data included in this work were generated under funding from the AusIndustry START Grant scheme. Some aspects of this work have been reproduced with permission from Elsevier Science B.V. (See Reference 32, Boyd, B.J. Characterization of drug release from cubosomes using the pressure ultrafiltration method. *Int J Pharm* 260: 239–247, 2003.)

REFERENCES

1. Allen, T.M., Mehra, T., Hansen, C., and Chin, Y.C. Stealth liposomes: an improved sustained release system for 1-beta-D-arabinofuranosyl-cytosine. *Cancer Res* 52: 2431–2439, 1992.

2. Engström, S., Nordén, T.P., and Nyquist, H. Cubic phases for studies of drug partition into lipid bilayers. *Eur J Pharm Sci* 8: 243–254, 1999.

3. Shah, J.C., Sadhale, Y., and Chilukuri, D.M. Cubic phase gels as drug delivery systems. *Adv Drug Deliv Rev* 47: 229–250, 2001.

4. Caboi, F., Amico, G.S., Pitzalis, P., Monduzzi, M., Nylander, T., and Larsson, K. Addition of hydrophilic and lipophilic compounds of biological relevance to the monoolein/water system I. Phase behaviour. *Chem Phys Lipids* 109: 47–62, 2001.

5. Burrows, R., Collett, J.H., and Attwood, D. The release of drugs from monoglyceride–water liquid crystalline phases. *Int J Pharm* 111: 283–293, 1994.

6. Engström, S. Drug delivery from cubic and other lipid–water phases. *Lipid Tech* 2: 42–45, 1990.

7. Wyatt, D.M. and Dorschel, D. A cubic phase delivery system composed of glyceryl monooleate and water for sustained release of water soluble drugs. *Pharm Tech* October: 116–130, 1992.

8. Chang, C.-M. and Bodmeier, R. Low viscosity monoglyceride-based drug delivery systems transforming into a highly viscous cubic phase. *Int J Pharm* 173: 51–60, 1998.

9. Chang, C.-M. and Bodmeier, R. Effect of dissolution media and additives on the drug release from cubic phase delivery systems. *J Controlled Release* 46: 215–222, 1997.

10. Helledi, L.S. and Schubert, L. Release kinetics of acyclovir from a suspension of acyclovir incorporated in a cubic phase delivery system. *Drug Dev Ind Pharm* 27: 1073–1081, 2001.

11. Ericsson, B., Eriksson, P.O., Löfroth, J.E., and Engström, S. Cubic phases as delivery systems for peptide drugs. *Am Chem Soc* 251–265, 1991.

12. Ericsson, B., Loander, S., and Ohlin, M. Liquid crystalline phases as delivery systems for drugs. III. *In vivo. Proc Int Symp Control Rel Bioact Mat* 15: 382–383, 1988.

13. Leslie, S.B., Puvvada, S., Ratna, B.R., and Rudolph, A.S. Encapsulation of haemoglobin in a bicontinuous cubic phase lipid. *Biochim Biophys Acta* 1285: 246–254, 1996.

14. Barauskas, J., Razumas, V., and Nylander, T. Entrapment of glucose oxidase into the cubic Q230 and Q224 phases of aqueous monoolein. *Prog Coll Polym Sci* 116: 16–20, 2000.

15. Drummond, C.J. and Fong, C. Surfactant self-assembly objects as novel drug delivery vehicles. *Curr Opin Coll Int Sci* 4: 449–456, 1999.

16. Larsson, K. Cubosomes and hexosomes for drug delivery. *Proc Int Symp Control Rel Bioact Mat* 24: 198–199, 1997.

17. Andersson, S., Jacob, M., Lidin, S., and Larsson, K. Structure of the cubosome — a closed lipid bilayer aggregate. *Z Kristallogr* 210: 315–318, 1995.

18. Gustafsson, J., Ljusberg-Wahren, H., Almgren, M., and Larsson, K. Cubic lipid–water phase dispersed into submicron particles. *Langmuir* 12: 4611–4613, 1996.

19. Siekmann, B., Bunjes, H., Koch, M.H.J., and Westesen, K. Preparation and structural investigations of colloidal dispersions prepared from cubic monoglyceride–water phases. *Int J Pharm* 244: 33–43, 2002.

20. Landh, T. and Larsson, K. Particles, method of preparing said particles and uses thereof. U.S. Patent 5,531,925. July 2, 1996.

21. Gustafsson, J., Ljusberg-Wahren, H., Almgren, M., and Larsson, K. Submicron particles of reversed lipid phases in water stabilized by a nonionic amphiphilic polymer. Langmuir 13: 6964–6971, 1997.

22. Spicer, P.T., Hayden, K.L., Lynch, M.L., Ofori-Boateng, A., and Burns, J.L. Novel process for producing cubic liquid-crystalline nanoparticles (cubosomes). *Langmuir* 17: 5748–5756, 2001.

23. Nielsen, L. S., Schubert L., and Hansen, J. Bioadhesive drug delivery systems I. Characterisation of mucoadhesive properties of systems based on glyceryl monooleate and glyceryl monolinoleate. *Eur J Pharm Sci* 6: 231–239, 1998.

24. Kim, J.S., Kim, H., Chung, H., Sohn, Y.T., Kwon, I.C., and Jeong, S.Y. Drug formulations that form a dispersed cubic phase when mixed with water. *Proc Int Symp Control Rel Bioact Mat* 27: 1118–1119, 2000.

25. Engström, S., Ericsson, B., and Landh, T. A cubosome formulation for intravenous administration of somatostatin. *Proc Int Symp Control Rel Bioact Mat* 23: 89–90, 1996.

26. Vadolas, J., Wijberg, O.L.C., and Strugnell, R.A. Liposomes as Systemic and Mucosal Delivery Vehicles. In *Antigen Delivery Systems*. 1st ed., Gander, B. and Merkle, H.P., Eds. Amsterdam: Harwood Academic Publishers, 1997, pp. 73–100.

27. Papahadjopoulos, D. Stealth liposomes: from steric stabilization to targeting. In *Stealth Liposomes*. Lasic, D. and Martin, F., Eds. Boca Raton, FL: CRC Press, 1995, pp. 1–6.

28. Chung, H., Kim, J., Um, J.Y., Kwon, I.C., and Jeong, S.Y. Self-assembled "nanocubicle" as a carrier for peroral insulin delivery. *Diabetologia* 45: 448–451, 2002.

29. Higuchi, W.I. Diffusional models useful in biopharmaceutics. *J Pharm Sci* 56: 315–324, 1967.

30. Nakano, M., Sugita, A., Matsuoka, H., and Handa, T. Small-angle x-ray scattering and ^{13}C NMR investigation on the internal structure of "cubosomes." *Langmuir* 17: 3917–3922, 2001.

31. Burgess, D.J., Hussain, A.S., Ingallinera, T.S., and Chen, M.-L. Assuring quality and performance of sustained and controlled release parenterals: AAPS workshop report, co-sponsored by FDA and USP. *Pharm Res* 19: 1761–1768, 2002.

32. Boyd, B.J. Characterisation of drug release from cubosomes using the pressure ultrafiltration method. *Int J Pharm* 260: 239–247, 2003.

33. Boyd, B.J. An *in vitro* method for the measurement of drug release from nanoparticulate drug delivery systems. "Nanoparticles, 2002," Orlando, Florida, Poster presentation #246. 2002.

34. Taillardat-Bertschinger, A., Marca Martinet, C.A., Carrupt, P.-A., Reist, M., Caron, G., Fruttero, R., and Testa, B. Effect of molecular size and charge on IAM retention in comparison to partitioning in liposomes and *n*-octanol. *Pharm Res* 19: 729–737, 2002.

35. Benita, S. and Levy, M.Y. Submicron emulsions as colloidal drug carriers for intravenous administration: comprehensive physicochemical characterisation. *J Pharm Sci* 82: 1069–1079, 1993.

36. Santos Magalhaes, N.S., Cave, G., Seiller, M., and Benita, S. The stability and *in vitro* release kinetics of a clofibride emulsion. *Int J Pharm* 76: 225–237, 1991.

37. Magenheim, B., Levy, M.Y., and Benita, S. A new *in vitro* technique for the evaluation of drug release profile from colloidal carriers — ultrafiltration technique at low pressure. *Int J Pharm* 94: 115–123, 1993.

38. Qui, H. and Caffrey, M. The phase diagram of the monoolein/ water system: metastability and equilibrium aspects. *Biomaterials* 21: 223–234, 2000.

39. Lynch, M.L. and Spicer, P.T. Functionalized cubic liquid crystalline phase materials and methods for their preparation and use. U.S. Patent Application 20020158226, October 31, 2002.

40. Nema, S., Washkuhn, R.J., and Brendel, R.J. Excipients and their use in injectable products. *PDA J Pharm Sci Technol* 51: 166–171, 1997.

41. Nema, S., Brendel, R.J., and Washkuhn, R.J. Excipients: their role in parenteral dosage forms. In *Encyclopedia of Pharmaceutical Technology*. Sworbrick, J. and Boylan, J.C., Eds. New York: Dekker, 2000, pp. 137–172.

42. Ribier, A. and Biatry, B. Cosmetic or dermatological composition in the form of an aqueous and stable dispersion of cubic gel particles based on phytanetriol and containing a surface active agent which has a fatty chain, as dispersing and stabilizing chain. US Patent 5,834,013 98.

43. Okumura, T., Suzuki, K., Kumada, K., Kobayashi, R., Fukuda, A., Fujii, C., and Kohama, A. Severe respiratory distress following sodium oleate ingestion. *J Toxicol Clin Toxicol* 36: 587–589, 1998.

44. Appel, L.E., Rajewski, L., Leppert, P.. and Zentner, G. *In vivo* evaluation of monoolein cubic phase gels in a mouse model. AAPS Annual Meeting 299: PDD 7286, 1996.

45. Leo, A., Hansch, C., and Elkins, D. Partition coefficients and their uses. *Chem Rev* 71: 525–616, 1971.

46. Hansch, C., Leo, A., and Hoekman, D.H. Exploring QSAR, Washington, D.C.: American Chemical Society, 1995.

11

Membrane Protein Crystallization in Lipidic Bicontinuous Liquid Crystals

MARTIN CAFFREY

CONTENTS

11.1 INTRODUCTION

One of the bottlenecks on the route that eventually leads to membrane protein structure through to activity and function is found at the crystal production stage. Diffraction-quality crystals, with which structure is determined, are particularly difficult to prepare currently when a membrane source is used. The reason for this is our limited ability to manipulate proteins with hydrophobic/amphipathic surfaces that are usually enveloped with membrane lipid. More often than not, the protein gets trapped as an intractable aggregate in its watery course from membrane to crystal. As a result, access to the structure and thus function of tens of thousands of membrane proteins is limited. In contrast, a veritable cornucopia of soluble proteins has offered up their structure and valuable insight into function, reflecting the relative ease with which they are crystallized. There exists therefore an enormous need for new ways of producing crystals of membrane proteins. One approach that looks promising involves overexpression as insoluble cytoplasmic inclusion bodies, refolding, and subsequent crystallization (Buchanan, 1999). Another makes use of antibodies to expand polar contacts in the crystal (Ostermeier and Michel, 1997). Very recently, a micelle-based method has been introduced that may share mechanistic similarities with the method described below (Faham and Bowie, 2002).

The more traditional methods involve solubilizing the membrane protein in surfactants and subsequent treatment with "precipitants" as for soluble proteins (Michel, 1991; Garavito et al., 1996; Rosenbusch, 1990). I refer to this as the *in surfo* method. The Rosenbusch stables at the Biozentrum in Basel has been a hotbed of activity in this area from the beginning and has contributed several membrane protein structures through the years (Garavito and Rosenbusch, 1986; Cowan et al., 1995; Dutzler et al., 1999). Little wonder then that it was this same group that introduced another method for growing crystals of membrane proteins that uses the lipidic mesophase as a host (Landau and Rosenbusch, 1996).

The cubic mesophase figures prominently in this, what is sometimes called the *in cubo*, method. We know very little about how and why the method works. Nor indeed do we know the identity of the structure or phase that feeds directly the growing crystal surface. Accordingly, I prefer to refer to it as the *in meso* method, where *meso* stands for the more general, noncommittal middle or liquid crystal phase. Mesophase and liquid crystal are used synonymously.

The basic recipe for growing crystals *in meso* follows: Combine two parts protein solution/dispersion with three parts lipid (monoolein). Mix. The cubic phase forms spontaneously. Add precipitant and incubate, and crystals form in hours to weeks. All operations are at 20°C. Having introduced the *in meso* method and having put it in its historical perspective, I will now set the stage for a discussion of how crystallization might come about by introducing a few fundamental issues regarding lipid phase science.

11.2 BACKGROUND

11.2.1 Form, Function, and Rational Design

I became interested in the *in meso* method because it relates to a project of my group that has to do with form and function as applied to lipidic systems. The view taken is that form, as in molecular structure, determines function. By function I am referring to the assorted activities ascribed to the lipid component of membranes through to the performance of lipid additives in pharmaceutical and food products. However, it is not structure alone that determines function. Rather it is structure, in concert with composition and environmental factors (temperature, pressure) that gives rise to a well defined mesophase and phase microstructure. These in turn dictate performance. The challenge is to understand the relationship between form and function such that the principles of rational design can be deciphered and implemented. The link between the two is phase behavior [as in the corresponding temperature–composition (T-C) phase diagram] and phase microstructure. Integral to realizing the principles of rational design is the need to establish the rules that tie together molecular structure, composition and environmental factors, and phase behavior. The final step involves establishing how phase properties impact on function.

With a view to establishing the rules referred to above, it is useful to work with a lipid whose molecular structure can be altered easily in a way that impacts sensibly on phase behavior. The lipid of choice is one that has a simple molecular structure and can access the full range of lyotropic (water-induced) and thermotropic (temperature-induced) mesophases in a biologically relevant composition (water content) and temperature range. The monounsaturated monoacylglycerols (MAGs) fit the bill reasonably well. Accordingly, the task of mapping the T-C phase diagrams for a host of MAGs where molecular form is varied systematically is in progress (Caffrey, 2000 and references therein).

Monoolein is one such MAG, and its phase diagram has been mapped out in great detail (Figure 11.1) (Qiu and Caffrey, 2000 and references therein). The *in meso* method is based on this system (Rummel et al., 1998). The original objective was to effect membrane protein crystallization from within a cubic phase at 20°C. Conditions for doing so in the monoolein/water system were arrived at by inspecting the 20°C isotherm where the cubic-Pn3m phase is accessed at an overall composition of 40% water and 60% monoolein (A in Figure 11.1). At this temperature, additional water creates a fully hydrated cubic phase in equilibrium with bulk water (B in Figure 11.1). Lesser amounts cause the cubic-Pn3m phase to transform to a cubic-Ia3d phase (C in Figure 11.1). The lamellar liquid crystal (L_α) phase forms at lower hydration levels (D in Figure 11.1). Within any single- (as opposed to two-) phase region, the lattice size, which scales with water-layer thickness in the mesophase, increases as hydration level rises. Lattice size (as well as phase identity) is determined accurately by low-angle x-ray diffraction (Qiu and Caffrey, 2000).

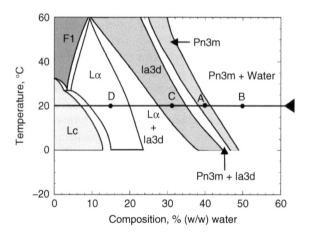

Figure 11.1 Temperature-composition phase diagram for the monoolein/water system. Points along the 20°C isotherm identified by letters are referred to in the text. (From Briggs J., Chung H., and Caffrey M. (1996) *J Phys II France* 6, 723–751 and Qiu H. and Caffrey M. (2000) *Biomaterials* 21, 223–234.)

11.3 HOW CRYSTALS GROW *IN MESO*

We comprehend little about how *in meso* crystallization comes about (Caffrey, 2000; Nollert et al., 2001). However, crystals of several membrane proteins have been grown from what starts out as a doped cubic phase. Given that we know the initial and end states, it is reasonable to speculate as to how we get from one to the other (Figure 11.2). If bacteriorhodopsin (BR) is used as an example, the end state is a crystal with proteins arranged in sheets. The starting condition has BR dispersed in an aqueous micellar solution to which residual purple membrane (PM) lipid remains bound. The protein presumably is cummerbunded with solubilizing alkyl glycoside (AG) detergent around its hydrophobic midsection. The dispersion is then combined with monoolein by mechanical mixing.

Let us now contemplate the fate of the different components in the system (neglecting buffer, salt, etc.) up to crystallization. The detergent has high aqueous solubility. But it is also amphipathic with a proclivity for hydrophobic spaces and surfaces. At the monoolein-to-water ratio used typically [40% (w/w) water], the cubic phase is stable (A in Figure 11.1). It incorporates a network of cubic membranes (lipid bilayers) that separate two interpenetrating but noncontacting aqueous networks (Figure 11.2). Both the polar aqueous and the apolar bilayer interior serve as compartments into which the detergent will partition. However, the detergent is also drawn to the hydrophobic surface of the protein on which it piggybacked into the mix in the first place.

Typically, the aqueous protein/detergent solution is combined with dry monoolein at 20°C. Upon initial contact, water will migrate from the solution into the monoolein. In so doing, it will establish a water activity gradient along which a series of phases will form (Figure 11.2A). The sequence of phases is the same as found on the 20°C isotherm in the monoolein/water phase diagram (Figure 11.1). At low hydration levels, the first liquid crystal phase to form is of the lamellar type (D in Figure 11.1). With increasing hydration, this gives way to the cubic phase (C in Figure 11.1).

As water leaves the aqueous protein solution for the dry lipid, there is a corresponding increase in the concentration of detergent and protein (etc.) in the residual aqueous solution. This will favor the formation of a lamellar phase by reference to the AG/water phase diagram (Warr et al., 1986). At the same time, the slightly soluble monoolein will partition into it, facilitated no doubt by the detergent. In this way, the mixed detergent/protein micelle will acquire monoolein.

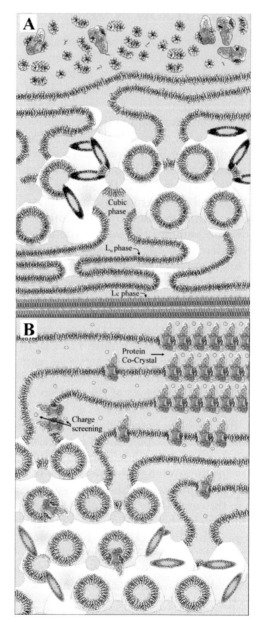

Figure 11.2   Reconstitution (A) and *in meso* crystallization (B) of membrane proteins as described in the text. With reference to the monoolein/AG/water phase diagram (Ai X. and Caffrey M. (2000)

With reference to the monoolein/AG/water phase diagram (Ai and Caffrey, 2000), we see that high concentrations of AG again favor the lamellar phase, and it is likely that the protein now finds itself in a monoolein-enriched micelle with strong lamellar tendencies (Figure 11.2A). Close by, bulk monoolein is giving way to a local lamellar phase and it is possible that the two fuse. This produces a bilayer that is continuous into the bulk lipid and in which the protein is now reconstituted. Vectoral orientation and oligomerization of the protein within a given layer may be imposed at this stage in the process. As the system approaches equilibrium, defined by the overall composition of the sample, the lamellar phase transforms to the bulk cubic phase. Since the lipid bilayer is continuous throughout and the protein and detergent can diffuse within it, both are likely to distribute in the cubic phase. Consistent with this is the observation that in the case of BR, the phase adopts a homogenous purple hue at this stage in the process.

In this state, the protein has been removed from the potentially hostile environment of a detergent micelle. It is stabilized in a lipid bilayer with physicochemical properties (lateral pressure, hydrophobic matching, etc.) more akin to those of the native membrane. We now turn our attention to the protein crystallization process. For nuclei to form and for crystals to grow requires that the system be perturbed in some way. Not unlike crystallization protocols for soluble proteins, typically, this involves adding salt (McPherson, 1999).

Figure 11.2 (Continued) *Biophys J* 79, 394–405), we see that high concentrations of AG again favor the lamellar phase, and it is likely that the protein now finds itself in a monoolein-enriched micelle with strong lamellar tendencies (Figure 11.2A). Close by, bulk monoolein is giving way to a local lamellar phase and it is possible that the two fuse. This produces a bilayer that is continuous into the bulk lipid and in which the protein is now reconstituted. Vectoral orientation and oligomerization of the protein within a given layer may be imposed at this stage in the process. As the system approaches equilibrium, defined by the overall composition of the sample, the lamellar phase transforms to the bulk cubic phase. Since the lipid bilayer is continuous throughout, and the protein and detergent can diffuse within it, both are likely to distribute in the cubic phase. Consistent with this is the observation that in the case of BR, the phase adopts a homogenous purple hue at this stage in the process.

What does salt do to trigger crystal nucleation and growth? Again, we have no hard answers here although a program is underway to obtain them. We know that the salt will compete with the lipid (and the detergent and the protein) in the cubic phase for available water. This is accompanied on occasion by deliquescence and the creation of a salt-saturated liquid (Caffrey, 2000). The water-withdrawing effect of the salt causes the cubic lattice to contract and bilayer curvature to rise (Chung and Caffrey, 1994a; 1994b). This may be perturbation enough to herd the proteins into a lipidic corral where they crowd together, associate, and eventually organize into a crystal lattice. All the while, the salt dissolves in the aqueous medium. By shielding charges on the protein, the elevated ionic strength should also facilitate close protein contact, nucleation, and crystal growth. The dehydrating effect of the salt, if sufficient in degree, can induce transient and local lamellar phase formation (D in Figure 11.1). Whether nucleation and growth take place directly in/from the cubic phase and without the involvement of any other, possibly local, intermediate structure or phase is not known. This is what I refer to as the postulated portal question.

As crystals grow, the mesophase loses color. This is a relatively slow process happening over a period of days to weeks in the case of the purple-colored BR. Presumably, during the course of growth the crystal face is being fed by protein diffusing in the lipid bilayer through the mesophase up to the crystal face to be ratcheted into place. This suggests the existence of a tethering portal structure between the bulk mesophase and the crystal surface. We are actively pursuing the identity of such an umbilicus using, among other things, x-ray diffraction with submicrometer-sized beams (Cherezov et al., 2000).

11.4 ADDITIONAL CONSIDERATIONS

11.4.1 Salt

It may be that the initial extreme and local salt-induced dehydration is what provides the impetus and driving force for nucleation and subsequent crystal growth. In addition to the lipid and detergent, the protein too must give up some of its bound water. Protein–protein and other types of contacts may be favored to compensate for the lost water. These serve to recruit more proteins from within the mesophase. Further, since the protein is designed to span a bilayer and does not normally encounter regions of high curvature, the natural inclination is for the accreting domain to grow laterally to produce a layered structure (Figure 11.2B). If the same is happening in adjacent layers, the proteins will establish contacts between layers and in so doing, set up the three-dimensional lattice of the crystal. The stacking of sheets is likely favored by local dehydration since it frees up additional water for sequestration by the salt. Charge screening may serve a role too.

11.4.2 Detergents

Nonionic detergents have found extensive use in solubilizing membrane proteins for subsequent characterization, reconstitution, and crystallization studies (Michel, 1991). They are likely to accompany the protein into the *in meso* crystallization mix. As surfactants, they can wreak havoc on the lipidic mesophase, which we assume is integral to growing crystals. Molecular geometry considerations suggest that the complementary molecular shapes of the popular AG detergents and monoolein should lead to a destabilization of the highly curved cubic phase in favor of a lamellar-type structure in the presence of the detergent (Ai and Caffrey, 2000). Using x-ray diffraction, we have shown that relatively high concentrations of the AG are tolerated by the cubic phase, which is good news for the *in meso* method. However, further additions bring about a complete transformation from the cubic to the lamellar phase (Ai and Caffrey, 2000). As noted, this is relevant to the mechanism of crystallization in the presence of detergents.

Where detergents are not a part of the *in meso* crystallization protocol, as in the case of the purple membrane (Nollert et al., 1999), the monoacylglycerol (in concert with membrane lipids) may serve in this capacity. Thus, for example, monoolein, which has finite water solubility and is amphipathic, may facilitate the shuttling between the membrane rafts and the mesophase and eventually to the protein crystal via transient, soluble intermediates.

11.4.3 Screen Compatibility

Typically, a precipitant is added to trigger nucleation and growth of membrane protein crystals in the *in meso* method. The commercially available screen solution series is convenient for use in such crystallization trials (McPherson, 1999). However, they contain an array of components, many of which could destroy the cubic phase. We have determined which of the Hampton Screen (50 solutions) and Hampton Screen 2 (48 solutions) series of solutions support cubic phase formation by means of x-ray diffraction (Cherezov et al., 2000). The data show that over 90% of the screen solutions produced the cubic phase at 20°C. In contrast, only half of the screens were cubic phase compatible at 4°C.

11.4.4 Colorless Proteins

The *in meso* approach is particularly suited to pigmented proteins. Their crystallization leads to obvious colored crystals in a colorless background. With colorless proteins the fear is that the crystals would go undetected. This has been tested by crystallizing lysozyme and thaumatin, "colorless" (water-soluble) proteins, *in meso*. Clearly visible crystals grew from the cubic phase (Caffrey, 2000). The exercise shows that the *in meso* method is not limited to colored proteins.

11.4.5 High Throughput

Crystallization trials involve vast numbers of samples. We have described the construction of an inexpensive mixing/delivery device mostly from commercially available parts that lends itself nicely to such high-throughput applications (Cheng et al., 1998). It was developed for working with milligram quantities of lipid (micrograms of protein), to facilitate the handling of highly viscous cubic mesophases and to have precise control over sample composition. The device is simple to use and is being implemented in *in meso* crystallization trials and other applications in labs worldwide. Large-volume syringes can be employed to prepare hundreds of milligrams of cubic phase/protein dispersions in a single step. Submicrogram quantities of the dispersion are then placed in wells containing screen solutions, and the samples are stored for crystallization. In the hands of an experienced user, hundreds of crystallization trials can be set up daily. With a few obvious and simple modifications, the device can be adapted for automation and robotization (Cherezov et al., 2004).

11.5   WHY LIPIDIC PHASES SHOULD PRODUCE GOOD CRYSTALS

11.5.1   The Microgravity Analogy

It is likely that growing crystals *in meso* is akin to growing them in microgravity (and in gels) for the following reasons. Upon nucleation and as the crystals begin to grow, a depletion zone is created in the surrounding lipid phase. However, *in meso*, the crystal presumably is tethered to and embedded in a highly viscous medium, which will not support convection as occurs in solution. Further, protein diffusion in the supporting phase is even slower than it is in solution (Cribier, 1993). Thus, a depletion zone is stabilized, and the crystal face is fed by slowly diffusing protein from the bulk. This increases the likelihood of growing larger and more ordered crystals. It is likely too that impurities are "filtered out" by the lipid bilayer, producing a higher grade of crystal. Settling of newly formed crystals onto others in the mix is commonly encountered in solution for earth-grown crystals (McPherson, 1999). It produces defects and limits crystal growth. Under conditions of microgravity and *in meso*, for obvious but different reasons, sedimentation-related impairments to crystal growth are not an issue.

11.6 QUO VADIS

The utility of *in meso* crystallization rests on its ability to support the production of diffraction-quality crystals. In this regard, the method has fared reasonably well. Thus far, crystals of bR, halorhodopsin, light harvesting complex 2 (LHC2), the reaction centers from *Rhodobacter sphaeroides* (RCS) and *Rhodopseudomonas viridis* (RCV), sensory rhodopsin II (SR II), and most recently, an SR II/transducer complex and Btub have been grown *in meso* (Luecke et al., 1998; Kolbe et al., 2000; Luecke et al., 2001; Gordeliy et al., 2002; Misquitta et al., 2004; M. Chiu, personal communication). bR, halorhodopsin, SR II, and the SR II/transducer complex represent new, high-resolution structures. The best published resolution has been obtained with bR at 1.55 Å (Luecke et al., 1998; 1.38 Å, unpublished data from H. Luecke, personal communication). In the case of bR and halorhodopsin, the process of deciphering the structure of photocycle intermediates is in progress (Lanyi and Luecke, 2001; Kolbe et al., 2000; Matsui et al., 2002; G. Bueldt (PDB 1CWQ) personal communication). As to the generality of the method, the community waits with bated breath for the *in meso* crystallization of nonbacterial proteins. Advances along these lines will be facilitated by an increased understanding of the fundamental mechanism of crystal nucleation and growth *in meso* and by making it a more user-friendly and accessible method. This is where the resources and effort should be directed. In essence, what we seek is a solution to what might be called the second "phase problem" in crystallography. Perutz resolved the first one half a century ago (Perutz, 1992). In the meantime, the fishing expedition continues for more membrane proteins willing to provide a close and personal view of their molecular innards and workings through the intermediate step of crystallization.

ACKNOWLEDGMENT

I thank V. Cherezov and Y. Misquitta for contributing to this work. This work was funded in part through grants from Science Foundation Ireland (02-IN1-B266), the National Institutes of Health (GM56969, GM61070) and the National Science Foundation (DBI9981990).

REFERENCES

Ai X. and Caffrey M. (2000) *Biophys J* 79, 394–405.

Briggs J., Chung H., and Caffrey M. (1996) *J Phys II France* 6, 723–751.

Buchanan S.K. (1999) *Curr Opin Struct Biol* 9, 455–461.

Caffrey M. (2000) *Curr Opin Struct Biol* 10, 486–497.

Cheng A., Hummel B., Qiu H., and Caffrey M. (1998) *Chem Phys Lipids* 95, 11–21.

Cherezov V., Fersi H., and Caffrey M. (2000) *Biophys J* 81, 225–242.

Cherezov V., Qiu H., Retsch C.C., McNulty I., and Caffrey M. (2000) *Biophys J* 78, 484A.

Cherezov V., Peddi A., Muthusubramaniam L., Zheng Y.F., Caffrey M. (2004) *Acta Cryst* D60: 1795–1807.

Chung H. and Caffrey M. (1994a) *Nature* 368, 224–226.

Chung H. and Caffrey M. (1994b) *Biophys J* 66, 377–381.

Cowan S.W., Garavito R.M., Jansonius J.N., Jenkins J., Karlsson R., Koenig N., Pai E.F., Pauptit R.A., Rizkallah P.J., Rosenbusch J.P., Rummel G., and Schirmer T. (1995) *Structure* 3, 1041–1050.

Cribier S., Gulik A., Fellmann P., Vargas R., Devaux P.F., and Luzzati, V. (1993) *J Mol Biol* 229, 517–525.

Dutzler R., Rummel G., Albert S., Benedi V.J., Rosenbusch J.P., and Schirmer T. (1999) *Structure* 7, 425–434.

Edman K., Nollert P., Royant A., Belrhali H., Pebay-Peyroula E., Hajdu J., Neutze R., and Landau E.M. (1999) *Nature* 401, 822–826.

Essen L-O., Siegert R., Lehmann W.D., and Oesterhelt, D. (1998) *Proc Natl Acad Sci USA* 95, 11,673–11,678.

Faham S. and Bowie J.U. (2002) *J Mol Biol* 316, 1–6.

Garavito R.M. and Rosenbusch J.P. (1986) *Methods Enzymol* 125, 309–328.

Garavito R.M., Picot D., and Loll P.J. (1996) *J Bioenerg Biomembr* 28, 13–27.

Kolbe M., Besir H., Essen L.O., and Oesterhelt D. (2000) *Science* 288, 1390–1396.

Landau E.M. and Rosenbusch J.P. (1996) *Proc Natl Acad Sci USA* 93, 14,532–14,535.

Lanyi J.K. and Luecke H. (2001) *Curr Opin Struct Biol* 11, 415–419.

Luecke H., Richter H-T., and Lanyi J. K. (1998) *Science* 280, 1934–1937.

Luecke H., Schobert B., Lanyi J.K., Spudich E.N., and Spudich J.L. (2001) *Science* 293, 1499–1503.

Luecke H., Schobert B., Richter H-T., Cartailler J-P., and Lanyi J.K. (1999a) *J Mol Biol* 291, 899–911.

Luecke H., Schobert B., Richter H-T., Cartailler J-P., and Lanyi J.K. (1999b) *Science* 286, 255–260.

Matsui Y., Sakai K., Murakami M., Shiro Y., Adachi S., Okumura H., and Kouyama T. (2002) *J Mol Biol* 324, 469–481.

McPherson A. (1999) *Crystallization of Biological Macromolecules.* CSHL Press, Cold Spring Harbor, NY.

Michel H. (Ed). (1991) *Crystallization of Membrane Proteins.* CRC Press, Boca Raton, FL.

Misquitta L.V., Misquitta Y., Cherezov V., Slattery O., Mohan J.M., Hart D., Zhalnina M., Cramer W.A., and Caffrey M. (2004) *Structure* 12: 2113–2124.

Ostermeier C. and Michel H. (1997) *Curr Opin Struct Biol* 7, 699–701.

Nollert P., Qiu H., Caffrey M., Rosenbusch J.P., and Landau E. (2001) *FEBS Lett* 504, 179–186.

Nollert P., Royant A., Pebay-Peyroula E., and Landau E.M. (1999) *FEBS Lett* 457, 205–208.

Perutz M. (1992) *Faraday Discuss* 93, 1–11.

Rosenbusch J.P. (1990) *J Struct Biol* 104, 134–138.

Rummel G., Hardmeyer A., Widmer C., Chiu M., Nollert P., Locher P., Pedruzzi I., Landau E.M., and Rosenbusch J.P. (1998) *J Struct Biol* 121, 82–91.

Qiu H. and Caffrey M. (2000) *Biomaterials* 21, 223–234.

Sass H.J., Buldt G., Gessenich R., Hehn D., Neff D., Schlesinger R., Berendzen J., and Ormos P. *Nature* 406, 649–653.

Warr G., Drummond C., and Greiser F. (1986) *J Phys Chem* 90, 4581–4586.

12

Application of Monoglyceride-Based Liquid Crystals as Extended-Release Drug Delivery Systems

CHIN-MING CHANG AND ROLAND BODMEIER

CONTENTS

12.1 BACKGROUND

Monoglycerides are esters of glycerol and fatty acids. Long-chain monoglycerides such as glyceryl monooleate (GMO) and monolinoleate (GML) are water insoluble but water swellable polar lipids. Upon swelling in water, monoglycerides form several mesophases. This property was first systematically studied by Lutton in 1965 [1]. GMO and GML swell into an inverted micellar phase (L_2), a lamellar phase (L), and a viscous isotropic phase with increasing water content. Later, Hyde et al. reported that the viscous isotropic phase is a cubic phase, which consists of a curved bilayer extending in three dimensions separating two congruent networks of water channels [2]. Extensive studies have been reported by Larsson on the structures of GMO cubic phases [3].

Due to the unique properties of the cubic phase, which include extremely large interfacial areas, containing both polar and nonpolar parts, and plasticity, monoglycerides have a great potential as carrier material in extended-release drug delivery systems. In this review, the swelling and drug release properties, as well as the potential applications of monoglyceride-based systems in extended-release drug delivery, are discussed.

12.2 SWELLING BEHAVIOR OF MONOGLYCERIDES

Upon immersion of GMO and GML in water, water molecules spontaneously penetrate into the matrix and interact with the hydrophilic head groups of monoglyceride molecules. The initial swelling self-associates GMO and GML molecules into an inverted micellar structure, with water molecules solubilized in the core of the inverted micelles. The inverted micelles further transform into a lamellar phase and then into the cubic phase upon further increase in water content.

Since both the swelling rate and the swollen mesophase structures of monoglycerides may affect the drug release, the swelling isotherms of the technical grades of GML (Myverol® 18-92) and GMO (Myverol 18-99) from Eastman Chemicals were studied by Chang and Bodmeier [4]. The water uptake increased rapidly initially and then leveled off and approached the equilibrium water content (Figure 12.1).

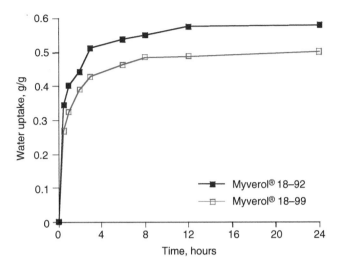

Figure 12.1 Swelling isotherms of Myverol 18-92 and Myverol 18-99 matrices (2 g) in 0.1 *M* pH 7.4 phosphate buffer at 37°C. (From CM Chang, R Bodmeier. *J Pharm Sci* 86: 747–752, 1997.)

The maximum water uptake at 37°C was approximately 0.5 g per gram of monoglyceride. The swelling of these monoglycerides into a cubic phase followed second-order kinetics [5], and the swelling capacity and the swelling rate of monoglycerides could be obtained by plotting the swelling data using the second-order equation below.

$$\frac{t}{W} = \frac{1}{kW_{\infty}^2} + \frac{t}{W_{\infty}}$$

The estimated maximal swelling capacity and the swelling rate were 0.6 g/g and 4.8 1/h*g for Myverol 18-92, and 0.5 g/g and 4.3 1/h*g for Myverol 18-99, respectively [4].

12.3 DRUG RELEASE FROM MONOGLYCERIDES

Due to the unique structure of monoglyceride liquid crystals, lipophilic, hydrophilic, and amphiphilic drugs can be incorporated into the monoglyceride-based systems. A drug can be either dissolved or dispersed in the system dependent upon its solubility. Extended-release applications of drugs with varying molecular weights and solubilities have been published [6–8]. The drug release kinetics followed the square-root of time diffusion model. As an example, the release of chlorpheniramine maleate and pseudoephedrine hydrochloride from GMO rose steeply initially and then leveled off gradually (Figure 12.2). Pseudoephedrine hydrochloride was completely released within 24 h, while only 60% of chlorpheniramine maleate was released within 48 h. No further release of chlorpheniramine maleate was observed after 48 h. The drug release followed the square-root of time relationship before reaching a plateau (Figure 12.3).

The release profile of a lipophilic drug such as ibuprofen from GMO was similar to those of the hydrophilic drugs. Approximately 80% of ibuprofen was released within 48 h, and the release profile also followed the square-root of time relationship (Figure 12.4).

The release of oligopeptides from GMO-based liquid-crystalline phases was reported by Ericsson et al. [9]. They found that the drug release also followed the square-root of time diffusion model. The *in vivo* release data from intramuscular and subcutaneous depots of cubic phase containing desmopressin and somatostatin showed a constant desmopressin-like and somatostatin-like immunoreactivity over 6 h. They suggested that the GMO liquid-crystalline phases are

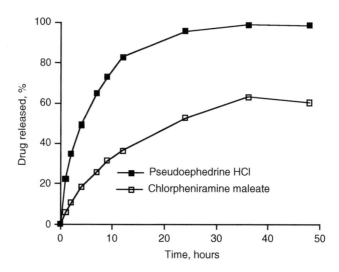

Figure 12.2 Release of chlorpheniramine maleate and pseudo-ephedrine hydrochloride from GMO matrices in 0.1 *M* pH 7.4 buffer at 37°C. (From CM Chang, R Bodmeier. *J Pharm Sci* 86: 747–752, 1997.)

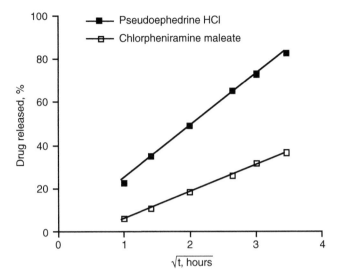

Figure 12.3 Drug release profiles of chlorpheniramine maleate and pseudoephedrine hydrochloride from GMO matrices in 0.1 M pH 7.4 buffer at 37°C according to the square-root of time relationship. (From CM Chang, R Bodmeier. *J Pharm Sci* 86: 747–752, 1997.)

not only able to extend the release of peptides but are also able to protect peptides against enzymatic degradation.

12.4 INFLUENCE OF pH ON SWELLING AND DRUG RELEASE

The influence of the pH of the dissolution medium on monoglyceride swelling and drug release in simulated gastric (pH 1.2) and intestinal (pH 7.4) fluids were investigated by Chang and Bodmeier [8]. The swelling of GMO was significantly influenced by pH. Approximately 10% less water was absorbed by GMO in the pH 1.2 medium. Under a polarized microscope, an isotropic viscous phase was observed for swollen GMO in both media. This indicated that although the pH of the dissolution medium did not affect the type of the swollen mesophases, it affected the swelling capacity of GMO. This phenomenon was not expected because monoglycerides are nonionic lipids. It was later confirmed that the free fatty acid impurities in the technical-grade monoglycerides play a significant role.

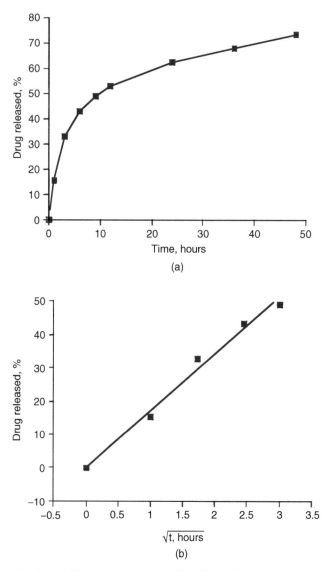

Figure 12.4 (a) Release of ibuprofen from GMO matrix in 0.1 *M* pH 7.4 buffer at 37°C. (b) Drug release profile of ibuprofen from GMO matrices in 0.1 *M* pH 7.4 buffer at 37°C according to the square-root of time relationship. (From CM Chang. Application of Monoglyceride-Based Materials as Sustained-Release Drug Carriers. Ph.D. dissertation, The University of Texas at Austin, Austin, 1995.)

The fatty acids are ionized in pH 7.4 buffer and unionized in pH 1.2 buffer. The ionized fatty acids readily interact with water molecules in pH 7.4 buffer medium; thus, more water can be absorbed in pH 7.4 buffer. On the other hand, fatty acids are unionized and solubilized in the lipid domain of monoglyceride mesophases at pH 1.2; thus, less water is absorbed. The swelling profiles of GMO containing various levels of oleic acid in pH 1.2 and 7.4 buffer media are shown in Figure 12.5. The water uptake of GMO at pH 1.2 decreased as the oleic acid content increased. The maximum water uptake dropped from 0.4 to 0.2 g/g with the presence of 5% oleic acid in GMO. The transformation of the cubic phase into an inverted hexagonal phase in the presence of oleic acid was observed by polarized microscopy. At pH 1.2, oleic acid is solubilized in the lipid domain of GMO mesophases, which may have affected the monoglyceride molecular packing and furthermore transformed the mesophase from the cubic phase into the inverted hexagonal phase. The swelling profile of matrices containing 10% or higher oleic acid could not be obtained because those matrices broke into irregular-shaped pieces upon hydration. In pH 7.4 buffer medium, an opposite result was obtained. The swelling capacity of monoglyceride increased from 0.5 to 1.2 g/g as 10% oleic acid was incorporated into the GMO matrix. These results suggest that the influence of dissolution media on the GMO swelling was mainly due to the free fatty acid impurities in the technical-grade GMO.

A significant influence of dissolution medium pH on drug release was observed for propranolol hydrochloride (Figure 12.6). Comparing the results from the swelling of the monoglyceride, which indicated that the swelling of monoglyceride at pH 7.4 was higher than that at pH 1.2, drug release was opposite to the monoglyceride swelling. Approximately 20% more propranolol hydrochloride was released at pH 1.2 than at pH 7.4. This is also attributed to the free fatty acids in the technical-grade monoglyceride. Propranolol hydrochloride is a basic drug; it is positively charged in both pH media. The ionic interaction between fatty acids and propranolol hydrochloride at pH 7.4 may have decreased the drug release. At pH 1.2, fatty acids were nonionized, and there was no ionic interaction between propranolol hydrochloride and the fatty acids. Therefore, more drug was released in pH 1.2 medium than in pH 7.4 medium. The influence of incorporated oleic acid on the propranolol hydrochloride release is shown in Figure 12.7. As expected, oleic acid has no influence on propranolol hydrochloride release in pH 1.2 dissolution medium; however, due to ionic interactions between propranolol hydrochloride and oleic acid

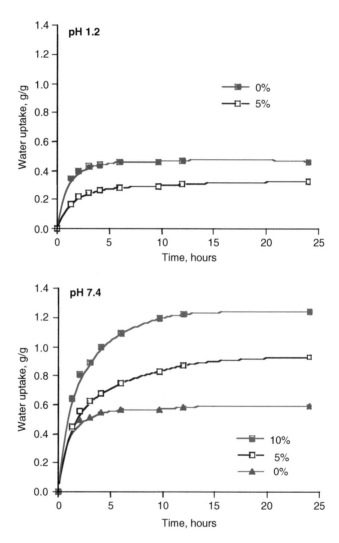

Figure 12.5 Effect of oleic acid concentration on the swelling capacity of GMO in 0.1 M HCl and 0.1 M pH 7.4 phosphate buffer at 37°C. (From CM Chang, R Bodmeier. *J Controlled Release* 46: 215–222, 1997.)

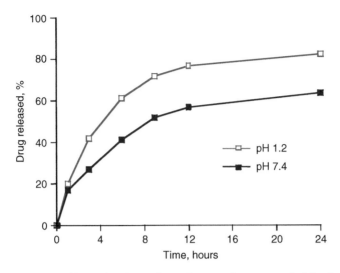

Figure 12.6 Effect of pH on the release of propranolol hydrochloride (2% drug content) from GML matrices at 37°C. (From CM Chang. Application of Monoglyceride-Based Materials as Sustained-Release Drug Carriers. Ph.D. dissertation, The University of Texas at Austin, Austin, 1995.)

in pH 7.4 medium, the release of propranolol hydrochloride decreased with increasing oleic acid content. It was not possible to investigate the release of propranolol hydrochloride from matrices containing 0 and 5% oleic acid at pH 1.2 because those matrices broke into irregular-shaped pieces in the dissolution medium.

For nonionizable drugs, a pH effect on drug release is not expected due to the lack of ionic interaction between drug and free fatty acids. The release of the nonionizable drug guaifenesin from GMO was investigated in both pH 1.2 and 7.4 media (Figure 12.8). As expected, no significant influence of oleic acid content on the drug release was observed.

The release of drugs from monoglyceride matrices was mainly affected by the type of mesophase and the interaction between the monoglyceride (especially ionizable ingredients) and drug molecules. Therefore, factors affecting mesophase formation and the ionization of lipid components could influence drug release. The elimination of free fatty acids may circumvent pH-dependent drug release behavior.

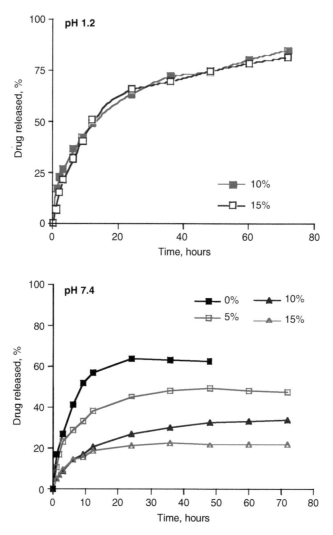

Figure 12.7 Effect of oleic acid concentration on the release of propranolol hydrochloride (10% drug content) from GMO in 0.1 *M* HCl and 0.1 *M* pH 7.4 phosphate buffer at 37°C. (From CM Chang, R Bodmeier. *J Controlled Release* 46: 215–222, 1997.)

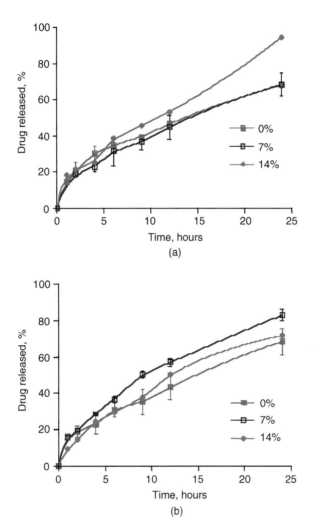

Figure 12.8 Effect of oleic acid concentration on the release of guaifenesin in (a) 0.1 M HCl and (b) 0.1 M pH 7.4 phosphate buffer at 37°C. (From CM Chang, R Bodmeier. *J Controlled Release* 46: 215–222, 1997.)

12.5 FORMULATION FACTORS AFFECTING DRUG RELEASE FROM MONOGLYCERIDES

The structure of monoglyceride mesophases is dependent upon various factors such as water content and the presence and concentration of additional solutes. The effect of formulation factors, including drug type, drug content, and the source of monoglycerides, on drug release from monoglyceride-based systems has been evaluated by a number of researchers [4,6,8].

The release of chlorpheniramine maleate from GML–water matrices with different initial water contents (10 to 30%) is shown in Figure 12.9. The drug release increased with increasing initial water content. Although these systems were eventually transformed into a viscous cubic phase during the dissolution study, the preexisting water channels in the initial water-containing matrices increased the drug release. The nearly parallel drug release curves, at longer time periods, indicated that the differences in drug release occurred during the transformation stage of monoglyceride to cubic phase within the initial release period. Once the cubic phase formed for all samples, no difference in the drug release rate was observed.

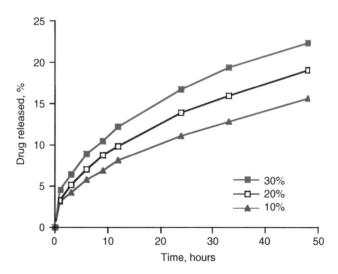

Figure 12.9 Effect of initial water content of the GML matrix (10 to 30%) on the chlorpheniramine maleate release in 0.1 *M* pH 7.4 buffer at 37°C. From CM Chang, R Bodmeier. *J Pharm Sci* 86: 747–752, 1997.)

Monoglycerides are produced by an esterification reaction of triglycerides (fat) and glycerol using calcium hydroxide as a catalyst. The reaction typically yields about 40 to 60% of monoglycerides and thus, a distillation process is required to increase the content of monoglyceride to over 90%. The final composition of monoglycerides is highly dependent upon the source of the starting materials and the process conditions. Chang and Bodmeier evaluated the release of propranolol hydrochloride from three technical-grade GMO (Myverol 18-99, GMOrphic-80®, and Dimodan DGMO®). The principal composition of these monoglycerides is shown in Table 12.1. Dimodan DGMO has the highest GMO content, at approximately 90%, while GMOrphic-80 and Myverol 18-99 have GMO contents of approximately 75 and 61%, respectively. Although the compositions of these technical-grade GMO varied, the drug release rate and the extent of drug release were not affected (Figure 12.10). Differences in GMO composition in these three products did not significantly affect the drug release. The drug release from mono-glyceride was primarily determined by the type of swollen meso-phase, and the GMO content, from 60 to 92%, did not affect drug release.

The effect of drug loading on drug release was investigated with drug loadings of 5, 10, 20, and 30% pseudoephedrine hydrochloride in GMO (Figure 12.11). Pseudoephedrine hydrochloride release increased with increasing drug loading; however, the percent drug released remained at the same level for all drug loadings. Pseudoephedrine hydrochloride was completely released within 36 h. This indicated that the same portion of drug was released in the same time interval regardless of the drug loading. The drug release rate was directly proportional to the drug loading (Figure 12.12).

Table 12.1 Main Composition of Three Technical-Grade GMO Products

Monoglyceride	Myverol 18-99	GMOrphic-80	Dimodan DGMO
Monoolein	~60.9%	~75%	~92%
Monolinolein	~21.0%	<15%	~4.3%
Monopalmitate	~4.1%	<10%	~0.5%

Source: CM Chang. Application of Monoglyceride-Based Materials as Sustained-Release Drug Carriers. Ph.D. dissertation, The University of Texas at Austin, Austin, 1995.

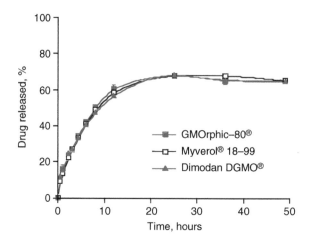

Figure 12.10 Release profile of propranolol hydrochloride from technical g rades of GMO at 37°C. (From CM Chang, R Bodmeier. *J Pharm Sci* 86: 747–752, 1997.)

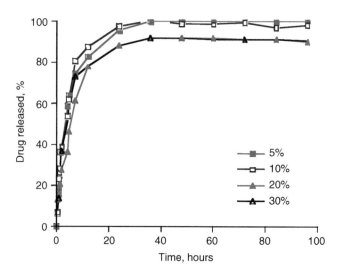

Figure 12.11 Effect of pseudoephedrine hydrochloride loading on the percent drug release in 0.1 *M* pH 7.4 buffer at 37°C. (From CM Chang, R Bodmeier. *J Pharm Sci* 86: 747–752, 1997.)

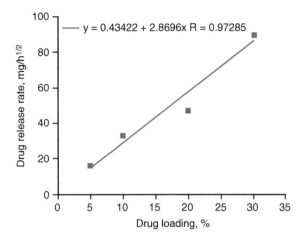

Figure 12.12 Relationship between pseudoephedrine hydrochloride release rate and percent drug loading. (From CM Chang. Application of Monoglyceride-Based Materials as Sustained-Release Drug Carriers. Ph.D. dissertation, The University of Texas at Austin, Austin, 1995.)

12.6 BINDING OF VARIOUS DRUGS TO MONOGLYCERIDES

The binding of drug molecules to monoglycerides was observed during the dissolution study of chlorpheniramine maleate and propranolol hydrochloride from monoglyceride matrices [4]. For example, the release of chlorpheniramine maleate was highly extended; however, only up to 60% of the drug was released from the GMO cubic phase. Reaching the equilibrium point before complete drug release suggested that a fraction of the drug might be bound to the swollen liquid-crystalline phase. Since the dissolution study was conducted under sink conditions, the incomplete release was not due to the limited solubility of the drug in the dissolution medium. The binding of several model drugs including chlorpheniramine maleate, diltiazem hydrochloride, metoclopramide hydrochloride, phenylpropanolamine hydrochloride, propranolol hydrochloride, pseudoephedrine hydrochloride, and theophylline anhydrous to GMO was investigated by dissolution studies at pH 7.4 (Figure 12.13) [10]. A nearly complete release of pseudoephedrine hydrochloride, phenylpropanolamine hydrochloride, metoclopramide hydrochloride, and theophylline anhydrous was observed, while chlorpheniramine maleate, propranolol hydrochloride, and diltiazem hydrochloride were not completely released.

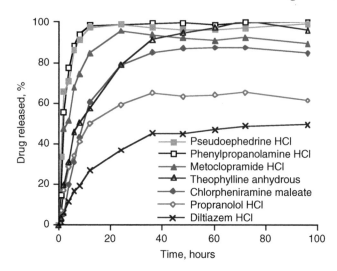

Figure 12.13   Release of various drugs from GMO in 0.1 *M* pH 7.4 buffer at 37°C. (From CM Chang, R Bodmeier. *Int J Pharmaceutics* 147: 135–142, 1997.)

This indicated that these three drugs were bound to the GMO matrices. The solubility of a drug did not seem to be a factor in incomplete drug release (Table 12.2). For example, the solubility of theophylline anhydrous in water is 12 mg/mL and is the lowest among the drugs measured. Theophylline anhydrous, however, was almost entirely released within 60 h. The aqueous solubility of chlorpheniramine maleate is higher than that of phenylpropanolamine hydrochloride; however, phenylpropanolamine hydrochloride was totally released within 24 h. Thus, it was clear that factors other than the solubility of the drug determined the drug release from GMO. Since GMO is an amphiphilic polar lipid, the solubilization of the drug in the amphiphilic GMO or mesophases may play an important role.

Metoclopramide hydrochloride, phenylpropanolamine hydrochloride, and pseudoephedrine hydrochloride are nonamphiphilic molecules. In contrast, chlorpheniramine maleate, propranolol hydrochloride, and diltiazem hydrochloride are amphiphilic in nature and can associate at the lipid–water interface. The high interfacial region of the monoglyceride mesophases can solubilize significant amounts of amphiphilic drug molecules. The association of drug molecules to monoglyceride mesophases, other than the ionic interaction with free fatty acid, caused this incomplete drug release. This was confirmed

Table 12.2   Solubility of Drugs in Various Media

	Solubility, mg/mL		
Drug	Water	pH 7.4 Buffer	0.1 M HCl
Chlorpheniramine maleate	705	640	661
Diltiazem HCl	511	487	633
Guaifenesin	620	652	659
Metoclopramide HCl	800	825	803
Phenylpropanolamine HCl	486	476	497
Propranolol HCl	291	301	249
Pseudoephedrine HCl	807	785	805
Theophylline anhydrous	12	13	14

Source: CM Chang. Application of Monoglyceride-Based Materials as Sustained-Release Drug Carriers. Ph.D. dissertation, The University of Texas at Austin, Austin, 1995.

with the absorption study conducted on equilibrating propranolol hydrochloride drug solution with high purity (99%) GMO and GML (Figure 12.14). A significant amount of propranolol hydrochloride absorbed to both monoglycerides.

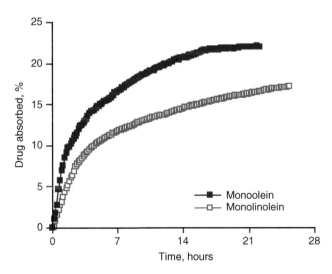

Figure 12.14   Absorption of propranolol hydrochloride to pure GMO and GML (22°C). (From CM Chang, R Bodmeier. *Int J Pharmaceutics* 147: 135–142, 1997.)

12.7 EFFECT OF DRUGS ON THE PHASE TRANSITION OF MONOGLYCERIDES

The addition of a third component such as a drug substance to monoglyceride–water binary mixtures may disturb molecular aggregation and mesophases. This depends mainly upon the physicochemical characteristics of the third component. Monoglycerides form liquid crystals that consist of three parts: the hydrophilic domain, the hydrophobic domain, and the interface to the two separate domains. Depending upon the type of incorporated substance, this substance can be dissolved in the hydrophilic domain (hydrophilic substances), in the hydrophobic domain (lipophilic substances), or at the interface (amphiphilic substances). The solubilization of the substance in the different domains affects the mesophase differently. This effect has been qualitatively described by Ninham's ratio,

$$\text{Ninham's ratio} = \frac{V_h}{a_0 l_c}$$

where V_h is the volume of the hydrocarbon chain, a_0 is the cross-sectional area of the polar group, and l_c is the length of the amphiphile [11]. Since l_c is a constant for a given amphiphile, the transformation of mesophases is affected by V_h and a_0. The location of the solubilized component influences these two parameters. If V_h is roughly equal to a_0, a lamellar phase may form. If $V_h > a_0$, an inverted cubic phase, hexagonal phase, or micellar phase may form. On the contrary, if $V_h < a_0$, a normal hexagonal phase or a micellar phase may form.

Theoretically, the unsaturated monoglyceride cubic phase has a larger V_h than a_0. If the added substance stays in the hydrophilic portion, a_0 will increase; hence, it may transform the mesophase from a cubic phase into a lamellar phase. Continually increasing the amount of substance may further transform the phase into a hexagonal and even into a micellar phase. For lipophilic substances, increasing the amount of the substance will increase V_h and therefore transform the cubic phase into an inverted hexagonal phase or an inverted micellar phase. If the substance is surface active and favorably remains at the interface, both V_h and a_0 may increase, but to different extents.

Chlorpheniramine maleate is a highly water-soluble drug. Adding more than 2% of this drug into the cubic phase transformed the cubic phase into a lamellar phase. The formation of anisotropic layers was observed by polarized microscopy. This type of optical appearance was observed at all drug loadings. The viscosity of the lamellar phase decreased with increasing drug loading. The same results were obtained with other hydrophilic drugs such as diltiazem hydrochloride, propranolol hydrochloride, and pseudoephedrine hydrochloride. On the contrary, the incorporation of the lipophilic drug ibuprofen into the cubic phase transformed the cubic phase into an inverted hexagonal phase. The inverted hexagonal phase was more viscous than the lamellar phase but less viscous than the cubic phase. Increasing the drug loading to 20% transformed the mesophase into an inverted micellar phase; therefore, the system became flowable. Propranolol also transformed the cubic phase into an inverted hexagonal phase. At 15% propranolol loading, the inverted hexagonal phase was transformed into a dispersed system; this suggested that propranolol was highly surface active and significantly changed the liquid crystalline structure. Griseofulvin could not be dissolved in the monoglyceride cubic phase; therefore, the addition of griseofulvin in the cubic phase did not transform the cubic phase into other phases. Dispersed drug crystals were observed under a polarized microscope. The influence of drug loading on mesophase formation and the fluidity is summarized in Table 12.3.

12.8 PHARMACEUTICAL APPLICATION OF MONOGLYCERIDE-BASED EXTENDED-RELEASE DOSAGE FORMS

The application of GMO-based extended-release dosage forms for various administration routes has been investigated by a number of researchers. The unique characteristics of GMO liquid crystals make it possible to be utilized as extended-release carrier material for oral, topical, parenteral depot, dental, ophthalmic, and buccal deliveries. Examples from the published literature are summarized in this section.

12.8.1 Oral Delivery

For oral application, GMO-based extended-release systems could be prepared by filling the molten GMO or GMO–water mixture with drugs into hard or soft gelatin capsules. The release of the water-soluble drugs pseudoephedrine hydrochloride and chlorpheniramine maleate from hard gelatin capsules containing the drug powders alone and GMO is shown in Figure 12.15. For the powder-filled capsules, the drug was released immediately after the hard gelatin capsule

Table 12.3 Influence of Drug Substances on the Phase Transformation of the Cubic Phase

Loading, %	Chlorpheniramine Maleate	Diltiazem HCl	Propranolol HCl	Pseudoephedrine HCl	Ibuprofen	Propranolol	Griseofulvin
5	Lamellar	Lamellar	Lamellar	Lamellar	Hexagonal	Hexagonal	No phase transformation
10	Lamellar	Lamellar	Lamellar	Lamellar	Hexagonal	Dispersion	No phase transformation
15	Lamellar	Lamellar	Lamellar	Lamellar	Hexagonal	Dispersion	No phase transformation
20	Lamellar	Lamellar	Lamellar	Lamellar	Hexagonal	Dispersion	No phase transformation

Source: CM Chang. Application of Monoglyceride-Based Materials as Sustained-Release Drug Carriers. Ph.D. dissertation, The University of Texas at Austin, Austin, 1995.

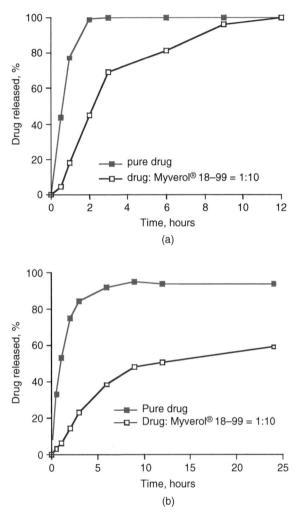

Figure 12.15 Release of (a) pseudoephedrine HCl and (b) chlorpheniramine maleate from hard gelatin capsules with and without Myverol 18-99 in 0.1 M pH 7.4 buffer at 37°C. (From CM Chang. Application of Monoglyceride-Based Materials as Sustained-Release Drug Carriers. Ph.D. dissertation, The University of Texas at Austin, Austin, 1995.)

shell was dissolved. On the contrary, the release of the drug from the capsule containing GMO was extended [12].

Sallam et al. evaluated the feasibility of developing extended-release furosemide capsules using a GMO-based system to enhance gastric residence time in order to improve the bioavailability of the drug and at the same time extend the action [13]. The release of furosemide in both simulated gastric and simulated intestinal fluids was found to be too slow to be considered for oral application. The slow release rate was caused by the low solubility and high lipid partitioning of the drug to the cubic phase bilayer. The release rate, however, could be enhanced by the addition of PEG400 and trisodium phosphate to the formulation. Although the influence of bile salts in the intestine on drug release was not mentioned in this study, bile salts may enhance the drug release and absorption [8].

12.8.2   Topical Delivery

Using GMO mesophases as topical drug delivery vehicles has several advantages:

1. A larger amount of drug can be solubilized in GMO–water systems when compared to creams and ointments.
2. The GMO mesophases have a stronger skin adhesive property and thus resist wash-off.
3. GMO is nonionic and therefore less irritating to the skin.

These properties are ideal for a topical formulation.

Hydrocortisone-containing GMO–water topical formulations were evaluated by Chang and Bodmeier [14]. They compared the release of hydrocortisone from GMO cubic phase against the commercially available hydrocortisone cream and ointment using Franz diffusion cells (Figure 12.16). The release of hydrocortisone from these formulations was in the order of: cream > cubic phase > ointment. Approximately 50% hydrocortisone was released from the cream within 20 min. The release of the drug from the cubic phase was significantly slower than that from the cream. This can be explained by the high solubility of hydrocortisone in the cubic phase compared to that in the cream. Garti et al. reported that more than 1% hydrocortisone can be solubilized in the GMO cubic phase [15]. The release of hydrocortisone from the ointment was far slower than from the other two vehicles because of the low diffusion coefficient of the drug in the ointment vehicle.

Another study published by Helledi and Schubert utilized a GMO–water system to deliver acyclovir topically [16]. In their study,

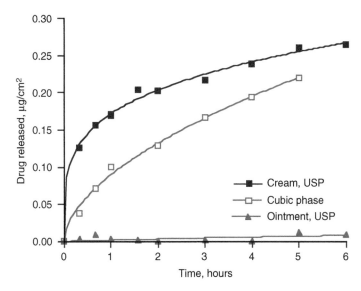

Figure 12.16 Release of hydrocortisone from various topical vehicles in 0.1 *M* pH 7.4 buffer at 37°C. (From CM Chang. Application of Monoglyceride-Based Materials as Sustained-Release Drug Carriers. Ph.D. dissertation, The University of Texas at Austin, Austin, 1995.)

acyclovir was suspended instead of solubilized in the cubic phase, and the release of acyclovir (1 to 5%) was investigated using Franz diffusion cells. Acyclovir was readily released from the system, and the release increased with increasing initial drug loading. The release rate of acyclovir from the cubic phase was sixfold faster than that from the commercial product, Zovir® cream.

The adhesive property of a preparation on the applied skin surface is another important factor to be considered for topical application. Formulations that can be washed off easily by water or detergents require more frequent applications on the skin surface. The washing off of a topical formulation from the skin surface delivers incomplete amounts of drug during the therapeutic period. Since the penetration of the drug from topical vehicles through the skin is typically slow, a long residence time of a topical vehicle on the skin surface is important. The ease of the wash-off of monoglyceride-based vehicles, cream, and ointment was evaluated by Chang and Bodmeier [14] by investigating the resistance of the vehicles applied on a cellulose acetate membrane to washing off in sodium dodecyl sulfate solution. The resistance was in the order GMO > ointment > cream.

12.8.3   Depot Drug Delivery Systems

GMO-based systems have a potential as carrier material in biode-gradable depot systems because monoglycerides are subject to lipol-ysis by esterases present in muscular and conjunctive tissues [17]. However, due to the highly viscous nature of the GMO cubic phase, the fully swollen GMO cannot be injected. The viscosity of a lamellar phase is lower, and it can be injected through syringe needles. The lamellar phase will absorb body fluids from the surrounding tissues and transform into a cubic phase upon injection. This process could potentially cause tissue irritation. The tissue response to mesophases upon subcutaneous and intramuscular injection of the cubic phase and the lamellar phase into rabbits were investigated by Ericsson et al. [9]. Irritation was found in a few animals, and a slightly greater irritating response was observed upon injection of a lamellar phase.

An ideal GMO-based depot formulation should contain enough water so that it does not absorb additional water from the sur-rounding tissues after injection. Furthermore, the viscosity of the formulation should have an acceptable syringability. This type of formulation may be developed by adding a third component, drugs or injectable solvents, to transform the cubic phase into lower-viscosity mesophases (lamellar or inversed micellar phases). After a portion of the drug is released or the solvent diffused away, the high-viscosity cubic phase then forms *in situ* without further hydration. These approaches have been evaluated by Chang and Bodmeier [18].

A GMO mesophase change upon the addition of drug sub-stances to the cubic phase was observed by Engstroem [19]. Surface-active drugs such as lidocaine transformed the monoglyceride cubic phase into a lamellar liquid-crystalline phase, a reversed hexagonal liquid-crystalline phase, or a reversed micellar phase when the base form of lidocaine was added. A similar observation was noticed by Chang and Bodmeier when chlorpheniramine maleate and propra-nolol hydrochloride were added to a GMO cubic phase [10]. Chlor-pheniramine maleate transformed the cubic phase into the lamellar phase, then into an isotropic solution phase as the drug content continuously increased. A different but similar phase transformation was noticed with propranolol hydrochloride. With increasing propra-nolol hydrochloride content, the lamellar phase was induced from the cubic phase; however, no isotropic solution phase was induced with increasing drug content.

In order to develop an injectable depot system using chlorphe-niramine maleate as a model drug, Chang and Bodmeier systemically studied the influence of chlorpheniramine maleate content on the

Table 12.4 Compositions of Chlorpheniramine
Maleate–Induced Low-Viscosity Preparations
for Dissolution Study

Formulation	Myverol, %	Water, %	Drug, %
A	43.48	43.48	13.04
B	60.87	26.09	13.04
C	78.26	8.70	13.04
D	34.48	34.48	31.03
E	48.28	20.69	31.03
F	26.32	26.32	47.37

Source: CM Chang, R Bodmeier. *Int J Pharmaceutics* 173: 51–60, 1998.

GMO mesophases using a triangular phase diagram [18]. The release
of chlorpheniramine maleate from low-viscosity solution and lamellar phases was evaluated in simulated intestinal fluid. Compositions
A, B, and C listed in Table 12.4 had the same amount of drug loading
of 13%, while compositions D and E had a drug loading of 31%. The
ratio of GMO and water was fixed at 1:1 for preparations A, D, and
F, and at 0.7:0.3 for B and E, respectively. The release of chlorpheniramine maleate from formulations with a 1:1 ratio of GMO:water,
but varied percentage of drug loading, is shown in Figure 12.17.

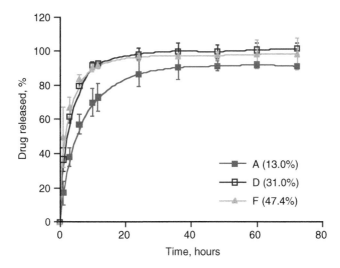

Figure 12.17 Effect of chlorpheniramine maleate loading on the
drug release from drug-induced low-viscosity formulations in 0.1 *M*
pH 7.4 buffer (GMO:water = 1:1) at 37°C. (From CM Chang, R
Bodmeier. *Int J Pharmaceutics* 173: 51–60, 1998.)

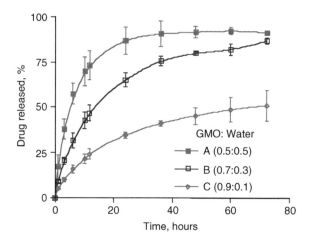

Figure 12.18   Effect of GMO:water ratio on the chlorpheniramine maleate release from drug-induced low-viscosity formulations in 0.1 *M* pH 7.4 buffer at 37°C (13% drug loading). (From CM Chang, R Bodmeier. *Int J Pharmaceutics* 173: 51–60, 1998.)

The drug release rate increased with increasing drug loading because the increase in drug content transformed a mesophase into an isotropic solution phase. The low viscosity and solution-like isotropic phase increased the drug release rate. The drug release significantly decreased with increasing GMO:water ratio (Figure 12.18). Although all preparations were in a lamellar mesophase, the increase in water content increased the diffusivity of the water-soluble drugs and therefore the drug release. In the solution phase, no such differences were present.

The influence of ethanol on GML mesophases was systematically studied by Chang and Bodmeier using a triangular phase diagram [18,20]. The cubic phase was observed at an ethanol content of less than 20%. Increasing the ethanol content in GML/water mixtures required more water to form the cubic phase. A hexagonal region was obtained in the low water and ethanol content region. A large area of a clear solution phase was obtained at high ethanol contents. This region started from the over 20% ethanol–containing monoglyceride/ water mixtures. This solution phase was clear and of low viscosity. Upon injection of formulations from this region in water, the cubic phase formed *in situ* with diffusion of ethanol away from the formulation. Thus, formulations from this region are capable

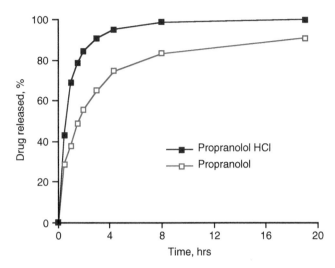

Figure 12.19 Release of propranolol and propranolol hydrochloride from an ethanol-induced low-viscosity formulation with a ratio of ethanol:water:GML equaling 20:16:64 in 0.1 *M* pH 7.4 buffer at 37°C. (From CM Chang, R Bodmeier. *Int J Pharmaceutics* 173: 51–60, 1998.)

of being utilized as injectable formulations. The release of propranolol and propranolol hydrochloride from the isotropic solution phase containing ethanol:water:GML in a ratio of 20:16:64 is shown in Figure 12.19. An initial burst release occurred and the drug was almost completely released within 12 h. Although a lower viscosity monoglyceride-based system could be obtained by combining drugs and solvent, the *in vitro* release rate, however, was too fast for a extended release depot system. The *in vivo* performance of these systems needs to be further evaluated in order to understand the performance of this type of depot system.

The application of a fully swollen GMO cubic phase as an implantable local delivery system for the antitumor compound, carmustine, for the treatment of malignant neutral gliomas after surgical debulking of tumors was demonstrated by Jones et al. [21]. The GMO cubic phase with extended release may offer the same advantages of the presently available polymer-based disk system by providing a product that allows local delivery of chemotherapeutic agents with decreased dose-related systemic toxicity. In addition, the cubic system fits well into the tumor pocket allowing greater intimate contact between the pocket walls and the delivery system and perhaps greater clinical efficacy.

12.8.4 Dental Application

Damani has demonstrated the use of a mixture of GMO and triglycerides for the treatment of periodontal disease [22]. The incorporation of the triglyceride sesame oil decreased the melting point of GMO, thus allowing injection at room temperature. The injection of a mixture of GMO and sesame oil with antibiotics into periodontal pockets was able to treat periodontal disease.

Another dental application of a monoglyceride-based system was demonstrated by Esposito et al. [23]. They prepared tetracycline HCl containing monoglyceride-based dental gels and studied them in patients with periodontitis. The formulation prepared with up to 10% water content maintained the fluidity necessary for administration with a syringe needle to the periodontal cavity. After application, the formulation absorbs biological fluids and swells into a viscous phase for extended drug release. The gel was easily administered by syringe needles appropriate for intrapocket delivery. Both the *in vitro* drug release profile and *in vivo* gel persistence data suggest that the gel had good bioadhesiveness at the application site and that the drug release rate was significantly extended. After subgingival application, the gel produced a significantly improved outcome in moderate to deep periodontal pockets.

12.8.5 Ophthalmic Application

An example of using a monoglyceride liquid-crystalline system as an extended-release ophthalmic dosage form was demonstrated by Engstroem et al. with a system that underwent a thermoreversible lamellar-to-cubic phase transition [24]. The system, containing GMO and Brij 96, formed a relatively low-viscosity lamellar phase below 30°C; it transformed into a stiff cubic phase with an increasing temperature. The authors suggested that the choice of Brij 96 is one of many possibilities, and other polar lipids such as phospholipids and soaps of fatty acids may be used to obtain the thermoreversible property. The *in vitro* release behavior of a system containing timolol maleate was much more extended compared with that of a polymer-based gel system.

Despite the potential application of GMO for ophthalmic use, the compatibility of GMO with ocular tissues has not been explored in detail.

12.8.6 Buccal Delivery

Due to the mucoadhesive property of GMO liquid-crystalline phases, it could be considered as an attractive buccal delivery carrier. The mucoadhesive properties of GMO and GML mesophases were demonstrated *in vitro* by Nielsen et al. [25]. He suggested that the cubic phase was mucoadhesive when formed on wet mucosa. The unswollen monoglycerides had the greatest mucoadhesion, followed by the partly swollen lamellar phase and the fully swollen cubic phase. The mechanism of mucoadhesion is unspecific and probably involves dehydration of the mucosa. Lee et al. investigated the mucoadhesive property of the GMO-based system using an *in vitro* tensile strength technique [26]. They observed that the mucoadhesion of the liquid-crystalline phases of GMO occurred after the uptake of water. A good linear relationship between initial water content of the liquid-crystalline phases and mucoadhesive force led to the conclusion that the mucoadhesive force increased with decreasing initial water concentration. Lee et al. also investigated the feasibility of using GMO-based systems for the buccal delivery of peptides [27]. They incorporated oleic acid as a lipophilic permeation enhancer and polyethylene glycol 200 (PEG 200) as a coenhancer into the cubic liquid-crystalline phase of GMO. The *ex vivo* buccal permeation of leucine-2-alanine enkephalin (DADLE) through porcine buccal mucosa mounted in a Franz cell showed that the buccal permeation flux of DADLE significantly increased when 5 or 10% of PEG 200 was coadministered with 1% oleic acid compared with the cubic phase containing 1% oleic acid alone. These results suggested that PEG 200 enhanced the action of the lipophilic permeation enhancer oleic acid, and that the combination of oleic acid and PEG 200 as a coenhancer could be a useful tool to improve the membrane permeability in the buccal delivery of peptide drugs using a GMO cubic liquid-crystalline phase.

12.9 CONCLUSION

Monoglyceride-based systems have unique properties and can be applied as extended-release drug delivery systems for both small and large molecules. Depending upon the solubility of the drug, it can be either solubilized or dispersed in the GMO-based vehicle. In some instances, the stability of the incorporated drug against chemical, physical, or enzymatic degradation may even be enhanced [28–30]. However, some hurdles remain to be overcome before GMO-based extended-release drug delivery systems will be commercializable.

Although GMO is a natural material and generally recognized as safe, very limited studies have been published on the safety of GMO for parenteral or ophthalmic use. In addition, the purity of the commercially available technical grades of GMO is not very high. Potential impurities such as free fatty acids and other monoglycerides may affect both the performance and the safety of GMO-based systems. Furthermore, most of the drug release studies have been performed *in vitro*, and very limited studies have been performed *in vivo*.

REFERENCES

1. ES Lutton. *J Am Oil Chem Soc* 42: 1068–1070, 1965.

2. ST Hyde, S Andersson, B Ericsson, K Larsson. *Z Kristallogr* 168(1–4): 213–219, 1984.

3. K Larsson. *J Phys Chem* 93: 7304–7314, 1989.

4. CM Chang, R Bodmeier. *J Pharm Sci* 86: 747–752, 1997.

5. H Schott. *J Pharm Sci* 81: 467–470, 1992.

6. R Burrows, JH Collett, D Attwood. *Int J Pharmaceutics* 111: 283–293, 1994.

7. DM Wyatt, D Dorschel. *Pharm Technol* 16: 116,118,120,122,130, 1992.

8. CM Chang, R Bodmeier. *J Controlled Release* 46: 215–222, 1997.

9. B Ericsson, PO Eriksson, JE Loefroth, S Engstroem. ACS Symposium Series 469 (Polym. Drugs Drug Delivery Syst.): 251–265, 1991.

10. CM Chang, R Bodmeier. *Int J Pharmaceutics* 147: 135–142, 1997.

11. BW Ninham, DJ Mitchell. *J Am Chem Soc Faraday Trans II* 76: 201, 1981.

12. CM Chang. Application of Monoglyceride-Based Materials as Sustained-Release Drug Carriers. Ph.D. dissertation, The University of Texas at Austin, Austin, 1995.

13. AS Sallam, E Khalil, H Ibrahim, I Freij. *Eur J Pharmaceutics Biopharmaceutics* 53: 343–352, 2002.

14. CM Chang, R Bodmeier. *Pharmaceutical Res* 11(10): S-185, 1994.

15. N Garti, D Ostfeld, R Goubran, EJ Wachtel. *J Dispersion Sci Technol* 12: 321–335, 1991.

16. LS Helledi, L Schubert. *Drug Develop Indust Pharm* 27: 1073–1081, 2001.

17. H Malonne, AJ Moes, J Fontaine. *Acta Technol Legis Medicamenti* 7(3): 165–170, 1996.

18. CM Chang, R Bodmeier. *Int J Pharmaceutics* 173: 51–60, 1998.

19. S Engstroem, L Engstroem. *Int J Pharmaceutics* 79: 113–122, 1992.

20. R Bodmeier, CM Chang. *Eur J Pharm Biopharm* 42 (Suppl): 32S, 1996.

21. C Jones, M LaPoint, S Patel, J Shah. Proceedings of the International Symposium on Controlled Release of Bioactive Materials 26th: 605–606, 1999.

22. NC Damani. European patent application EP 429224 A1, 1991.

23. E Esposito, V Carotta, A Scabbia, L Trombelli, P D'Antona, E Menegatti, C Nastruzzi. *Int J Pharmaceutics* 142: 9–23, 1996.

24. S Engstroem, L Lindahl, R Wallin, J Engblom. *Int J Pharmaceutics* 86: 137–145, 1992.

25. LS Nielsen, L Schubert, J Hansen. *Eur J Pharm Sci* 6(3): 231–239, 1998.

26. J Lee, SA Young, IW Kellaway. *J Pharm Pharmacol* 53: 629–636, 2001.

27. J Lee, IW Kellaway. *Int J Pharmaceutics* 204: 137–144, 2000.

28. Y Sadhale, JC Shah. *Int J Pharmaceutics* 191: 65–74, 1999.

29. Y Sadhale, JC Shah. *Int J Pharmaceutics* 191: 51–64, 1999.

30. Y Sadhale, JC Shah. *Pharm Devel Technol* 3: 549–556, 1998.

13

Cubic Liquid-Crystalline Particles as Protein and Insoluble Drug Delivery Systems

HESSON CHUNG, SEO YOUNG JEONG,
AND ICK CHAN KWON

CONTENTS

13.1 INTRODUCTION

Based on the scientific understanding of lipid particulate systems, many scientists have pursued the development of better lipid formulations for their own application fields. Numerous innovative lipid particulate systems have been formulated in recent years. One particularly interesting system is Cubosome® (1). Cubosome refers to a submicron-sized dispersed lipid particle of the bicontinuous cubic liquid-crystalline phases in an aqueous environment. A great number of studies have been performed to investigate the structure (2,3) and the mathematical description of Cubosome particles (4,5). One of the main application fields for Cubosome is drug delivery (6,7). Cubosomes have distinctly different characteristics from other conventional lipid-particulate systems including liposomes and lipid emulsions. The interior of the particles is considered the hydrated cubic phase, which contains hydrophilic water channels, the hydrophobic hydrocarbon chain region, and the interfacial headgroup region (6). This somewhat complex internal structure is an ideal setting for many drugs of hydrophobic, hydrophilic, and amphiphilic nature.

A Cubosome can be formed by first forming a very viscous lipid cubic phase by adding water and an emulsifier to monoglyceride, and then by dispersing the mixture in water. A Cubosome can have an average particle size of as large as several micrometers in diameter. Since it is preferable to have submicron-sized Cubosomes to solubilize pharmaceutical compounds, submicron-sized particles have been formulated by applying mechanical forces, such as microfluidizing the coarse dispersion (6). Preparing submicron-sized Cubosome particles by means of applying mechanical force, however, may result in physicochemical instability of the constituting ingredients or the enclosed materials due to high energy and high temperature accompanying the mechanical process.

To overcome some of the drawbacks in the conventional preparation method, we developed a homogeneous liquid formulation that can be readily dispersed in water (8,9). The liquid formulation comprises monoolein, an emulsifier, and a biocompatible organic solvent but contains no or negligible amounts of water, which can degrade drug or other components by hydrolysis; thus, the formulation is remarkably stable at room temperature. The formulation can be prepared without heat or mechanical force and can be sterilized easily by filtration. The liquid formulation can be dispersed easily in excess water by simple shaking or vortexing. By mixing the liquid formulation with an excess amount of water or phosphate-buffered saline (PBS), submicron-sized particles ranging from 200 to 500 nm in diameter can be formed. Drug encapsulation efficiency inside the dispersed particles varies mainly depending on the nature of the drug itself. While a hydrophobic or protein drug is enclosed inside the particles at a high ratio, a small hydrophilic drug resides mostly in the bulk aqueous phase.

In this chapter, we describe the preparation procedure, stability, and characteristics of our liquid formulation. One of the main application fields of our liquid formulation and its dispersion is drug delivery systems. Due to the toxicity of monoolein in blood and muscle, the dispersion could not be administered intravenously or intramuscularly (see below). The formulation, however, is an efficient oral drug delivery system without any toxicity. Oral administration of hydrophobic or protein drug will be described.

13.2 PREPARATION OF THE LIQUID FORMULATION AND SIZE DISTRIBUTION OF THE DISPERSED PARTICLES

Distilled monoglyceride (RYLO™ MG 19, NF, Danisco A/S, Denmark) consists of monoolein (ca. 77%) and other monoglycerides, diglycerides, and triglycerides as the minor components. Monoolein refers to this distilled monoglyceride in this chapter. Pure monoolein (>99%, Nu Chek Prep, Elysian, MN) can also be used without changing most of the physical characteristics of the liquid formulation and its dispersion with or without drugs. One hundred milligrams of monoolein and 15 to 30 mg Pluronic F-127 (Lutrol F127, Poloxamer 407, BASF Corporation, Parsippany, NJ) were dissolved completely in 1 to 5 ml of absolute ethanol (Table 13.1).

Table 13.1   Size Distribution of the Particles in the Dispersions
of the Liquid Formulations at Different Emulsifier
Compositions in Various Solvents[a]

Weight of PF-127 per 100 mg Monoolein (mg)	Average Particle Diameter (nm)/Polydispersity in Various Solvents			
	No Solvent	Ethanol	PEG_{400}	Propylene Glycol
10	ND[b]	ND	ND	ND
15	ND	258.0/0.123	278.0/0.290	312.1/0.257
20	ND	254.4/0.069	249.3/0.219	399.1/0.996
25	ND	259.3/0.345	277.9/0.428	351.3/0.320
30	ND	262.7/0.376	246.1/0.611	336.8/0.356

[a] The liquid formulation contains 50% (w/w) of ethanol or PEG400 or 70% (w/w) of propylene glycol.
[b] Not dispersed.

In some cases, the mixture was heated to ca. 40°C to accelerate the dissolution process. Ethanol was evaporated under a stream of oxygen-free dry nitrogen for a few hours and subsequently at a reduced pressure overnight to remove any traces of the solvent. The mixture was dissolved again in absolute ethanol, 1,2-propanediol, or a low-molecular-weight polyethyleneglycol (mol. wt. 400, PEG_{400}) to form the liquid formulation, which was stored at room temperature or at 4°C for further experiments.

These liquid formations were dispersed *in situ* by adding 100 μl of the liquid formulations in 2 ml distilled water and by vortexing the mixture for 1 min. The particle size distribution was determined by quasielastic laser light scattering with a Malvern Zetasizer® (Malvern Instruments Limited, England).

The liquid formulations in water formed dispersions of lipid particles ranging from 250 to 400 nm in diameter depending on the concentrations of Pluronic F-127 and the organic solvents as shown in Table 13.1. The dispersion was the most stable with small-sized particles when more than 15 mg of Pluronic F-127 were added to disperse 100 mg of monoolein in the liquid formulation. The particle sizes did not decrease significantly when the amount of Pluronic F-127 was increased beyond 15 mg per 100 mg monoolein. When less than 15 mg of Pluronic F-127 was used, the mean particle size of the lipid dispersions was large and not reproducible. The use of Pluronic F-68 or Tween 80 instead of Pluronic F-127 also yielded a liquid formulation that dispersed well when more than 15 mg was used per 100 mg monoolein (data not shown).

Table 13.2 Size Distribution of the Particles in the Dispersions of the Liquid Formulations at Different Solvent Compositions in Various Solvents[a]

Solvent Content in liquid formulation [% (w/w)]	Average Particle Diameter (nm)/Polydispersity in Various Solvents		
	Ethanol	PEG_{400}	Propylene Glycol
0	ND[b]	ND	ND
50	253.6/0.191	278.0/0.290	ND
60	254.4/0.105	288.4/0.377	ND
70	271.9/0.175	294.2/0.393	312.1/0.257
80	296.7/0.246	268.0/0.378	295.3/0.222
90	296.8/0.191	310.3/0.394	284.8/0.224

[a] The liquid formulation contains monoolein and PF-127 at 100:15 by weight.
[b] Not dispersed.

To determine an adequate content of solvents in the liquid precursor formulations, formulations containing various amounts of solvents were also prepared (Table 13.2). Liquid formulations containing 100 mg monoolein, 15 mg Pluronic F-127, and different amounts of the solvent selected from ethanol, 1,2-propanediol, or PEG_{400} were prepared. One hundred microliters of the liquid formulations were dispersed in 2 ml distilled water to determine the mean particle size in the dispersion. Liquid formulations containing 0 to 40% ethanol or PEG_{400}, or 0 to 60% 1,2-propanediol did not disperse but formed viscous gel-like aggregates in water. With higher solvent contents, the liquid formulations dispersed easily in water. The mean particle size did not decrease significantly by increasing the amount of solvents beyond 50% (w/w) for ethanol and PEG_{400} and 70% (w/w) for 1,2-propanediol. It is worthy to mention that the liquid precursor formulations containing PEG_{400} underwent gelation in a few minutes at ambient temperature and therefore were difficult to handle. It is possible that gelation is caused by the phase transformation to the sponge or L_3 phase (10,11). The L_3 phase has been observed when the cubic phase of the monoolein/water system encounters a third component, a hydrophilic solvent such as dimethylsulfoxide, ethanol, *N*-methylpyrrolidine, PEG_{400} or 1,2-propanediol. Pluronic F-127, a nonionic amphiphile, acting as the fourth component, could also alter the gross phase behavior of the system (12).

Liquid formulations containing a drug were also prepared by mixing monoolein, emulsifier, organic solvent and the drug (Table 13.3). In case of protein drugs, monoolein was mixed homogeneously with Pluronic F-127, ethanol, and propylene glycol first.

Table 13.3 Particle Size Distribution, Drug Loading Efficiency, and Structural Characteristics of the Dispersions of the Liquid Formulations at Different Solvent Compositions in Various Solvents

Encapsulated Molecules	Particle Size (nm)/Polydispersity	Weight of Drug/100 mg Monoolein	Loading Efficiency (%)	Space Group	Unit Cell Parameter (Å)
Hydrophobic Drugs (50% Ethanol[a])					
Cresol red	348.0/0.351	4	86	Im3m	121[b]
Rifampicin	272.6/0.247	0.5	75	Im3m	138
Paclitaxel	196.6/0.329	0.42	100	Im3m	132
Pyrene	275.0/0.045	6	100	Im3m	123
NBD-PE[c]	290.0/0.393	1	100	ND[d]	—
Rhodamine-PE	264.4/0.306	2	100	ND	—
Hydrophilic Molecules (50% Ethanol)					
Bromocresol green	283.1/0.583	0.5	20	Im3m	130
FITC[e]	277.8/0.194	1	3	Im3m	130
Rhodamine B	290.9/0.229	1	3.9	Im3m	130
HPTS[f]	289.8/0.312	1	5	Im3m	130
Methylene blue	282.2/0.231	0.5	0	Im3m	130
Proteins (70% Propylene Glycol[g])					
FITC-BSA[h]	314.5/0.165	8	85	Pn3m	134
Insulin	350.7/0.170	15	87	Im3m	129
Cholera toxin B	475/0.356	2	50 to 60	Ia3d	152

Tetanus toxoid	412.1/0.143	0.23	100	Spotty[i]	—
Cyclosporin	424/0.133	40	100	ND	—
Lysozyme	345.8/0.182	5	85	Spotty	—
Calcitonin	319.0.260	0.6	0	Im3m	130
Other Macromolecules (70% Propylene Glycol)					
HPTS-dextran[j]	332.0/0.455	3	67	ND	—
Blue dextran[k]	398.0/0.482	1	75	ND	—

Note: The liquid formulation contains monoolein and PF-127 at 100:20 by weight.

[a] The composition of ethanol in the liquid formulation was 50% (w/w).

[b] The error in the unit cell parameter was ca. ± 5 Å.

[c] 1,2-Dioleyl-sn-glycero-3-phosphoethanolamine-*N*-(7-nitro-2-1,3-benzoxadiazol-4-yl).

[d] Not determined.

[e] Fluorescein isothiocyanate.

[f] Pyranine, 1,3,6-pyrenetrisulfonic acid, 8-hydroxy-, trisodium salt.

[g] The composition of propylene glycol in the liquid formulation was 70% (w/w).

[h] Fluorescein isothiocyanate conjugated bovine serum albumin.

[i] Spotty diffraction patterns indicating the coexistence of different cubic phases.

[j] HPTS conjugated dextran (molecular weight 10,000).

[k] Molecular weight 600,000.

Concentrated protein aqueous solution was added to the liquid mixture and stirred until clear. Ethanol in the mixture was evaporated completely under low vacuum to prepare the liquid formulation containing a protein drug. Since propylene glycol does not evaporate under this pressure, the final formulation contains monoolein, emulsifier, propylene glycol, protein and a trace of water. Liquid formulations containing different model drugs and proteins are listed in Table 13.3. All of the liquid formulations containing drugs were clear homogeneous liquids when freshly prepared. In case of rifampicin and paclitaxel, however, drugs precipitated out irreversibly with time. Precipitation of rifampicin could be prevented by adding antioxidants such as ascorbic acid to the dispersion. When dispersed in water, most of the liquid formulations produced dispersions containing finely dispersed particles. In case of paclitaxel, precipitation occurred within a few hours after preparation.

13.3 ENCAPSULATION EFFICIENCY

Two hundred microliters of the liquid formulation were dispersed in 1 to 2 ml water. The dispersion was transferred into a retentate vial of Centricon® (MWCO 100,000 or 300,000 depending on the molecular weight of the encapsulated molecules) and centrifuged at $1000 \times g$ for 30 min. After centrifugation, only the solution (filtrate) passed through the membrane, leaving the particles in the retentate vial. The concentration of the drug in the filtrate was determined spectrophotometrically (UV-VIS or fluorescence) or by ELISA. Aqueous drug solutions were prepared to examine whether the drug in the bulk aqueous phase releases out completely through the membrane. Also, aqueous drug solution was mixed with the dispersion of the liquid formulation without drug to examine the interaction between the drug molecules and the particles. In the cases of hydrophobic drugs including paclitaxel, pyrene, and phosphatidylethanolamine derivatives, control experiments could not be performed since aqueous drug solutions could not be prepared due to the low solubility of these molecules in water. Low-molecular-weight drugs solubilized in the aqueous solution passed through the membrane completely in the absence of dispersed particles proving the validity of the measurements. Approximately 5% of the protein in the aqueous solution remained inside the retentate vial after centrifugation, probably since it is adsorbed on the membrane of the retentate vial. The presence of dispersed particles had no influence on the release of drugs into the bulk water phase.

Cresol red and rifampicin, which are slightly soluble in water, were not encapsulated completely in the particles. More hydrophobic molecules, paclitaxel or pyrene, and phosphatidylethanolamine derivatives, NBD-PE or Rhodamine-PE, were encapsulated completely inside the particles. As mentioned above, paclitaxel precipitated out with time even though the drug was completely loaded in the particles when freshly prepared. The encapsulation efficiency of hydrophilic drugs was lower than that of hydrophobic drugs. Most of the proteins were encapsulated inside the particles at high ratios. Calcitonin, a small hydrophilic peptide, was not encapsulated inside the particles at all, however. Dextrans, which are soluble macromolecules, were loaded efficiently in the particles.

13.4 SMALL-ANGLE X-RAY DIFFRACTION

Since the dispersed particles were composed of monoolein, emulsifier, organic solvent, and water, it is expected that the structure of the particles would be different from that of conventional lipid particulate systems such as liposomes or oil-in-water–type lipid emulsions. The structure of the dispersed particles was investigated by small angle x-ray diffraction. One hundred microliters of the liquid formulation was dispersed in 1 ml distilled water and transferred into 1-mm diameter quartz x-ray capillaries using an 18-gauge needle. To obtain a clear x-ray diffraction pattern, the particles in the dispersion were concentrated by centrifugation at 3000 r/min for 10 min. We confirmed that the centrifugation process did not alter the physical nature of the dispersion by redispersing the concentrated lipid particles in water. Redispersed particles retained the original size distribution. The capillary was flame-sealed and glued to prevent water leakage. X-ray diffraction data was obtained by using an x-ray diffractometer with general area detector diffraction system (GADDS, Bruker, Karlsruhe, Germany). The CuK radiation (1.542 Å) was provided by an x-ray generator (FL CU 4 KE, Bruker, Karlsruhe, Germany) operating at 40 kV and 45 mA. The sample-to-detector distance was 300 mm, and exposure time was 3 h. To avoid air scattering, the beam path was filled with helium.

Fully hydrated bulk mixtures containing monoolein, Pluronic F-127, and water were prepared by mixing the contents in two syringes, each containing lipids and aqueous phases, coupled with a three-way stopcock (Discofix®, B. Braun, Emmenbrücke, Switzerland) using a method slightly modified from that described previously (13,14). Without organic solvents, the bulk mixture of monoolein and

PF-127 at 5:1 by weight formed an Im3m cubic phase with the lattice parameter of 120 Å in excess water. The small-angle diffraction pattern was also collected for the concentrated dispersion of the liquid formulations. The diffraction peaks observed at spacing ratios of 2:4:6:10:12:14 indicate that the particles of the dispersion, regardless of the amount or kind of the organic solvent, have the internal structure of the Im3m cubic phase (15). The lattice parameter, however, was 130 Å, which was different from that of the bulk cubic phase formed by the mixture of the same composition in excess water (120 Å) but identical to that of Cubosome (1). It is highly likely, therefore, that our liquid formulation is a precursor or a preconcentrate of Cubosome (6). The fact that the lattice parameter of the dispersed particles differs from that of the bulk cubic liquid-crystalline phase may also be indirect evidence that there exist interior and exterior of the particles whose compositions are different in a single particle. It is highly probable that PF-127 acts as an emulsifier and distributes mainly on the exterior similar or identical to Cubosome.

When a hydrophobic drug was encapsulated, the type of the cubic phase did not change, but the lattice parameter was lower for pyrene and cresol red while it was higher for rifampicin. For hydrophilic drugs, the structure and the lattice parameter were virtually identical to those of Cubosomes without any drugs, probably due to low drug encapsulation efficiency. The structure and the lattice parameter changed to a great extent when proteins were encapsulated inside the particles. Dispersion containing insulin remained as an Im3m cubic phase with a lattice parameter of 129 Å. Pn3m and Ia3d cubic phases were observed for dispersions containing FITC-BSA and cholera toxin B, respectively. Dispersions with tetanus toxoid or lysozyme yielded spotty patterns that could not be indexed as a single cubic phase.

13.5 CUBOSOMES AND LIPID CUBIC PARTICLES

Since the components and the composition of the dispersion of our liquid formulation were similar to the Cubosome dispersion except for the presence of the organic solvent, it was highly likely that the particles in the dispersion are also similar to Cubosomes. If they are different from Cubosomes, the difference could be instigated by the different preparation procedures and/or the presence of the organic solvent in our system. The preparation procedure of the original Cubosome includes a step to cool the microfluidized particles slowly to room temperature to form a well-defined internal cubic lattice (1,16) The preparation procedure of our lipid cubic particles is lacking this step (17). We can make a scientific guess on how the particles are formed from our liquid formulation by thinking through the sequence of events that may have happened when the liquid formulation was submerged in water.

When the liquid formulation is placed in water, the lipid, emulsifier, and organic solvents will feel the presence of water suddenly and will try to form an equilibrium phase, which may be a homogeneous viscous cubic phase. The system, however, will have to undergo a dramatic change since we either shake or vortex it. Since the bulk of the liquid formulation still remains as free-flowing liquid at this time, it can be broken easily into smaller pieces, whose surface will subsequently face surrounding water. Many events may happen in and on these small particles. Water-friendly polyethyleneoxide chains of PF-127 are highly likely to be located at the interface between bulk water and the particles and form the exterior of the particles. Some of the monoolein molecules, which are more hydrophobic than PF-127, may still be mixed with PF-127 on the interface. Most of monoolein and some PF-127 molecules will comprise the internal structure of the particles along with the organic solvent that was not able to escape to the bulk aqueous phase. The organic solvent, however, would not stay inside the particles. Since solvent molecules are small in size and can be mixed easily with water, most of them will diffuse out of the particles into the bulk water phase. Also, water will rush into the particles. The rate of the molecular exchange between the organic solvent and water may be a factor to control the formation of the well-defined cubic lattice. Whether the events happen as narrated above has yet to be investigated experimentally.

The liquid formulation contains 50 to 70% (w/w) of organic solvent. Since the dispersion contains 5 to 20% (w/w) of the liquid formulation, the content of the organic solvent becomes 2.5 to 14% (w/w) in the final dispersion prepared from our liquid formulation. Since the organic solvents used in the liquid formulation are miscible with monoolein and PF-127, they can partition into the particles as well as into the bulk aqueous phase. How the organic solvent changes the characteristics of the dispersed particles needs further investigation. Despite the possible factors that may create differences between the original Cubosome and our lipid cubic particles, our particles will be referred to as Cubosomes hereafter since they are lipid particulate systems with the internal structure of the cubic liquid-crystalline phase satisfying the necessary and sufficient conditions for the definition of Cubosome.

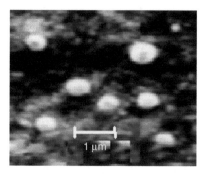

Figure 13.1   Low-temperature scanning electron microscope image of the dispersion of the liquid formulation.

13.6 LOW-TEMPERATURE SCANNING ELECTRON MICROSCOPY

Cubosomes prepared from the liquid dispersion were observed by using a low-temperature scanning electron microscope. To prepare a dispersion of Cubosomes, 10 µl of the liquid formulation was dispersed in 1 ml water by shaking. One drop (ca. 3 µl) of the Cubosome dispersion was mounted on a stub and immersed rapidly into nitrogen slushing chamber of the cryo-transfer system (CT 150 Cryotrans, Oxford Instruments Ltd., U.K.). The sample was transferred into the cryo-preparation chamber and cooled to –170°C under vacuum. The sample was transferred onto the cryo-stage of a scanning electron microscope (JSM 5410 LV, JEOL, Japan), and the temperature was raised to –70°C for 5 min to sublimate water. The sample was withdrawn to the cryo-preparation chamber again, coated with gold, and imaged at an accelerating voltage of 15 kV. Discrete submicron-sized particles were observed, but accurate structure was not visible due to the low resolution of the microscope (Figure 13.1).

13.7 PHYSICAL AND CHEMICAL STABILITY OF THE LIQUID FORMULATION

Immediately after preparation, the liquid formulation was a clear single-phase liquid. When stored at 4°C, the formulation became an opaque semisolid. Depending on the temperature, the bulk viscosity of the formulation changed. For instance, the formulation was a free-flowing liquid at 30°C whereas it was fairly viscous at 10 to 15°C.

The liquid formulations containing ethanol had a tendency to become more viscous than those with propylene glycol. Since PF127 alone precipitates in ethanol with time, the high viscosity of the liquid formulation made with ethanol may originate from the decreased solubility of the pluronics at lower temperature. The formulation containing PEG_{400} underwent gelation with time as described above. Sometimes, a small amount of white semitransparent aggregation formed in the ethanol-based liquid formulation with time, especially at or below 25°C. The heterogeneous formulation, however, became transparent liquid again when heated to ca. 35 to 40°C for a few minutes. There were no apparent changes in the physicochemical properties of the reheated sample when compared to the freshly prepared liquid formulation. There were no chemical degradations of the components in the formulation for at least 2 years when stored at 4 or –70°C. At room temperature, however, monoolein had a tendency to dissociate into glycerol and oleic acid (data not shown).

Liquid formulations containing drugs were also stable physically for at least 1 year at 4°C. For formulations containing rifampicin, however, an aggregation formed irreversibly within a few days after preparation. Pyrene and paclitaxel in the liquid formulation did not degrade for at least one year at 4°C. In the case of protein drugs, the situation was a little different. Since the liquid formulation contains a small amount of water, it is possible that the proteins can undergo a variety of destabilization processes including the conformational change, hydrolysis, or oxidation. The liquid formulation containing insulin was physically and chemically stable for at least 1 month when stored at 4°C. However, prolonged storage was not possible since the protein degraded with time. The stability was improved greatly when the liquid formulation was prepared in the absence of oxygen and stored at –70°C subsequently.

13.8 STABILITY OF CUBOSOME DISPERSION PREPARED FROM LIQUID FORMULATION

Good storage stability is crucial to a successful drug delivery system. To have commercial value, the formulation must be stable for more than 1 year at room temperature (17,18). The original Cubosome is known to be very stable at room temperature (1). We have tested the stability of Cubosomes freshly prepared from our liquid formulation and compared it with the stability of Cubosome prepared by microfluidization. Since our Cubosome dispersion has a wider size distribution function (see below), we expected that our Cubosome system would be less stable than the original Cubosome, which has a narrow size distribution. As anticipated, the Cubosome prepared from the liquid formulation was stable for only a few days to a few weeks depending on the composition. Therefore, Cubosome prepared from our liquid formulation would not be suitable for commercialization in the form of the dispersion. Our liquid formulation, a precursor to Cubosome, however, has greater advantages that can more than compensate for the stability problems of the dispersion. Unlike its dispersion, the liquid formulation was very stable for a long period of time. Many hydrophobic drugs can be dissolved in the liquid formulation and be spontaneously encapsulated at a high rate in the Cubosome particles when dispersed in water without using mechanical devices. When mechanical force was used to disperse particles, many drugs would be chemically destabilized. In many systems, water is often the cause of physical instability and chemical degradation. Water can host reactive oxygen species, proton or hydroxide ions, and free radicals. Since our liquid formulation is a thermodynamically stable system containing no or a minimum amount of water, the sources of physicochemical destabilization were completely taken away. Even when the liquid formulation was prepared in the presence of oxygen, it was stable for more than a year without degradation of components including the drugs when stored at low temperature. For protein drugs, it was important to remove air from the precursor to prevent inactivation or conformational change of the protein since the liquid formulation contains water.

13.9 ABILITY TO FORM CUBOSOME FROM LIQUID FORMULATION AFTER STORAGE

As mentioned above, the liquid precursor formulation can be dispersed readily in excess water by mere shaking. As a drug delivery system, the liquid formulation is an ideal environment for drugs that might be oxidized and hydrolyzed in the presence of water. When the particulate dispersion system is needed, one just has to disperse the liquid formulation in water for immediate use. With this point of view, therefore, it is important to evaluate the physicochemical storage stability and the ability to form dispersion after storage of the liquid precursor formulation. To this end, liquid formulations comprising 100 mg monoolein, 20 mg Pluronic F-127, and 120 mg ethanol or PEG 400 or 280 mg 1,2-propanediol were prepared

Figure 13.2 Photograph of the dispersion of the liquid formulation after storage at 4°C for 3 days.

and stored at ambient temperature (10 to 30°C) or at 4°C for more than 500 days. There were no apparent visual changes in these liquid formulations. As mentioned earlier, white aggregates were sometimes found in the formulation but disappeared again when it was warmed to ca. 40°C. An important characteristic of the liquid formulation is the ability to form a fine dispersion of lipid particles in water. The particle size distribution was measured immediately after dispersing the stored liquid precursor formulation at different time points. These liquid formulations formed lipid dispersions of small mean particle sizes (<300 nm) and polydispersity (<0.3) for 18 months.

One of the advantages of the precursor system is that it can be stored at low temperature. The liquid formulation can be frozen or refrigerated for a long period of time. When frozen rapidly, the solution solidified without undergoing phase separation. If the liquid precursor formulation was refrigerated, however, phase separation was observed. Pluronic F-127 solidified before other components turned solid. We have to bear in mind that Cubosome is a thermodynamically unstable system, unlike the liquid formulation. The dispersion of the precursor, for instance, did undergo irreversible phase separation in a few days and generated bulk cubic phase when stored at 4°C (Figure 13.2).

13.10 *IN VITRO* DRUG RELEASE

The *in vitro* release experiments were performed with Cubosome dispersions containing four model drugs: methylene blue, bromocresol green, rifampicin, and insulin. Lipid formulations containing model drugs were dispersed in excess water. One milliliter of the dispersion was put into a dialysis membrane with molecular weight cutoff values of 6000 for methylene blue, bromocresol green, and rifampicin and 100,000 for insulin. The bags containing the dispersion were sealed and immersed in 10 ml of 0.1 M Na_2HPO_4/citric acid buffer solution at pH 7.4 prewarmed at 37°C. In case of rifampicin, we were cautious to prevent oxidation of the drug by replacing the aqueous medium containing 0.5% sodium ascorbate at pH 7.4 during the experiment (19). The fact that rifampicin was not oxidized was confirmed by identical UV-VIS spectra of the released rifampicin to the freshly made one (the spectrum changes upon oxidation). If rifampicin was oxidized, it precipitated out and could not be released through the membrane. Aqueous drug solutions were used for controls. The samples in triplicate were placed in a shaking incubator at 37°C with a shaking frequency of 2 Hz. The dissolution media were replaced completely with 10 ml of fresh media at preset time intervals.

The release patterns of methylene blue, bromocresol green, rifampicin, and insulin from Cubosome are shown in Figure 13.3 (open circles). Aqueous drug solutions were used as controls (solid circles).

The release patterns of methylene blue from Cubosome and from the aqueous solution were virtually identical. Methylene blue was released completely in ca. 3 h.

An initial burst of bromocresol green release was followed by a plateau. Fifty percent of the drug was released during the first 5 h from the Cubosome dispersion. Bromocresol green in the aqueous solution was released quickly and completely in 2 h.

Rifampicin from Cubosome was released in a more sustained manner than bromocresol green from Cubosome. About 10% of the drug was released during the first 2 h, and about 30% was released during the following 20 h at zero-order rate. Complete release of the drug was not observed for the duration of the experiment. Rifampicin from the aqueous solution was released completely in 3 h. The difference in the release rate between bromocresol green and rifampicin may originate from the differences in the hydrophobicity of the drugs.

In the case of insulin, ca. 20% of the protein was released in 5 h. Insulin from the aqueous solution also showed an initial burst of release for the first 5 h. Therefore, it is possible that the initially released insulin from the Cubosome may represent those located in bulk water, not inside the Cubosome particles in the beginning. After the initial burst, insulin was not released at all from Cubosome, indicating the difficulty of diffusing a bulky protein out of the

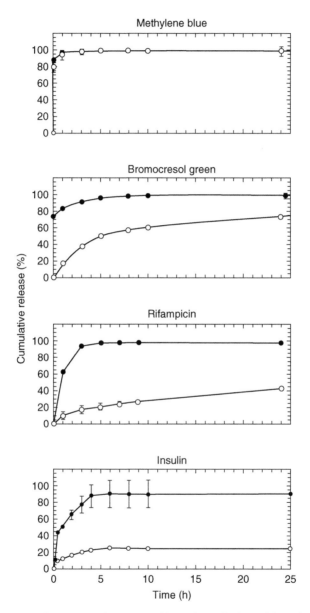

Figure 13.3 *In vitro* release profiles of methylene blue, bromocresol green, rifampicin, and insulin from aqueous solution (solid circles) and Cubosomes (open circles).

Cubosome particles to the buffer solution. Also, complete release was not observed even for the aqueous insulin solution, probably since some of the insulin molecules bind to the interior of the semipermeable membrane (MWCO 100,000).

For small and relatively hydrophilic drugs such as methylene blue, Cubosome does not act as a release barrier as it cannot encapsulate these drugs. In the case of hydrophobic drugs, however, Cubosome can help sustain the release of the encapsulated drug. Encapsulated protein, a macromolecule, was not released at all from Cubosomes. The resistance against release can be advantageous in some aspects. For instance, Cubosome can protect the loaded proteins from proteases until they reach the target cells when administered into the body. For this reason, the Cubosome system has a potential to become an effective drug delivery system for hydrophobic or protein drugs.

13.11 *IN VITRO* AND *IN VIVO* TOXICITY OF CUBOSOME PREPARED FROM A LIQUID FORMULATION

Since Cubosome prepared from the liquid formulation is a submicron-sized particle, it is natural to consider using it as a drug delivery system via the intravenous route. To be injected through a vein, however, Cubosomes must not be toxic or hemolytic. Since monoolein causes severe inflammation in muscle and subcutaneous tissues (Unpublished data, personal communication with T. Landh and M. Caffrey, 2002), it cannot be used as an injectable formula.

Hemolytic activity was evaluated for Cubosome prepared from the liquid formulation and the oil-in-water–type emulsion of the egg phosphatidylcholine (PC)/soybean oil system. The egg PC/soybean oil emulsion, prepared by sonicating a mixture of soybean oil and egg PC at 5 : 1 by weight in water, did not cause hemolysis even at 50 µg of total lipids/ml, while Cubosome was hemolytic at 6 µg of total lipids/ml (Figure 13.4). Cubosome, therefore, is not suitable to be injected intravenously.

The *in vitro* cytotoxicity of Cubosome in Caco-2 cells was evaluated by incubating the cells with various concentrations of Cubosomes for 1 day at 37°C, as shown in Figure 13.5 (9). Cubosome was highly toxic at concentrations equal to or greater than 500 µg of total lipids in 1 ml culture medium. At concentrations of up to 100 µg of total lipids/ml, however, cell viability did not decrease for 3 days.

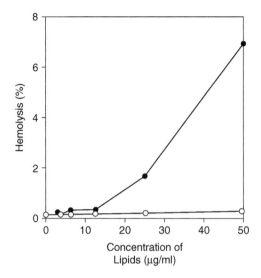

Figure 13.4 Hemolytic activity of Cubosomes (solid circles) and o/w emulsion of egg phosphatidylcholine/soybean oil (open circles). The concentration represents the total weight of the lipids and oil in the system per unit volume.

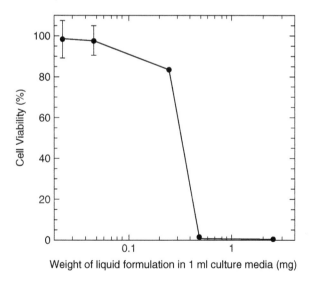

Figure 13.5 *In vitro* cell viability of Caco-2 cells upon incubation with Cubosomes made by dispersing the liquid formulation in culture media. The cells were incubated with Cubosomes for 1 day at 37°C.

To evaluate the *ex vivo* cytotoxicity of the Cubosome dispersion, rat everted jejunum was incubated with Cubosomes at a concentration of 10 mg of total lipid in 1 ml Krebs-Ringer bicarbonate buffer solution at 37°C for 1 h and observed with light microscopy and transmission electron microscopy (Figure 13.6).

There was no evidence of damage in the cells when observed by light microscopy (Figure 13.6a). Although many droplet structures were formed inside the periphery of the cells upon incubation with Cubosomes (arrows in Figure 13.6b), the cells functioned well as demonstrated by their ability to exclude trypan blue. Caco-2 cells in the *in vitro* culture system were much more susceptible to the toxic effects of Cubosomes, while the morphology of rat jejunum did not change even when a concentration of the Cubosome dispersion 50 times higher than the concentration that had shown *in vitro* toxicity was applied to the tissue. The vacuole-like structures formed after incubation with Cubosome may be similar to those formed after consuming a large amount of neutral lipids. These droplets seem to be analogous to the large lipid bodies formed during lipid digestion and absorption in intestinal absorptive cells (20). Since monoolein can enter the enterocyte by simple diffusion with or without the aid

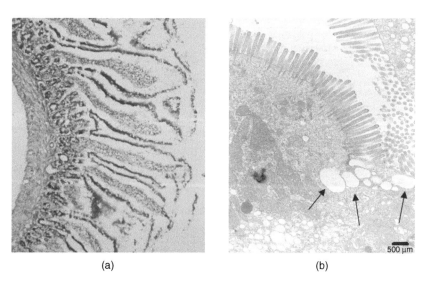

(a)   (b)

Figure 13.6 Light microscopy (a) and transmission electron microscopy (b) of rat small intestine after incubation of Cubosome made by dispersing 20 mg of the liquid formulation in 1 ml KBR solution for 1 h at 37°C.

of bile salts, constituents of these structures could be triglycerides synthesized from monoolein after being absorbed by the cells.

One of the main components of Cubosome is 1-monoolein, and one of the major digestion products of triglycerides in intestine is 2-monoolein. 2-Monoolein can convert to 1-monoolein and vice versa with an equilibrium composition of ca. 90% 1-monoolein and ca. 10% 2-monoolein at room temperature (21). Also, the equilibrium mixture of monoolein is in the GRAS (Generally Recognized As Safe) category for oral consumption (22). Single or repeated oral feeding every day for 2 weeks also showed that Cubosomes did not have any toxicity in mice (in preparation). Therefore, while Cubosome may not be a good drug delivery system for intravenous injection, it can be taken orally without exhibiting toxicity at high concentrations.

13.12 *IN VITRO* CELLULAR ASSOCIATION OF DRUGS FROM CUBOSOME

Since the oral tract is one of the promising administration routes for Cubosomes, *in vitro* and *ex vivo* uptake experiments were performed with Caco-2 cells and rat everted sac, respectively (9). A hydrophobic fluorescence probe, pyrene, was encapsulated in Cubosomes to visualize the location of lipid absorption (23,24). The encapsulation efficiency of pyrene in Cubosome was ca. 100% (Table 13.3). When the liquid formulation was dispersed in PBS or DMEM, the particle size was 278.9 nm (polydispersity 0.267) and 267.8 nm (polydispersity 0.261), respectively. Pyrene was selected as a probe to follow the location of the lipids, particularly monoolein due to its hydrophobic nature. Pyrene was not released at all into the aqueous phase over 24 h in PBS at 37°C, indicating that the dye is located in the Cubosome, probably due to the hydrophobicity of pyrene and the stability of cubic particles.

The localization of pyrene and the morphology of the cells were visualized by fluorescence and phase contrast microscopy, respectively (Figure 13.7a and Figure 13.7b). We did not observe any blue fluorescence in and on the untreated cells (data not shown). When cells were incubated with Cubosome containing pyrene for 3 h at 37°C, the blue fluorescence of pyrene was observed clearly in the cells (Figure 13.7a). Intense punctual fluorescence as well as diffused blue fluorescence were observed in most of the fields. The number of dotted structures and the intensity of the diffused fluorescence increased with time, analogous to the increase of the vacuole-like structures by phase contrast microscopy (Figure 13.7b).

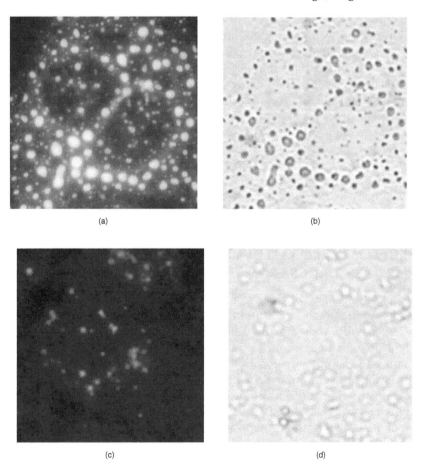

(a)   (b)

(c)   (d)

Figure 13.7 Fluorescence (a, c) and phase contrast microscopy (b, d) of Caco-2 cells incubated with Cubosomes encapsulating pyrene (a, b) and FITC-BSA (c, d).

Cubosome containing FITC conjugated bovine serum albumin (FITC-BSA) was incubated with Caco-2 cells for 3 h at 37°C. The protein was visualized by fluorescence microscopy and phase contrast microscopy (Figure 13.7c and Figure 13.7d). When FITC-BSA–loaded cubic particles were incubated with cells, droplet structures were observed by fluorescence microscopy (Figure 13.7c). Diffused fluorescence was not observed. The level of green fluorescence of FITC was similar in cells that were incubated with FITC-BSA aqueous solution or with Cubosome containing FITC-BSA. The fact that the droplet structure inside the cells is not fluorescent may indicate that molecular diffusion of the components in the Cubosome, not endocytosis, would be the main uptake mechanism under *in vitro* conditions. We note that the lack of protein uptake *in vitro* may not imply that the same phenomenon would be observed under *in vivo* conditions; it would be risky to extrapolate the *in vitro* data to *in vivo* conditions.

To study the time-dependent cellular association of Cubosome prepared from liquid formulation, Caco-2 cells were incubated for up to 8 h with Cubosomes encapsulating pyrene in DMEM containing 10% fetal bovine serum at 37°C (Figure 13.8).

An emulsion comprising pyrene, egg phosphatidylcholine, and soybean oil (1:12:100 by weight) was also prepared for comparison. The particle size of the emulsion was ca. 250 nm. The amount of cellular-associated pyrene increased smoothly with time at 37°C upon incubation with Cubosome. After an 8-h incubation at 37°C, the cellular-associated pyrene was ca. 20% for Cubosome (Figure 13.8).

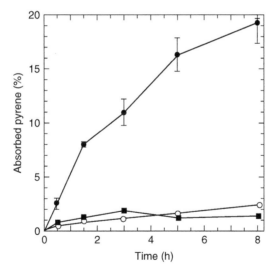

Figure 13.8 Absorption of Cubosomes by 1×10^6 Caco-2 cells as a function of incubation time. Caco-2 cells were incubated at 37°C (closed circles, n = 3) or at 4°C (open circles, n = 3) for up to 8 h with Cubosomes and at 37°C with egg PC emulsion (closed squares, n = 2). Data are represented as mean ± S.E.M. (Reproduced from JY Um, H Chung, KS Kim, IC Kwon, SY Jeong. *Int J Pharm* 253:71–80, 2003 with slight modifications with permission from Elsevier Science B.V.)

The absorption of pyrene for the emulsion formulation was only ca. 2% after 8 h of incubation. The number of associated pyrene molecules was approximately ten times higher for Cubosomes than for the emulsion. When Caco-2 cells were incubated with Cubosomes (ca. 250 nm) at 4°C, only 2% of the pyrene in the medium was absorbed on the cells after incubation. This result coincides well with that observed by phase contrast microscopy, where the interaction between Cubosomes and cells at 4°C seems to be much weaker than at 37°C (unpublished data). Since it is highly probable that Cubosomes do not retain their internal structure at 4°C, the phase behavior of the particles must be studied further to understand the differences in the absorption efficiency at the two temperatures.

13.13 SOLUBILIZATION OF CUBOSOME BY BILE SALT AND ABSORPTION OF MIXED MICELLES

In the gastrointestinal tract, lipid particles can be digested by many enzymes and solubilized by bile salts. Since the main component of Cubosome is monoolein, bile salts secreted to the intestine can become a good solubilizer of Cubosome. Various concentrations of sodium taurodeoxycholate micelle systems were prepared to test the solubilizing power of the bile salt. Coarsely dispersed Cubosomes with an average size of ca. 495 nm were prepared. Since the concentration of the bile salts in human intestine is ca. 5 and 10 to 20 mM in the fasted and at fed states, respectively, taurocholate solutions ranging from 0 to 20 mM were prepared to encompass the physiological range (25). The size distribution of the particles was measured using dynamic light scattering (Figure 13.9).

As taurodeoxycholate solution was added to Cubosomes, the size distribution of the particles changed from a unimodal peak at 495 nm to bimodal peaks at 114 nm and 34 nm at 10 mM taurocholate (Figure 13.9A and Figure 13.9B). When the concentration of taurodeoxycholate was 15 mM, mixed micelles of the taurocholate/monoolein/Pluronic system with a size of 8 nm were produced (Figure 13.9C). The size of the taurocholate micelle (15 mM) without any other components was ca. 3 nm (Figure 13.9D).

Since the bile salt can destroy the Cubosome structure completely under physiological conditions, the absorption of the loaded drug, pyrene, by cells can be altered greatly in the presence of bile salts. To study how the addition of the bile salt changes the absorption of pyrene by intestinal absorptive cells, the mixtures of Cubosome encapsulating pyrene and sodium taurocholate at different

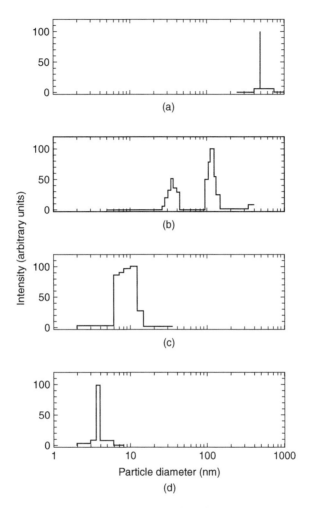

Figure 13.9 The diameter change of Cubosome upon mixing with sodium taurodeoxycholate (Na-TDC) measured with dynamic light scattering. Cubosomes were mixed with 0 mM (a), 10 mM (b) and 15 mM (c) taurodeoxycholate solution. The final concentration of Cubosomes was 20 mg of the liquid formulation in 1 ml bile salt solution. The diameter of bile salt micelle at 15 mM was measured in the absence of Cubosomes (d). (Reproduced from JY Um, H Chung, KS Kim, IC Kwon, SY Jeong. *Int J Pharm* 253:71–80, 2003 with slight modifications with permission from Elsevier Science B.V.)

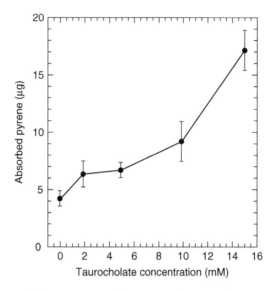

Figure 13.10 The amount of pyrene absorbed by 100 mg everted rat jejunum as a function of sodium taurodeoxycholate concentration. One milliliter of the medium contains 0.5 mg pyrene and 20 mg of the dispersed liquid formulation, Cubosomes. Vertical bars denote ± S.E. of the mean value for three separate determinations. (Reproduced from JY Um, H Chung, KS Kim, IC Kwon, SY Jeong. *Int J Pharm* 253:71–80, 2003 with permission from Elsevier Science B.V.)

concentrations were added to media containing the rat jejunum everted sac (Figure 13.10). After a 1-h incubation, pyrene concentration in the jejunum was quantified by extraction. There was a small increase in pyrene absorption when bile salt solutions at low concentrations (2 to 5 m*M*) were added to the Cubosome dispersion. As the bile salt concentration increased, the concentration of absorbed pyrene inside the cells increased by at least three times. Therefore, the bile salt can help the absorption of the drug encapsulated in Cubosome via the formation of mixed micelles.

Cubosome can be solubilized by the bile salts in the small intestine into mixed micelles and absorbed into the intestinal absorptive cells when administered orally. It is possible that ethanol used to prepare the precursor liquid formulation can also enhance the absorption. The monoolein–bile salt mixed micelle is known to penetrate into cells better than the bile salt micelle (26). Absorption of

Cubosome taken by oral administration can be further enhanced by the bile salts secreted into the intestine.

13.14 ORAL DELIVERY OF CUBOSOME ENCAPSULATING PYRENE

Monoolein and Cubosome containing 2 mg/ml of pyrene were prepared for an oral delivery experiment. Balb/c mice were fed 0.5 ml of monoolein or Cubosome corresponding to a dose of 1 mg pyrene per mouse. Monoolein containing pyrene was melted, super-cooled to room temperature, and fed to the animals as a liquid. Two or four hours after the feeding, the mice were sacrificed to obtain blood and the organs to determine the concentrations of pyrene (Figure 13.11). For mice fed with monoolein, pyrene was detected mainly in liver, kidney, and blood. In case of Cubosomes, pyrene concentration in each tissue was higher than that for monoolein. The concentration of pyrene was ca. 300 μg/g blood, which was at least 10 times higher than that of the monoolein group. Since monoolein forms the cubic phase upon contact with body fluid, the results indicate that micronization of the bulk cubic phase into cubic particles can indeed help the absorption of drugs in the oral tract.

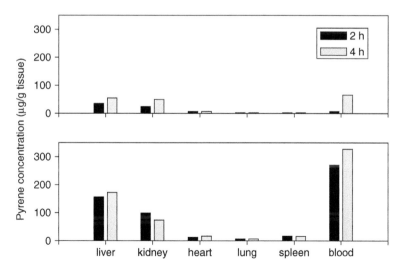

Figure 13.11 *In vivo* organ distribution profile of pyrene after oral administration of 1 mg pyrene in monoolein (A) and in Cubosome (B).

13.15 ORAL DELIVERY OF CUBOSOME ENCAPSULATING INSULIN

Since our liquid formulation or Cubosome can encapsulate protein drugs at a high ratio and therefore could be a good oral drug delivery system, a liquid formulation containing monoolein, PF-127, propylene glycol, and insulin was prepared as a peroral insulin formulation (8).

Male Wistar rats, age 5 weeks, weighing 120 to 150 g, obtained from Charles River Japan Inc. (Yokohama, Japan), were housed in groups of three in separate cages with free access to food and water for 1 week in a temperature-controlled room under a 12-h light/dark cycle prior to the experiments. Diabetes was induced by three consecutive intraperitoneal injections of streptozotocin (45 mg/kg). After 2 weeks, rats were considered diabetic if their blood glucose level was above 300 mg/dL in the fasting state. No insulin treatment was performed until the experiment.

Baseline blood glucose level was determined in normal and diabetic rats fasted for 4 h. During the experiment the rats were kept in a fasting state but allowed water *ad libitum*. Normal (Figure 13.12a) or streptozotocin-induced diabetic (Figure 13.12b)

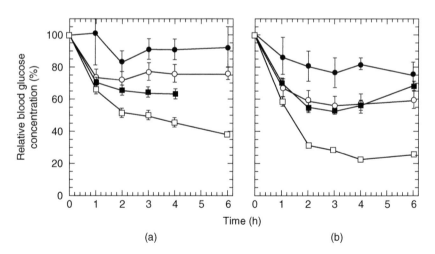

(a)   (b)

Figure 13.12 Relative blood glucose concentration after oral administration of 30 (open circles), 50 (closed squares), and 100 IU/kg (open squares) encapsulated in Cubosome in normal (a) and diabetic (b) rats. Relative blood glucose concentration of untreated rats (closed circles) was also measured as a control.

rats were fed 30 (open circles), 50 (solid squares), and 100 IU/kg (open squares) insulin-encapsulated Cubosome dispersion. Untreated rats were used as controls (solid circles).

Relative blood glucose level decreased for up to 2 h for the normal mice fed 30 or 50 IU/kg, but decreased continuously in those fed 100 IU/kg. The reduction of blood glucose concentration in normal rats was dose dependent. There was essentially no change in blood glucose concentration in the untreated rats. In diabetic rats, blood glucose concentration decreased slightly for untreated animals since fasting can lower the blood glucose concentration rather effectively. The blood glucose level was similar when 30 and 50 IU/kg of insulin in Cubosome were administered. When 100 IU/kg of insulin in Cubosome was administered, low glucose level was reached rapidly and maintained for at least 6 h.

The fact that insulin loaded inside Cubosome has a superior effect to control hyperglycemia indicates that Cubosome could be widely applied in oral protein delivery systems. It would be interesting to examine whether the particles themselves increase intestinal penetration or merely protect insulin from the attack of proteolytic enzymes.

13.16 ORAL VACCINATION BY ANTIGEN LOADED IN CUBOSOME

Another class of proteins that can be administered orally is antigens for vaccination. Oral vaccination is an attractive and convenient alternative to injection. However, oral immunization has not been widely used since antigen uptake is inefficient in the intestine and oral tolerance could also be induced. Antigens could also be degraded by proteolytic enzymes before they reach the immune cells. To overcome these problems, many scientists have developed polymer- and lipid-based delivery systems to encapsulate antigens for oral delivery. To test whether Cubosomes can be used as an oral vaccine carrier, we have encapsulated cholera toxin B subunit (CTB) inside Cubosome to administer orally into Balb/c mice three times at intervals of 2 weeks. The sera were collected 2, 4, and 6 weeks after the first immunization to follow the increase in the systemic antibody response (Figure 13.13).

Sera collected from the untreated mice and mice fed with Cubosome without CTB were used as negative controls. CTB alone or mixed with Cubosomes showed elevated systemic IgG and IgA concentrations when compared to negative controls. IgG and IgA

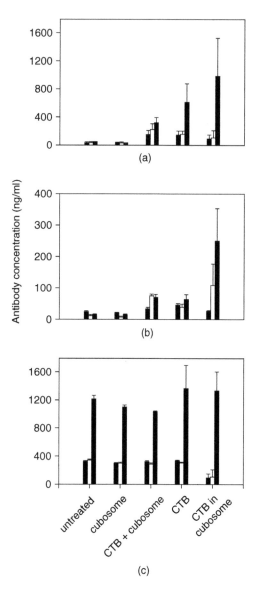

Figure 13.13 CTB-specific IgG (a), IgA (b) and IgM (c) antibody response in serum 2, 4, and 6 weeks after first immunization. Groups of 5- to 6-week-old Balb/c mice were orally immunized three times at intervals of 2 weeks. Values represent geometric mean ± S.E. (n = 5 to 6).

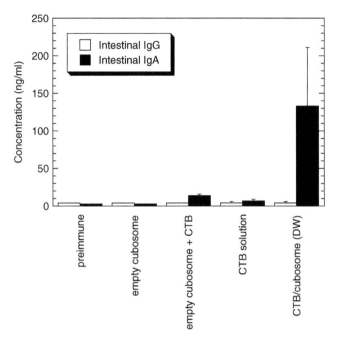

Figure 13.14 CTB-specific IgG and IgA responses in intestinal lavage samples 2, 4, and 6 weeks after first immunization. Groups of 5- to 6-week-old Balb/c mice were orally immunized three times at intervals of 2 weeks. Values represent geometric mean ± S.E. (n = 5 to 6).

levels of the mice fed CTB loaded inside Cubosomes were the highest among the experimental groups, showing that the oral vaccination was more efficient when the antigen was encapsulated inside Cubosome. Local immune reaction, intestinal IgA, was even higher for the group of mice orally administered CTB inside Cubosomes (Figure 13.14). The results imply that CTB inside the Cubosome was indeed taken up by the immune cells and elicited immune reactions efficiently.

13.17 CONCLUSIONS

Self-dispersing homogeneous liquid formulations containing monoolein, emulsifiers, and organic solvents have been prepared. The liquid formulation was a single anhydrous liquid and was stable for more than 6 months at room temperature and at 4°C. The homogeneous liquid formulation can be dispersed easily in an excess amount of water by mere shaking or vortexing to form Cubosomes. Our liquid formulation overcame some of the stability problems associated with the Cubosomes especially when protein drugs were used. Since our formulation is an anhydrous homogeneous mixture, it does not undergo oxidation or hydrolysis, which can potentially destabilize the system. The liquid formulation can be used as a drug delivery system since many drugs can be mixed easily in the formulation. The interaction between Cubosomes and Caco-2 cells was studied by various microscopic techniques. Lipid droplets were observed in the cytosol of enterocytes after incubation with Cubosomes. The amount of pyrene absorbed by Caco-2 cells was ca. 20% of the total at 37°C after an 8-h incubation. Cubosomes were easily solubilized by bile salts to produce mixed micelles. As the bile salt concentration increased, pyrene absorption into the jejunum of rat everted sac *ex vivo* increased.

Cubosomes were used as a peroral insulin delivery system. Serum glucose levels could be controlled for more than 6 h after oral insulin administration for normal and diabetic rats. Cubosome can also deliver antigens efficiently into the immune cells in the intestine resulting in prominent systemic and local immunity. Even though we have limited the delivery route of Cubosome to the oral tract in this chapter, other routes including those with mucosal tissues can be excellent targets for drug delivery using Cubosomes.

REFERENCES

1. J Gustafsson, H Ljusberg-Wahren, M Almgren, K Larsson. *Langmuir* 13:6964–6971, 1997.

2. B Siekmann, H Bunjes, MHJ Koch, K Westesen. *Int J Pharm* 244:33–43, 2002.

3. M Almgren, K Edwards, G Karlsson. *Colloid Surf A: Physicochem Eng Aspects* 174:3–21, 2000.

4. K Larsson. *Curr Opin Coll Interface Sci* 5:64–69, 2000.

5. K Larsson. *Chem Phys Lipids* 88:15–20, 1997.

6. T Landh, K Larsson. US Patent No. 5531925, 1996.

7. CJ Drummond, C Fong. *Curr Opin Colloid Interface Sci* 4:449–456, 2000.

8. H Chung, JS Kim, JY Um, IC Kwon, SY Jeong. *Diabetologia* 45:448–451, 2002.

9. JY Um, H Chung, KS Kim, IC Kwon, SY Jeong. *Int J Pharm* 253:71–80, 2003.

10. S Engström, K Alfons, M Rasmusson, H Ljusberg-Wahren. *Progr Colloid Polym Sci* 108:93–98, 1998.

11. K Alfons, S Engström. *J Pharm Sci* 87:1527–1530, 1998.

12. A Ridell, K Ekelund, H Evertsson, S Engström. *Colloid Surf* 228:17–24, 2003.

13. H Chung, M Caffrey. *Biophys J* 66:377–381, 1994.

14. A Cheng, B Hummel, H Qiu, M Caffrey. *Chem Phys Lipids* 95:11–21, 1998.

15. CH Macgillavry, G.D Rieck. *International Tables for X-Ray Crystallography.* Birmingham, England: The Kynoch Press, 1962.

16. J Gustafsson, H Ljusberg-Wahren, M Almgren, K Larsson. *Langmuir* 12:4611–4613, 1996.

17. SS Davis, J Hadgraft, KJ Palin. In: P. Becher, Ed. *Encyclopedia of Emulsion Technology,* Vol. 2, New York: Marcel Dekker, 1985, pp. 159–238.

18. AG Floyd. *Pharm Sci Technol Today* 2:134–143, 1999.

19. H Chung, TW Kim, M Kwon, IC Kwon, SY Jeong. *J Cont Rel* 71:339–350, 2001.

20. H Shen, P Howles, P Tso. *Adv Drug Deliv Rev* 50: S103–S125, 2001.

21. J Martin. *J Am Chem Soc* 75:L5482–L5486, 1953.

22. Cosmetic Ingredient Review Expert Panel. *J Am Coll Toxicol* 5:391–413, 1986.

23. AL Plant, DM Benson, LC Smith. *J Cell Biol* 100:1295–1308, 1985.

24. AL Plant, RD Knapp, LC Smith. *J Biol Chem* 262:2514–2519, 1987.

25. A Lindahl, AL Ungell, L Knutson, H Lennernas. *Pharm Res* 14:497–502, 1997.

26. S Muranishi. *Pharm Res* 2:108–118, 1985.

14

Bicontinuous Liquid Crystalline Mesophases—Solubilization Reactivity and Interfacial Reactions

NISSIM GARTI

CONTENTS

ABSTRACT

Lyotropic liquid crystalline mesophases, and in particular cubic phases, have been of major interest in a number of areas ranging from controlled uptake and release of solubilized molecules (mainly drugs and enzymes) to an environment for growing crystals of membrane proteins.

The packing parameters along with the pore sizes, channeling in the bicontinuous areas, and the three-dimensional nanostructures have been widely studied. However, since these mesophases are semi-solids or gel-like, it is not surprising that as microreactors they were almost completely ignored.

On the other hand, the extremely high surface areas, along with the unusual curvatures derived from these structures, are good reasons (for some scientists) to believe that interfacial reactivity and certain organic reactions in particular should be explored on these condensed and highly concentrated liquid-crystal mesophases with the aim of enhancing regioselectivity, yields, and kinetic rates.

In this chapter we discuss some studies related to the solubilization of guest molecules in cubic phases with relation to their interfacial reactivity. We stress the relationship between solubilization capabilities, loci of guest molecules, and chemical, electrochemical, and enzymatic reactivity. We also review some emerging studies demonstrating the possible advantages of these systems. Some Maillard reaction pathways for generation of food flavors are discussed in detail.

The main conclusions from this review are that lyotropic liquid crystals and mainly cubic phases provide enormous available surface areas for reactions to occur at the interface, resulting in improved product selectivity, formation of new and unexpected products, and enhanced rate of reactions. In addition there are some promising results regarding polymerization reactions of polymerizable surfactants and some interesting reactions related to biosensors.

Research on lyotropic liquid-crystalline mesophases as microreactors is in its infancy. Many more studies on many more model reactions are needed before one can estimate their industrial potential as microreactors in food applications.

14.1 INTRODUCTION

Liquid-crystalline structures, because they are semisolid or gel-like, have some potential advantages over macroemulsions or microemulsions, offering high solubilization capacities of hydrophilic and hydrophobic compounds (including proteins and enzymes), sustained and controlled drug release from semisolid matrices, improved drug bioavailability, and high capacity reservoirs for inorganic complexes along with high surface areas for carrying out organic processes.

A common practice in the formation of surfactant self-assemblies is the need to construct phase diagrams, and defining the isotropic regions and mesophases reflecting best utilization of the interrelations between surfactant/cosurfactant, oil (solvent), and aqueous phases (water and/or water with polyols). Binary, pseudoternary, and multicomponent phase diagrams have been widely studied and described. Surfactants and cosurfactants at different mass ratios, at various HLB values with variety of cosolvents, were utilized to form microemulsions with an extended one-phase isotropic region or extended lyotropic liquid-crystalline mesophases (1–5). In most cases attempts were made to minimize the areas in the phase diagram that revealed formation of lyotropic liquid crystals and to maximize the isotropic regions representing microemulsions.

The extremely high surface areas, the oriented surfactant layers, and the extremely high capacity of the core of the lyotropic liquid crystals can provide, at least in theory, some advantages over other self-assembled liquid nanosized structures (such as microemulsions), for example as microreactors for organic, inorganic, and enzymatic processes, especially those based on two immiscible reagents, sensitive to temperature, dependent upon the environment, or requiring extreme reaction conditions.

The extremely interesting properties of monoglyceride-based cubic phases, including temperature stability, bicontinuous structure, high internal surface area, solidlike viscosity, health safety, and low-cost raw materials, make them desirable for consumer products and pharmaceutical industry applications; therefore, we will mainly deal with monoglycerides.

The most popular application of cubic phase (the most studied mesophase) is as a delivery vehicle for hydrophobic and/or hydrophilic materials that, after solubilization into the cubic gel, diffuse out in a controlled-release manner. However, some new and emerging applications recently seen in the literature relate to the utilization of the cubic phase interfaces and channels for specific regioselective reactions. We will discuss those new findings and will elaborate on the options that they open to the scientific community.

14.2 SOME RELEVANT COMMENTS ON THE CHARACTERISTICS OF CUBIC PHASES

Self-assembly of amphiphilic molecules into complex phases is a broad and active area of research. Examples of self-assembled amphiphilic molecules include not only surfactants but also some polymers, biopolymers, and polar lipids. The past decade has seen a renaissance in the study of self-assembled mesophase behavior, especially that of liquid-crystalline mesophases and microemulsion systems. One of the most fascinating liquid-crystalline mesophases is the cubic phase formed from polar bilayer bicontinuous lipids. Bicontinuous cubic phases were first documented by Luzzati et al. (6), and a model of their geometric structure was first supplied by Scriven (7). More detailed structure description of a bicontinuous cubic phase was derived from differential geometry theories, as summarized by Hyde et al. (8). Many important and interesting works have been published in recent years relating to the structural aspects of the different lyotropic liquid crystals (see other chapters of this book), but this chapter will concentrate on the cubic phase and structural aspects that are relevant to their activity as microreactors.

Different types of cubic structures formed in aqueous systems have been identified (9–12). They depend on the particular type of the lipophilic amphiphile. The only structures that will be considered in this review are bicontinuous phases comprised of curved nonintersecting bilayers organized to form two disjoint continuous water channels (13). If a plane is placed in the gap between the methyl end groups of the lipid bilayer of the cubic phase, the surface obtained can be described by an infinite periodic minimal surface (IPMS) (10,14). Two principal radii of curvature, R_1 and R_2, give the curvature of any surface. The average surface curvature $1/2(1/R_1 + 1/R_2)$ at any point of a minimal surface is zero by definition. The packing parameter, v/al, for a lipid in such a curved bilayer is close to or slightly larger than unity and can be connected to the Gaussian curvature ($1/R_1R_2$) of the IPMS (15).

Three types of IPMSs, describing different cubic space groups, are important in lipid systems (10,14): the diamond (D) type [primitive lattice ($Pn3$m)], the gyroid (G) type [body-centered lattice ($Ia3$d)], and the primitive (P) type [body-centered lattice ($Im3$m)]. With such an arrangement, at any place on the surface the average curvature is zero, similar to a saddle.

One of the best-characterized systems in which reactivity was explored is based on glycerol monounsaturated fatty acid ester and water. It exhibits two main forms of the bicontinuous cubic phase (Figure 14.1a): the gyroid $Ia3$d (low hydration), and the diamond $Pn3$m (high hydration) of an optically clear, solidlike bicontinuous cubic phase at water contents between 20 and 40 wt% at room temperature. Such phases are composed of porous matrices (pores of ca. 5 nm) built from bilayers' contours into infinite periodic minimal surfaces. Figure 14.1b demonstrates the calculated representation of the hierarchy of a subunit of the diamond bicontinuous cubic geometry and the combination of several subunits to form finite particles (16). The bicontinuous arrangement minimizes stress and free energy and produces bicontinuous water and oil domains with an extremely high surface area (on the order of 400 m²/g of cubic phase). The interconnectedness of the structure results in a clear viscous gel similar in appearance and rheology to crosslinked polymer hydrogels. It should also be stressed that monoglyceride-based cubic gels (as an example) possess significantly more long-range order than hydrogels do, and because of their composition (i.e., lipid and water), they exhibit excellent biocompatibility.

A cubic gel in equilibrium with water can be dispersed into particles called "cubosomes" (Figure 14.1b) (16). This review will not discuss cubosomes.

14.2.1 Mesomorphism

The rich mesomorphism (mesomorph transformation) of mono-acylglycerols, because of their biological importance, has been widely studied. The mesomorphism is temperature and hydration dependent. Accordingly, the temperature–composition phase diagrams for a homogenous series of monoacylglycerols have been constructed and characterized. Qiu and Caffrey (17) have reported on the phase diagrams and mesomorphisms of several members in the homologue series including monomyristolein (C14:1c9) (18), monopentadecenoin (C15:1c10) (19), monoolein (C18:1c9) (20), monoundecenoin

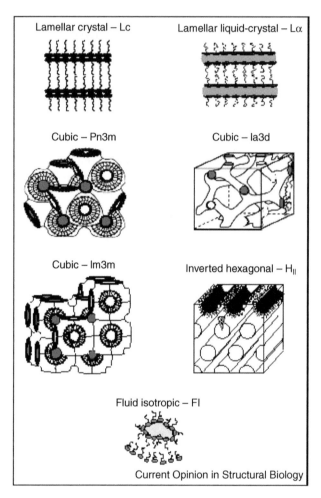

Figure 14.1a Representations of different solid (lamellar crystal phase, Lc), mesophase (lamellar liquid-crystal phase, L_α; cubic Pn3m phase [space group 224]; cubic Ia3d phase [space group 230]; cubic Im3m phase [space group 229]; and inverted hexagonal phase, H_{II}, and liquid [fluid isotropic phase, FI] states.

(C11:1c10), monotridecenoin (C13:1c12), monopalmitolein (C16:1c9), monoheptadecenoin (C17:1c10), monoeicosenoin (C20:1c11), mono-erucin (C22:1c13) (21), and monovaccenin (C18:1c11) (22).

The authors (17–22) discuss and compare an important member of these series, the monovaccenin with monoolein. Monovaccenin

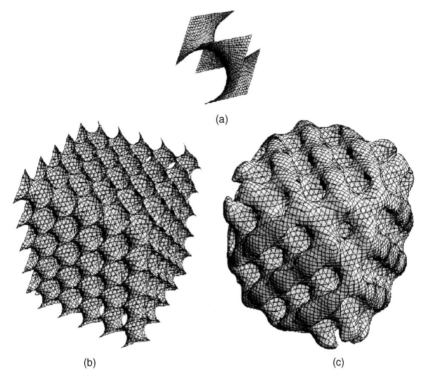

(a)

(b)   (c)

Figure 14.1b Calculated representation of the hierarchy of cubic phases and cubosomes, starting with the double diamond subunit (a), the lattice structure of a cubosome (b), and the lattice surface sealed off with a cube (c). In each case, the calculated surface represents a lipid bilayer that separates two continuous but nonintersecting aqueous (hydrophilic) regions. (From PT Spicer, WB Small, ML Lynch, JL Burns. *J Nanoparticles Res* 4:297–311, 2002.)

contains a fatty acid that is 18 carbon atoms long with a *cis* double bond at position 11 (C18:1c11) in ester linkage at C1 on the glycerol backbone, while the monoolein (C18:1c9) has a similar structure but the *cis* double bond is at C9. The structural differences (Figure 14.2 and Figure 14.3) that the authors describe are fascinating and help to enhance understanding of the self-assembly driving forces, characteristics, and mesomorphisms. The reader is referred to numerous papers that have been published recently by Caffrey et al. (17–23).

Important reports from the same authors (22,23) relate to the cubic phase's preparation conditions and protocols. The authors

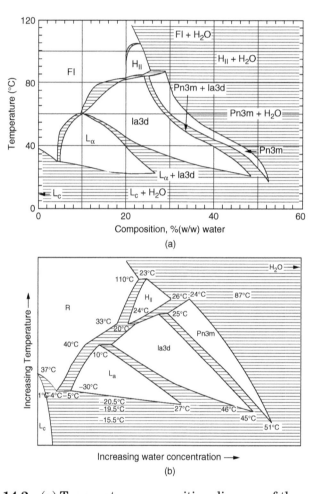

Figure 14.2 (a) Temperature–composition diagram of the monovaccenin/water system as determined by x-ray diffraction in the heating direction, based on diffraction data obtained from synchrotron measurements. (b) Same diagram, extending from 10 to 60% (w/w) of monovaccenin/water from 0 to 30°C showing the nonequilibrium diagram from undercooling of the L_α, Ia3d, and Pn3m phases, when compared with the equilibrium phase diagram shown in Figure 14.2(a). Schematized view of the diagram to reveal the nature of the assorted eutectic (at ca. 15.5, 19.5, 20.5, 30, and 83°C) and peritectic (at 60, 85, 87, and 105°C) transitions, and the temperatures and compositions at which they occur. The temperature and composition values included in this figure identify important transitions and phase coexistence regions.

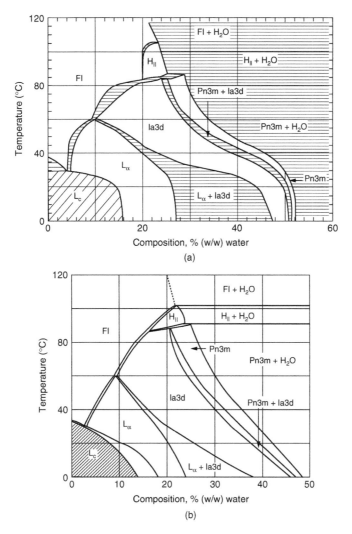

Figure 14.3 (a) Temperature–composition diagram of the monovaccenin/water system as determined by x-ray diffraction in the heating direction, based on diffraction data obtained from synchrotron measurements. The lower right-hand corner of the diagram [extending from 10 to 60% (w/w) water and from 0 to 30°C] shows undercooling of the L_α, *Ia3d*, and *Pn3m* phases when compared with the equilibrium phase diagram shown in Figure 14.2; (b) Temperature–composition diagram of the monoolein/water system in the composition range from 0 to 48% (w/w) water and in the temperature range from 0 to 102°C in the heating direction, as obtained by x-ray diffraction.

worked under various cooling protocols and have found that two different phase diagrams can be obtained for the same monoglycerides in water. Figure 14.2a and Figure 14.2b illustrate monovaccenin temperature–composition behavior including the assorted eutectic and peritectic points, and in Figure 14.3a and Figure 14.3b, one can see the phase diagrams for monovaccenin compared to monoolein

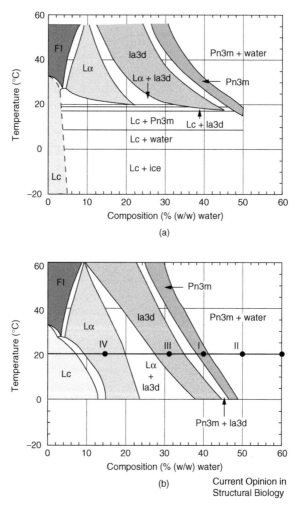

Figure 14.4 Temperature/composition phase diagram for the monoolein/water system: (a) Equilibrium phase diagram; (b) Metastable phase diagram. Points along the 20°C isotherm identified as I to IV.

(MO) under dynamic synchrotron measurements showing the under-cooling effect at temperatures below 30°C, which arise from an under-cooling of the liquid crystal phases. The phenomenon is better stressed in the phase diagrams in Figure 14.4 (23). Figure 14.4a represents equilibrium behavior at the low-temperature range of the phase diagram, while Figure 14.4b represents nonequilibrium behavior of the monoolein. The authors discuss the methods used to prove that under certain protocols the system can be at nonequilibrium conditions. Scientists working in this area should bear in mind that the mesophases, prepared at noncontrolled conditions (nonreproducible cooling protocols), might not be at "real" thermodynamic equilibrium conditions and might not represent the "true" mesophase boundaries. This work is extremely important since it also explains the difference between phase diagrams that were previously described by various authors versus those reported presently. The work also provides cooling condition protocols and guidelines for scientists who will prepare cubic phases in the future. The authors also established parameters such as pore size (diameters) of the water channel radii as a function of temperature and lipid length and the water carrying capacity in various lyotropic liquid crystals (LLC).

14.3 POTENTIAL APPLICATIONS OF CUBIC PHASES

Three major subjects (as we see it) related to cubic phases were extensively studied:

1. Solubilization (capacity and nature), reactivity, and release of drugs, nutraceuticals (vitamins), and enzymes
2. Structural aspects of biological protein membranes (crystallization and solubilization)
3. Chemical reaction properties of active molecules such as inorganic complexes, enzymes, vitamins, and drugs

This review covers only those aspects relevant to the interfacial reactivity.

14.3.1 Solubilization and Controlled Release

Knowledge of the phase behavior of monoacylglycerol/water systems can be used to design controlled drug release systems based on the lamellar, hexagonal, or cubic phases that they form (21–25). The unique bicontinuous structure of the cubic phases makes them especially attractive for use as a controlled release matrix.

14.3.1.1 Drugs

Much has been written about solubilization of drugs in cubic phases. Most of the studies are based on experimental attempts to solubilize certain (not maximal) amounts of drug into the cubic phase and to test their release profiles. In most cases the results show promising release patterns, which indicate slow or sustained release, as expected from gellike systems, in comparison to solutions or microemulsions.

Shah et al. (24) have reviewed the cubic phase gels as drug delivery systems and stressed the ability of cubic phases to deliver small drug molecules and large proteins by oral and parental intake and local delivery (in the vagina and in periodontal cavities). The authors have listed some novel applications of cubic phases and stressed the possible mechanistic routes of delivery from cubic gels. Drugs are released from the cubic phase typically by diffusion controlled from a matrix process as indicated by Higuchi's square root time-release kinetics. Incorporation of drug in the cubic phase can cause also phase transformation (26,27).

A number of different proteins have been shown to retain their native conformation and bioactivity and are protected from chemical and physical inactivation due to reduced water activity. The authors comment on the advantages (viscosity, biodegradability, and bioavailability) and disadvantages (regulation, health concerns) of cubic phases and stress the potential future and possible applications for cubic phases.

Our review will discuss only a few examples related to the solubilization of certain drugs, vitamins, and enzymes that are sensitive, chemically and enzymatically, more reactive, and more demanding as guest compounds than are regular small molecules of common drugs.

Studies have shown that small-molecule drugs, incorporated into the cubic phases formed by monoacylglycerols in water, are released slowly over a period of hours to days, thereby improving the performance of the drug (28–30). For certain drugs, sustained release is preferable to immediate release coupled with repeated administration. Since release rates are modulated by the size of the pores that permeate such materials (31), the ability to adjust pore size (at will) affords the opportunity to customize release rate. Studies show clearly that cubic phase pore size can be changed by selecting different monoacylglycerol lipids and/or by adjusting conditions such as temperature and composition (23) (Figure 14.5).

Monovaccenin and monoolein provide two excellent examples for studying the effect of water channel size on release properties

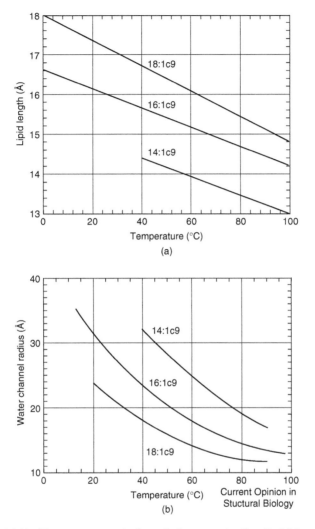

Figure 14.5 Temperature-induced changes in the lipid length and aqueous channel dimensions of the cubic phase. Temperature dependence of: (a) the lipid length in the cubic phase and (b) the radius of fully hydrated, cubic *Pn3m* phase water channels for three MAGs (monoacylglycerol). Lipid identity is given in shorthand, where the first two digits indicate the fatty acyl chain length in number of carbon atoms and the last digit refers to the position of the *cis* double bond along the chain, where the carbonyl carbon is number one. MO is represented by 18:1c9.

because of their close resemblance in molecular structure and indistinguishable effective molecular length, combined with a distinct difference in water channel pore sizes (23). Work along these lines is already in progress in more than one lab, with the aim of understanding and exploiting the mechanism of drug release in cubic phases. Other means of controlling the release profile from cubic phases make use of lipid mixtures and mesophases incorporating biodegradable polymers.

The two major advantages of the cubic phases — the high solubilization capacity and controlled release properties — were tested and verified for several drugs. One interesting study explores the various aspects determining the lipid bilayer/water partition coefficient of drugs in cubic phases along with the kinetic terms, such as time, needed for equilibrium. Effects of such variables as drug concentration, pH, agitation, sample preparation, and the monoglyceride phase structure were examined. Some examples include clomethiazole (CMZ), lidocaine, prilocaine, and 4-phenylbutylamine (4-PBA), which were chosen as model drug compounds. It was shown (32) that it is possible to determine a pH-dependent apparent partition coefficient, $K_{bl/w}$, of a drug compound using a $K_{bl/w}$ lipid bilayer expressed as a cubic liquid-crystalline structure. Good agreement was found when K vs. pH curves, for bilayer/water systems, were drawn. Clomethiazole, lidocaine, and prilocaine were fitted to a mathematical expression. This included the bilayer/water partition coefficient for the unionized and ionized drug, respectively, and the pK of the drug. The effects of different experimental conditions, such as amount of cubic phase, temperature, agitation, sample preparation, and interfacial area between the cubic phase and the aqueous bulk, on the partition kinetics were investigated and are presented in Figure 14.6. The studies revealed that the time needed to reach partition equilibrium was, as expected, substantially reduced (from days to hours) by decreasing the amount of cubic phase, increasing the interfacial area between the cubic phase and the aqueous phase, and increasing the temperature and the agitation of the sample. It was also shown that the bilayer affinity of 4-PBA was increased when a zwitterionic lipid (i.e. dioleoyl phosphatidylcholine, DOPC) was incorporated in the bilayer. The authors conclude, "it should be possible, in principle, to design a cubic system which will better mimic the chemical nature of biomembranes... and that cubic phases may play an important role as subcellular space organizers." Those of us interested in reactions and reactivity in structured membranes might consider the "mimic" structures as reaction media.

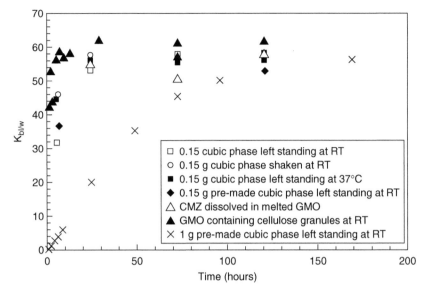

Figure 14.6 The effect of different experimental conditions on the kinetics of CMZ partitioning into a GMO/water cubic phase: 0.1 g GMO allowed to swell to ~0.15 g cubic phase in the presence of 1 ml CMZ (aq) at 20°C (□), at 20°C shaken (○), and at 37°C (■). One milliliter of CMZ (aq) allowed to equilibrate with 0.15 g (◆), and 1.0 g (×) ready-made cubic phase at 20°C. CMZ dissolved directly in 0.1 g GMO followed by the addition of 1 ml aqueous buffer at 20°C (△). GMO incorporated in cellulose granules (▲). Each experiment was performed in duplicate, and the mean values are presented in the figure.

Semisuccessful attempts were made to take advantage of the cubic phase as a solubilization medium and to link it to the bioadhesive properties for oral dosage (with increasing gastric resistance time) to improve bioavailability of a drug and at the same time to obtain a sustained action (33). The drug furosemide, a diuretic agent used for the treatment of hypertension associated with heart failure, had limited solubility in the monoolein, and therefore, additional "solubility modifiers" (trisodium phosphate and polyethylene glycol) (TSP/PEG) were added. Some significant improvement in dissolution kinetics of the release of the drug was obtained due to the solubility modifier (cosolvent or cosurfactant), but the mucoadhesive properties (advantages) were not well established.

14.3.1.2 Proteins

The effects of encapsulating bovine hemoglobin (BHb) in the bicontinuous cubic phase formed by monooleoylglycerol and water were investigated with Fourier Transform Infrared (FTIR) spectroscopy and x-ray diffraction (34). The cubic phase was formed in the presence of 1 to 10 wt% BHb. Studies using x-ray diffraction revealed that at 0.5 to 2.5 wt% BHb, the cubic phase structure is characterized by a double-diamond lattice (*Pn3m*). At 2.5 to 5 wt% BHb, the coexistence of two cubic phase structures, *Pn3m* and the gyroid lattice (*Ia3d*), was observed, while at BHb concentrations higher than 5 wt%, the gyroid structure persists. These structural transformations in the presence of the solubilizates are important events since the solubilization capacity, interfacial reactivity, and release profiles are expected to alter from one mesophase to the other.

The porous nature of the bulk cubic phase was further demonstrated by diffusion of $K_2Fe(CN)_6$ and conversion of 73% of the oxyhemoglobin to methemoglobin after 1 h (Figure 14.7) (34). *In vitro* release studies showed that 45% of the entrapped BHb was released after 144 h at 37°C (Figure 14.8) (34). The mobility within the porous mesophases suggests that the cubic phase may be a useful medium for encapsulation of Hb as a red cell substitute and for the encapsulation and delivery of other bioactive agents.

14.3.1.3 Enzymes

Enzymes (glucose oxidase, ceroplasmin) with a molecular weight of up to 590 kDa can be entrapped in cubic lyotropic liquid-crystalline phases. Both pure monoolein and monoolein with phosphatidylcholine mixtures have been used for the preparation of cubic phases (35). Electrochemical measurements of enzyme activity show that the entrapment is liable to stabilize the enzyme. Figure 14.9 shows the long-term stability of glucose oxidase in two types of cubic phases. It was also demonstrated that the composition of the lipid (zwitterionic and nonionic amphiphiles) might influence the stability of the enzyme, as it is manifested in an increase in the long-term stability of glucose oxidase with the introduction of the zwitterionic phosphatidylcholine (Figure 14.10). The enhanced stability is assigned both to the differences in the polar interface of the lipid bilayer and the changes in structure of the cubic phase. The biosensors consisting of solubilized glucose oxidase and ceruloplasmin in cubic phases were compared. Both enzymes have about the same molecular weight, but

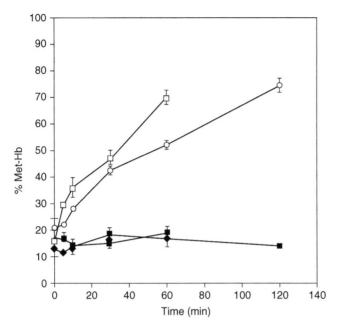

Figure 14.7 Conversion of cubic entrapped bovine hemoglobin from oxy- to methemoglobin at 25°C (○) and 37°C (□) in the presence of $K_2Fe(CN)_6$. Controls (●,■), are cubic entrapped bovine hemoglobin not exposed to $K_2Fe(CN)_6$. The amount of hemoglobin formed in the presence of $K_2Fe(CN)_6$ is significantly higher than that formed in the control at both 25 and 37°C.

they have different electrochemical reactions, which can be used for monitoring enzyme activity.

14.3.1.4 Vitamins

It was shown that vitamin K_1 (VK_1) can be solubilized in the different phases of the MO/W system to various extents, depending on the mesophase microstructure (36). Due to their hydrophobicity, the VK_1 molecules are distributed among the MO chains. It should be noted that the vitamin K_1 is practically solubilized in the hexagonal phase rather than in the cubic phase. A reverse hexagonal phase, absent at room temperature in the binary system, appears in the presence of a small amount of VK_1. The penetration of the vitamin into the lipid

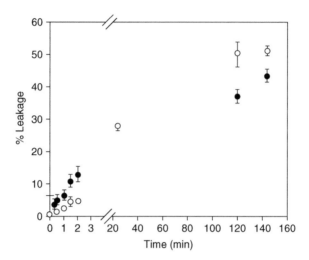

Figure 14.8 Leakage of bovine hemoglobin from cubic phase monoolein over time at 25°C (○) and 37°C (●), at incubation less than 24 h. The amount of hemoglobin released from samples at 37°C is was significantly higher than the amount released from samples at 25°C. At incubation times greater than 24 h, the trend is reversed.

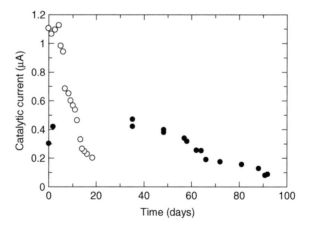

Figure 14.9 Catalytic steady-state current calibration curve for the glucose oxidase cubic phase–based platinum electrode at 600 mV vs. SCE (see text) in 10 mM potassium phosphate 0.01 M KCl/1 mM EDTA at pH 7 at 25°C. (○) MO/buffer/glucose oxidase (65:34.9:54: 0.046 wt/wt), (●) monoolein/phosphatidylcholine/buffer/glucose oxidase (48:12:39.956:0.044 wt/wt).

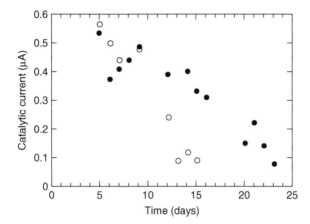

Figure 14.10 The long-term stability of ceruloplasmin entrapped in the cubic phase of MO/phosphatidylcholine/buffer ceruloplasmin at a weight ratio of 49.0:12.3:38.58:0.12. The supporting electrolyte was 0.05 M piperazine buffer, containing 0.1 M NaCl using 0.1 M HCl to adjust the pH to the desired value. The long-term stability is shown as the steady-state current versus storage time. The data from two different electrodes are shown, one investigated at pH 4.5 (●) and the other at pH 5.5 (○).

bilayer is expected to increase the apparent volume of the hydrocarbon chain of the monoolein, thus increasing the packing parameter (v/al). In essence, this will have the same effect as an increase of temperature: in the binary system, a transition of the cubic phase to the reverse hexagonal phase occurs at about 80°C. However, it was found that small amounts of VK_1 sufficient to induce a phase transition cannot possibly affect the packing parameter substantially enough. It is more likely that the VK_1 might not be uniformly distributed, in particular in the cubic phase. This produces a local change of the bilayer curvature in the cubic phase, which in turn induces a phase transition to the reverse hexagonal phase.

In an earlier study (37), the channeling effect of the cubic phase was demonstrated for a water-soluble protein, cytochrome C. The presence of cytochrome C leads to a decrease of the temperature of the cubic reverse hexagonal transition from 80°C in the binary system to about 56°C. However, the effect in this case is more likely achieved by the protein affecting the polar part of monoolein, which can reduce the size of the lipid polar head group. From the phase diagram of the ternary system MO/VK$_1$/W, it is clear that the gyroid type of cubic phase can accommodate more VK$_1$ than the diamond type can, before the transition to the reverse hexagonal phase occurs. This can be understood by considering the different structures of the two cubic phases. The diamond type is geometrically more restricted than the gyroid structure. It is more difficult to pack the lipid bilayer into the former type of cubic structure, as it contains circular necks. The gyroid type of cubic phase, because of fewer geometrical constraints, can occur over a larger concentration range and also at lower water content. Obviously, the lamellar phase has no geometrical constraints, and hence it can solubilize a much larger amount of VK$_1$ until the local curvature increases enough to form the reverse hexagonal phase. Interesting results, derived from the same study, stress the structure–reactivity relationships in cubic phases. The study clearly demonstrates that VK$_1$ molecules are electrochemically active in the cubic phase, being available for redox conversion. This agrees with the observation that VK$_1$ molecules display high mobilities within the bilayer of the cubic phase, even more than the MO molecules do. Finally, the results point to possible use of bicontinuous cubic structures (by virtue of their well-defined porosity) to construct electrochemical systems with large effective surface area.

It should also be mentioned that particles of such cubic phases, cubosomes, can be produced (38). These types of nanostructures could be used to improve the performance of bioanalytical systems based on enzymes entrapped in the cubic lipid/aqueous phases, as has been already demonstrated in our earlier studies (12,39).

14.3.2 Biological Systems

14.3.2.1 Membrane Protein Crystal Formation

Cubic structures have received increased attention with relation to biological systems (23,35,39–42). Several biological membrane lipids form cubic phases that possibly participate in various biological processes. For example, it has been suggested that acylglycerols assume a cubic structure in the intermediate stage of fat digestion (43–45). Recent studies have demonstrated also that cubic phases can accommodate a wide range of globular proteins (43) and that entrapped enzymes can be used to produce electrochemical biosensors (46,47).

Cubic phases have been used recently to procure crystals of a variety of materials, ranging from simple salts through to water-soluble and integral membrane proteins. The exact role played by the cubic phase medium in such applications, particularly membrane protein crystal formation, is not fully understood. However, the prospect exists that cubic phases provide the conduit to membrane protein crystals and thus to three-dimensional protein structures. Comparing monovaccenin and monoolein, monoacylglycerols with the same acyl chain length, it was found (22,23) that shifting the *cis* double bond away from the glycerol headgroup produces a "slimmer" molecular shape, which permitted stabilization of a less curved polar/apolar interface and an increase in the water-carrying capacity of the phase. Genomic sequencing has revealed that approximately one third of all genes encode membrane proteins having at least one membrane-spanning helix. A complete understanding of the function of integral membrane proteins requires knowledge of their three-dimensional structure at or near atomic resolution. The major bottleneck en route to obtaining high-resolution structures of membrane proteins is the preparation of diffraction-quality single crystals.

A paradigm of a membrane protein is the light-driven proton pump, bacteriorhodopsin (bR), with its seven transmembrane helices. bR has long been a target for conventional crystallization using mixed detergent–protein micelles (41). With one exception, this methodology has yielded bR crystals of comparably low diffraction quality, in which the protein is in a functional state markedly different from the native one. In contrast, the recently introduced concept of membrane protein crystallization in lipidic cubic phases has produced crystals of functional bR of exceptional diffraction quality. These have been instrumental in resolving the high-resolution structure of bR in the ground state at increasing resolution. The structures reveal the intricate interactions between bR, native purple membrane lipids, and water molecules in an environment closely resembling that of the cell membrane.

Lipidic cubic phase-grown bR crystals were instrumental in yielding high-resolution structures of photocycle intermediates, which should eventually allow a complete understanding of the proton pumping mechanism of the molecule at the atomic level. Moreover, it was recently demonstrated that bR crystals can also be grown in a lipidic cubic phase directly from the native membrane without exposure to any detergent.

Investigators have extended the crystallization method and demonstrated that the lipidic matrix used for the formation of a cubic phase allows the formation of three-dimensional crystals with membrane proteins differing with respect to the size of their membranous and extramembranous domains.

   Using the ability of bR to crystallize provided experimental evidence in support of a hypothesis for the molecular mechanism in cubo crystallization, which was originally proposed by Rummel et al. (48) and later developed further by others (41,49–53). The hypothesis describes how the protein, upon reconstitution into the continuous curved membrane of the cubic phase, traverses its convoluted and highly curved bilayer and eventually forms a lamellar-type packing arrangement within a crystal. Figure 14.11 illustrates the packing aspects with the three-dimensional crystal formation of the bR into

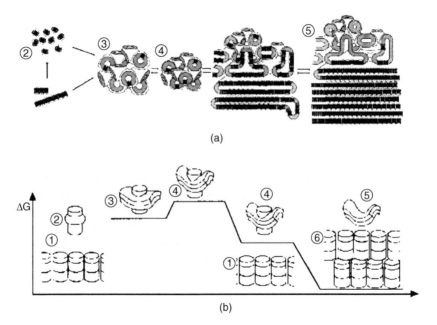

Figure 14.11   A hypothesis of the steps leading to the formation of three-dimensional crystals by incorporation of bR into curved lipid bilayers and subsequent phase separation into lamellar structures. The amphipathic species bR, purple membrane lipids, and detergent partition into the curved membrane of the cubic phase and may diffuse freely therein. Panel A: Schematic description of events occurring during the crystallization of bR in lipidic mesophases. Purple patches (1) or detergent-solubilized monomers (2) insert spontaneously into the curved bilayer (3) of the bicontinuous cubic-Pn3m phase. Addition of Sørensen salt increases the membrane curvature and reduces unit cell size [(3)→(4)].

curved lipid bilayers and subsequent phase separation into lamellar structures.

Significant cubic unit cell deformations due to the presence of the protein may occur at this stage. Separation of protein and purple membrane lipid from the highly curved cubic phase bilayer into growing planar domains favors crystal nucleation. Bending frustration is relieved by this process, which occurs by means of lateral diffusion analogously to defect migration and fusion described for similar systems (54). Diffusion of a misaligned membrane protein along the curved membrane to an adjacent layer may enable a crystallographic match with the growing crystal. Portal lamellae and a putative intermittent sponge phase connect the protein crystal with the surrounding cubic phase.

The crystallization process is completed when mature crystals coexist with a bR-depleted and highly curved cubic phase (54). Upon reswelling of the cubic phase by hydration, the crystallization process may be reversed, and protein molecules may diffuse back into the curved bilayer via the portal lamellae.

14.3.3 Chemical Reaction and Interfacial Reactivity

14.3.3.1 Maillard Reactions in Cubic Phases

Most studies discussed in this review are not directly concerned with the reactivity of cubic interfaces. Some of the major structure–reactivity relationships in this review are based on the effect of guest molecules on cubic phase transformations and only in some cases the effect of the host mesophase on certain reactions. To the best of our knowledge, very few studies are exploring cubic phases as microreactors for improved product selectivity or rates. Those examples will be further discussed.

For some of the monoglycerides, mainly the unsaturated, LLC coexist together with "fluid isotropic" regions at low water contents (up to 40 to 50 wt% water at elevated temperatures but only up to ca. 10 to 15 wt% water at room temperature). These regions are sometime termed "pseudo water-in-oil microemulsions." The isotropic regions are significantly larger at elevated temperatures, which is an advantage for organic reactions enhanced by high temperature.

For lyotropic liquid crystals, the cubic phase is of particular interest for organic reactions. Authors have reported the existence of cubic phases in a variety of other binary mixtures, such as sugar esters and water, and in many ternary systems that will not be further discussed here (55–58). Some recent reports have shown formation of cubic phases in ternary food and cosmetic systems based on monoolein/water in the presence of modified starch (16) or sodium oleate (55). The phase diagrams exhibit large areas of cubic phases. After more in-depth investigation, new opportunities for running organic reactions in these unique microreactors are likely to be found.

Vauthey et al. (59) used simple binary self-assembled structured fluids based on water and unsaturated monoglycerides as microreactor. The acyl chain of the monoglycerides in use differed in the degree of unsaturation.

Vauthey et al. (59) explored, on a quantitative basis, the reaction products of cysteine/furfural and cysteine/ribose model systems performed in the cubic phase and in a binary microemulsion phase. They compared the reaction products to those obtained from an aqueous medium under the same reaction conditions.

Maillard reactions are often studied at elevated temperatures of 140 to 180°C. In this study, 100°C was chosen due to the necessity to stay in the cubic phase region of the monoolein/water (80:20) (80 to 120°C). It is also known that cubic-to-hexagonal and hexagonalto-L_2 phase transition temperatures decrease for unsaturated monoglycerides.

The reaction between cysteine and furfural was particularly efficient both in L_2 microemulsions and cubic phases compared to aqueous systems. The reaction led to the formation of 2-furfurylthiol (FFT) and two new sulfur compounds, which were identified as 2-(2-furyl)-thiazolidine and N-(2-mercaptovinyl)-2-(2-furyl) thiazolidine. Similarly, generation of 2-furfurylthiol and 2-methyl-3-furanthiol was derived from the reaction between cysteine and ribose mixtures. The two reaction rates and yields were strongly enhanced, over the aqueous phase reactions, by the structured fluids, with some preference for the cubic phase over the pseudomicroemulsion.

The obtained products are interpreted in terms of a surface and curvature control of the reactions defined by the structural properties of the formed surfactant associates. Figure 14.12 and Figure 14.13

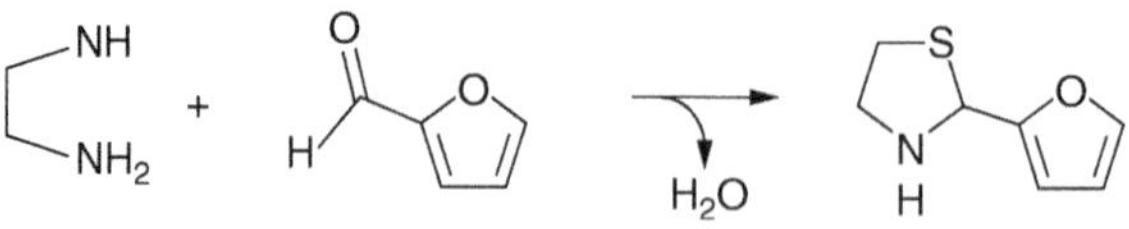

Figure 14.12 Synthesis of 2-(2-furyl)thiazolidine from cysteamine and furfural.

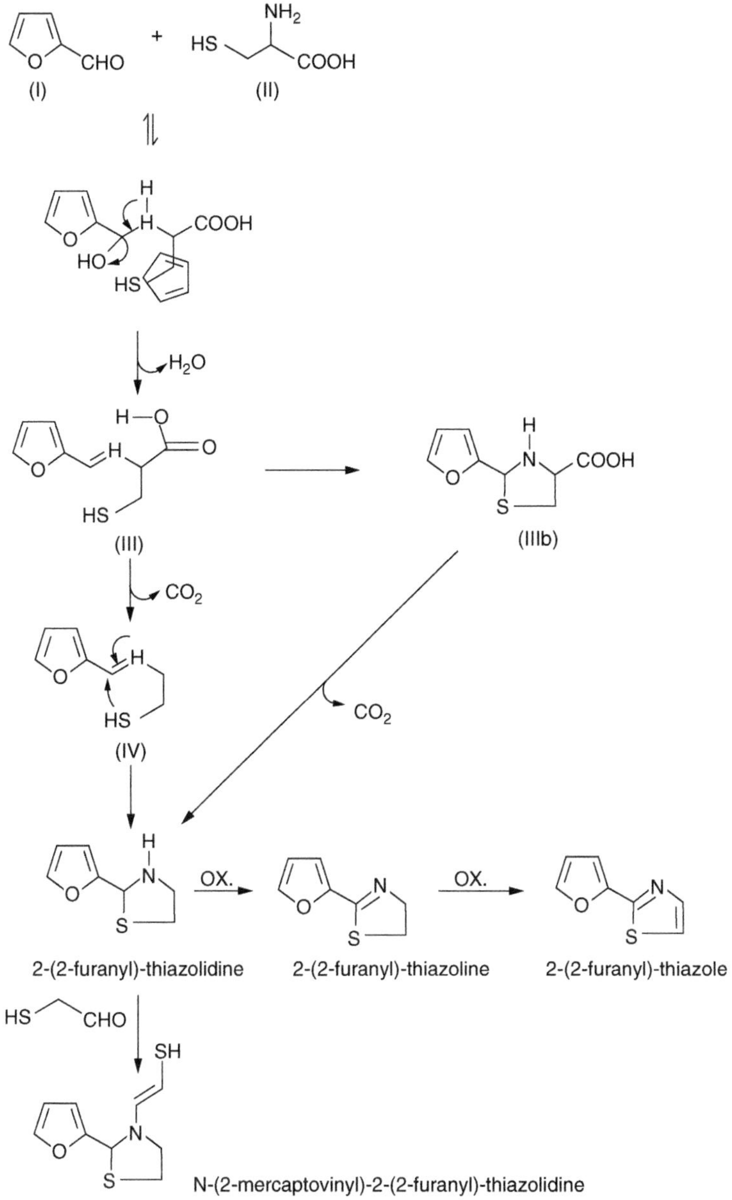

Figure 14.13 Hypothetical reaction pathway leading from cysteine and furfural to 2-(2-furyl)thiazolidine and *N*-(2-mercaptovinyl)-2-(2-furyl) thiazolidine.

Table 14.1 Amounts of Volatile Reaction Products Generated
from Furfural and Cysteine Solubilized In Different Matrices

Matrix	Amount[a] (*mg*)		
	2-Furfurylthiol	2-(2-Furyl)thiazolidine	*N*-(2-Mercaptovinyl)-2-(2-furyl)thiazolidine
Water	1.8	nd[b]	nd
Microemulsion	8.7	42	11.7
Cubic phase	11.6	240	123

[a] Mean values of duplicates (SD < 20%).
[b] Not detected.

demonstrate the speculated reaction pathways for FFT and the two
new products that were obtained during the thermal Maillard process
that was conducted in the cubic phase. Table 14.1 summarizes the
quantitative results in terms of selectivity and product ratios for reactions carried out in water solutions versus microemulsions and cubic
phases. In the aqueous phase, only two sulfur compounds were
detected in trace amounts. One compound was identified as FFT, which
is the expected product from the reaction of furfural with cysteine. In
the cubic phase, new "signature products" were obtained

These results indicate that the cubic structure is the most efficient matrix for the generation of the three volatile compounds under
the chosen reaction conditions. Moreover, the ratio of the amount for
each compound formed in the cubic phase and in the microemulsion
differed significantly. Compared to the microemulsion, the amount of
N-(2-mercaptovinyl)-2-(2-furyl)thiazolidine was 10 times higher in
the cubic phase, whereas ratios of about 1.3 and 6 were observed for
FFT and 2-(2-furyl)thiazolidine, respectively. A hypothetical pathway
for the generation of 2-(2-furyl)thiazolidine from furfural (**I**) and cysteine (**II**) is displayed in Figure 14.13. The first step in the reaction
is the addition of cysteine to the carbonyl group of furfural. The
subsequent elimination of a molecule of water gives rise to a Schiff
base (**III**). This compound may undergo decarboxylation, resulting in
a product (**IV**). This intermediate can finally form a five-member
ring by addition of the nucleophilic sulfur to the imine, which leads
to the formation of 2-(2-furyl)thiazolidine. Compound **III** may also
first cyclize to compound **IIIb** and subsequently lose CO_2. The 2-
(2-furyl)thiazolidine may then react with mer-captoacetaldehyde,
released from cysteine upon Strecker degradation, to finally give
rise to the tentatively identified *N*-(2-mercaptovinyl)-2-(2-furyl)
thiazolidine.

The second example is of the ribose/cysteine system, which was chosen because some parameters influencing flavor generation such as pH, water activity, and temperature have already been studied for this combination of reagents (60) and because of the importance of C5 sugars in the formation of meat-like and savory flavors. Some of the potent odorants were identified. The reaction carried out in the cubic phase was found to yield a more intense overall flavor. The cubic phase was shown, again, to be even more efficient in flavor generation than the L₂ microemulsion (Table 14.2). This cubic phase was denoted "cubic catalyst" or "cubic selective microreactor." The obtained results were also interpreted in terms of a surface and curvature control of the reactions defined by the structural properties of the formed surfactant associates.

14.3.3.2 Polymerization in Lyotropic Liquid Crystal Mesophases

The *in situ* polymerization in lyotropic mesophases on monomeric surfactants (*surfmers*) (61) is an interesting attempt to decisively stabilize the long-range order of the mesophases (in a similar way that *in situ* polymerization of organized surfmers has been successfully applied to stabilize monomers at the air/water interface (62,63). We chose to investigate, in short, some of the reactions

Table 14.2 Amount of Volatile Reaction Product Generated from Ribose and Cysteine Solubilized in Different Matrices

Compound[b]	Amount[a] (µg)	
	Cubic Phase	Water
2-Methyl-3-furanthiol (MFT)	18.4	nd[c]
2-Methyl-3(2H)-furanone	36.7	13.1
Furfural	874	351
2-Furfurylthiol (FFT)	12.0	tr[d]
3-Mercapto-2-pentanone	tr	nd[c]
Norfuraneol	698	291
MFT-MFT	8.3	tr
MFT-FFT	tr	tr

[a] Mean values of duplicates (standard deviation < 20%).
[b] Compounds were identified by comparison with the reference compound based on retention indices on DB-1701 and MS/EI spectra.
[c] Not detected.

even if carried out in lamellar or reverse hexagonal phase and not in cubic phase and even if some of the cases the structured mesophases are not fully maintained and possibly transformed or completely destroyed during the polymerization. The relevance of these reactions is in the fact that the oriented monomers stabilize the long-range order. Also, the reaction rates are significantly enhanced, and in some cases full monomer conversion is obtained. Several attempts were made to polymerize phospholipids (64,65) in reverse hexagonal mesophases or polymerizable mono- and diglycerides in cubic phases (66), *p*-styryloctadecanoate in reversed hexagonal phase (66), allyldodecyldimethylammonium bromide and allyldidodecylmethylammonium bromide in lamellar phases (67), and others. An interesting successful attempt was performed to carry out γ-ray polymerization of cationic surfactant methacrylates in lyotropic mesophases, where the phase was maintained intact throughout the polymerization, and full conversion with enhanced reaction rates was achieved (Figure 14.14) (62).

14.3.3.3 Electrochemical Sensor in the Cubic Phase

A very interesting and unique idea for utilizing cubic phases as hosting environment, immobilizing phase, and microreactor comes from a very original work by Rowinski and Bilewicz (68), who decided to use cubic phases for an immobilizing electrode based on a nickel complex. A cubic phase was used as a hosting layer for catalytic tetraazamacrocyclic Ni(II) complexes. The role of the cubic phase was to hold the catalyst close to the electrode surface to retain its catalytic activity and to provide a suitable environment for the catalytic reaction (Figure 14.15). Nickel complexes were incorporated into the bicontinuous cubic phase layer modifying the thin mercury film or glassy carbon electrode and were used for study of catalytic reduction of carbon dioxide. The results were very reproducible and useful for CO_2 determination in the gas phase with better linear dependence of the catalytic reduction than other modified electrodes.

14.3.3.4 Electrochemical Behavior of Ferrocene

The reactivity of the cubic phase interface can be illustrated in an additional example where no direct reaction between two reagents was examined, but rather the activity of solubilized ferrocene toward electrochemical reactivity was the focus of the investigation (47). The phase and electrochemical behavior of the aqueous mixtures of

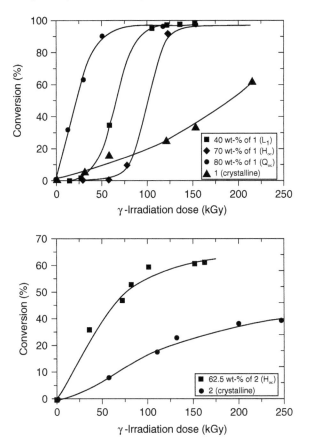

Figure 14.14   Dose vs. conversion curves of **1** (a) and **2** (b) in aqueous mixtures and in the crystalline state.

monoolein (MO) and synthetic ferrocene (Fc) derivatives containing long alkyl chains, (Z)-octadec-9-enoylferrocene (I), (Z)-octadecen-9-ylferrocene (II), and ferrocenylmethyl (Z)-octadec-9-enoate (III), were studied. At low hydration, the reversed micelles (L_2 phase) and cubic Q^{230} phase of MO can accommodate relatively high amounts (>6 wt%) of Fc derivative II, whereas at high hydration, the pseudoternary cubic phase Q^{224} is destabilized even at about 2 wt% of this Fc. Increasing the Fc-derivative content induces L_α to L_2 and L_α to reversed bicontinuous cubic phase (Q_{II} to H_{II}) transitions depending upon hydration. Compound II apparently has no effect on the lipid monolayer thickness in the pseudoternary L_α, H_{II}, and Q_{II} liquid-crystalline

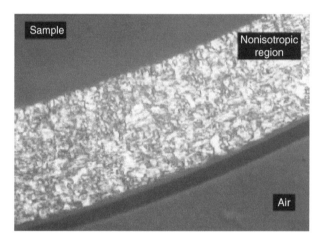

Figure 14.15 The photograph of cubic phase containing NiLC$_{16}$, observed in polarized light using a microscope with 200-fold magnification, $T = 30°C$.

phases of MO. Within a three-dimensional structure of the Q^{224} phase, the derivatives I and III exhibit electrochemical activity on a gold electrode. The values of the apparent diffusion coefficients and the heterogeneous electron-transfer rate constant of the ferrocenes are significantly lower in the cubic phase matrix when compared to the acetonitrile solution. By contrast the MO H$_{II}$ phase with entrapped ferrocene derivatives does not exhibit electrochemical activity on the electrode surface. It was suggested that the diffusional anisotropy and/or localized aggregation of the three compounds (Fc I to III) within a two-dimensional structure of the H$_{II}$ phase accounts for this feature.

14.3.3.5 Encapsulation and Diffusion of Dendrimers in Cubic Phases

Jeong et al. (69) described the synthesis of water-soluble poly(amido-amaine) (PAMAM) dendrimer derivatives labeled with fluorine and their diffusion in the water channels of a Q$_{II}$ phase with *Ia3d* symmetry. The phase was prepared by hydration of a 9:1 molar mixture of polymerizable monoacylglycerol and the corresponding 1,2-diacylglycerol. The dendrimers were synthesized by a Michael reaction of PAMAM dendrimers with a mixture of ethyl 4,4,4-trifluorocrotonate and methyl acrylate (Figure 14.16 and Figure 14.17). The authors evaluated the

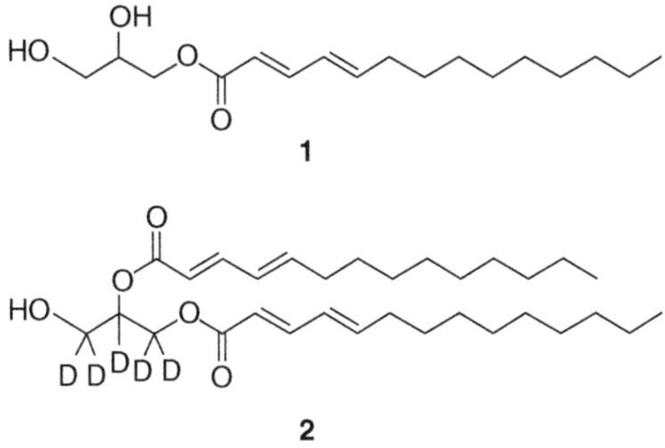
OH
HO
O
O
1
O
O
HO
O
D D D D D
O
2

Figure 14.16 Polymerizable monoacylglycerol (1) and 1,2-diacylglycerol (2).

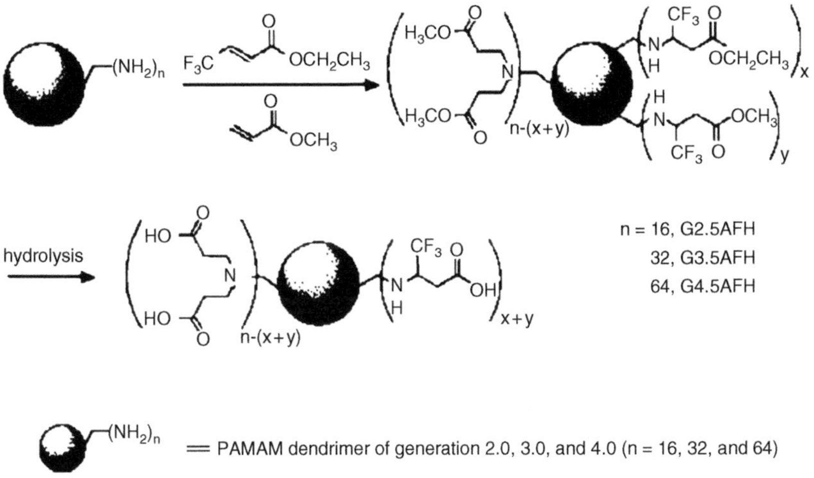
(NH2)n
F3C
OCH2CH3
OCH3
H3CO
N
n-(x+y)
CF3
N
H
x
y
hydrolysis
HO
OH
x+y
n = 16, G2.5AFH
32, G3.5AFH
64, G4.5AFH
= PAMAM dendrimer of generation 2.0, 3.0, and 4.0 (n = 16, 32, and 64)

Figure 14.17 Phase diagrams of the ternary (a) MO–W–NaO and (b) MO–W–OA systems. Composition is in wt%. Phase notations are: C, bicontinuous cubic phase; C_{mic}, cubic micellar phase; L_α, lamellar phase; L_2, reversed micellar solution phase; lyotropic liquid crystals, hydrated lipid crystals; H_{II}, reversed hexagonal phase; and H_I, normal hexagonal phase. The arrow, which starts at the used sample composition after lipase addition, indicates the expected path for hydrolysis provided that the water content does not change significantly.

hydrated diameter of the dendrimers (32.6 Å) and their diffusion coefficient (1×10^{-12} in comparison to 1.42×10^{-10} m²/sec in water), which indicate that small globular polymers (or proteins) can diffuse rapidly enough in stabilized cubic phases. The rapid diffusion phenomenon has both theoretical as practical significance. The ability to polymerize dendrimers *in situ* also has practical importance.

14.3.3.6 Lipase Action on the Interfacial Monoolein in a Monoolein/Sodium Oleate Cubic Phase (Auto Hydrolysis)

Lipase is known to hydrolyze monoolein. In order to understand the action of lipase on a "real" substrate that consists of lipid self-assembly structures, it was suggested to consider the lipolytic process that leads to monoolein, oleic acid, and sodium oleate, which monitors the lipolytic process in an intestinal-like environment (70). A three-component phase diagram was constructed (oleic acid, monoolein, water), and lipase was added to the aqueous phase (Figure 14.18). For the lipase to act, it must penetrate through the layers and solubilize at the interface (or at the core). It was demonstrated that the cubic phase favors the lipolysis process and enhances the reaction rate (larger effective reaction surface area). From various techniques (H-NMR, Self-Diffusion Nuclear Magnetic Resonance [SD-NMR], Small Angle X-ray Scattering [SAXS], etc.) it was also shown that the lipase changes the structure of the cubic phase and transforms it to reverse hexagonal.

The authors conclude that "lipase-catalyzed hydrolysis" affects the lipid organization in such a way that structures with increasing

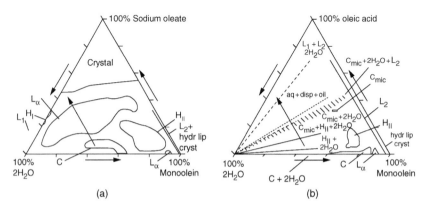

(a)   (b)

Figure 14.18 Preparation of the fluorine-labeled PAMAM dendrimers.

reverse mean curvature are promoted as the process progresses with time, that is bicontinuous cubic → reversed → hexagonal → reversed micellar cubic phase.

14.3.3.7 Activity of Lipases on Triglycerides in Cubic Phases

It is well known that lipase in the aqueous phase needs an oil/water interface in order to catalyze efficient hydrolysis of triglycerides. It is assumed that during the lipolysis, *in vivo* and *in vitro* lipid liquid-crystalline phases are in equilibrium with water and that a cubic lipid/water phase plays a dominant role. A preliminary study was carried out by Wallin and Arnebrant (71) to investigate whether the lipase is active at interfaces between water and LLC lipid phases and to compare it to that of the oil/water interface. The results show clear lipolysis in cubic phases prepared with distilled water (pH 5) and in cubic phases formed with buffered pH 5 and 10 aqueous solutions. The work is very preliminary and needs more experimental attention in order to quantify the results, but it seems clear that cubic phases are potential media for lipase activity.

14.4 CONCLUSIONS

Microemulsions have all the required characteristics to act as microreactors for special organic, inorganic, and enzymatic reactions with high regioselectivity, enhanced kinetic rates, and improved yields. However, they suffer from uncontrolled release and from health restrictions that are severe and difficult to overcome in real food systems. Lyotropic liquid-crystalline mesophases and the cubic phase in particular are semisolid, gel-like systems with slow- and controlled-release capabilities; they can be made from simple polar lipids (monoglycerides), which are known as GRAS (generally recognized as safe) compounds and can be safely utilized. It is therefore obvious that one should consider mesophases as microreactors both at low temperatures (gel-like mesophases) and at elevated temperatures (pseudoliquid mesophases).

Several examples demonstrate the potential and advantages of lyotropic liquid crystals in general and cubic phases in particular. In most cases, the authors have demonstrated interfacial reactivity combined with controlled diffusion within the channels and throughout, as well as controlled release. In other studies interfacial selectivity, due to orientation of the molecules, and enhanced reaction rates were demonstrated.

The feasibility to form by a simple procedure cubosomes that are easily dispersed in water opens some new options for scientists to explore them as microreactors for many organic reactions and products that are difficult to control or where regioselectivity is crucial.

Micellar catalysis can now be rejuvenated in condensed highly structured bicontinuous phases.

In a different possible interfacial reactivity, authors have demonstrated how the cubic phase can serve as a microreactor for membrane protein crystallization. The exact role played by the cubic phase medium in membrane protein crystal formation is not very clear, but investigators have extended the crystallization methods to cubic phases and demonstrated that the lipidic matrix used for the formation of a cubic phase allows the formation of three-dimensional crystals with membrane proteins differing with respect to the size of their membranous and extramembranous domains. The hypothesis describes how the protein, upon reconstitution into the continuous curved membrane of the cubic phase, traverses its convoluted and highly curved bilayer and eventually forms a lamellar-type packing arrangement within a crystal.

REFERENCES

1. C Solans, R Pons, H Kunieda. In C Solans, H Kunieda, Eds. *Industrial Applications of Microemulsions*. New York: Marcel Dekker, 1997, pp. 1–17.

2. J Sjöblom, R Lindberg, SE Friberg. *Adv Colloid Interface Sci* 65:125–287, 1996.

3. K Holmberg. In DO Shah, Ed. *Micelles, Microemulsions, and Monolayers*. New York: Marcel Dekker, 1998, pp. 161–192.

4. S Engström, K Larsson. In P Kumar, KL Mittal, Eds. *Handbook of Microemulsion Science and Technology*. New York: Marcel Dekker, 1999, pp. 789–796.

5. J Klier, CJ Tucker, TH Kalantar, DP Green. *Adv Mater* 12:1751–1757, 2000.

6. V Luzzati, A Tardieu, T Gulik-Kryzwicki, E Rivas, F Reiss-Husson. *Nature* 220:485, 1968.

7. LE Scriven. *Nature* 263:123–125, 1976.

8. S Hyde, A Andersson, K Larsson, Z Blum, T Landh, S Lidin, BW Ninham. In *The Language of Shape*. Elsevier: New York, 1997, Chapter 1.

9. M Lawrence. *J Chem Soc Rev* 23:417–423, 1994.

10. HG von Schnering, R Nesper. *Z Phys B: Condens Matter* 83:407–412, 1991.

11. JM Seddon, *Biochim Biophys Acta* 1031:1–69, 1990.

12. J Clogston, J Rathman, D Tomasko, H Walker, M Caffrey. *Chem Phys Lipids* 107:191–220, 2000.

13. G Lindblom, K Larsson, L Johansson, K Fontell, S Forsén. *J Am Chem Soc* 101:5465–5470, 1979.

14. S Andersson, ST Hyde, K Larsson, S Lidin. *Chem Rev* 88:221–242, 1988.

15. ST Hyde, *J Phys Chem* 93:1458–1464, 1989.

16. PT Spicer, WB Small, ML Lynch, JL Burns. *J Nanoparticles Res* 4:297–311, 2002.

17. H Qiu, M Caffrey. *J Phys Chem* 102:4819–4829, 1998.

18. J Briggs, M Caffrey. *Biophys J* 66:573–587, 1994.

19. J Briggs, M Caffrey, *Biophys J* 67:1594–1602, 1994.

20. J Briggs, H Chung, M Caffrey. *J Phys II* 6:723–751, 1996.

21. H Qiu, M Caffrey. *Biophys J* 68:A431, 1995.

22. H Qiu, M Caffrey. *Biomaterials* 21:223–234, 2000.

23. M Caffrey. *Curr Opin Struct Biol* 10:486–498, 2000.

24. JC Shah, Y Sadhale, DM Chilukuri. *Adv Drug Delivery Rev* 47:229–250, 2001.

25. K Larsson. *J Phys Chem* 93:7301–7314, 1989.

26. J Shah, M Maniar. *J Control Release* 23:261–270, 1993.

27. C Chang, R Bodmeier. *J Pharm* 147:135–142, 1997.

28. D Wyatt, D Dorschel. *Pharm Technol* 16:16,116,118,120,122, 130, 1992.

29. S Engstrom. *Polym Prepr* 31:157–158, 1990.

30. S Puvvada, SB Qadri, J Naciri, BR Ratna. *J Phys Chem* 97:11,103–11, 107, 1993.

31. WM Deen. *AIChE J* 33:1409–1424, 1987.

32. S Engstrom, T P Norden, H Nyquist. *Eur J Pharma Sci* 8:243–254, 1999.

33. A Sallam, E Khilil, H Ibrahim, I Freij. *Eur J Pharma Biopharma* 53:343–352, 2002.

34. SB Leslie, S Puvvada, BR Ratna, AS Rudolph. *Biochim Biophys Acta* 1285:246–254, 1996.

35. T Nylander, C Mattisson, V Razumas, Y Miezis, B Hakansson. *Colloids Surf A* 114:311–320, 1996.

36. F Caboi, T Nylander, V Razumas, Z Talaikyté, M Monduzzi, K Larsson. *Langmuir* 13:5476–5483, 1997.

37. V Razumas, K Larsson, Y Miezis, T Nylander. *J Phys Chem* 100:11, 766–11,774, 1996.

38. C. Ostermeier, H. Michel. *Curr Opin Struct Biol* 7:699–701, 1997.

39. J Gustafsson, H Ljusberg-Wahren, M Almgren, K Larsson. *Langmuir* 12:4611–4613, 1996.

40. J Born, T Nylander, A Khan. *J Colloid Interface Sci* 257:310–320, 2003.

41. P Nollert, H Qiu, M Caffrey, JP Rosenbusch, EM Landau. *FEBS Lett* 504, 179–186, 2001.

42. JH Collier, PB Messersmith, *Annu Rev Mater Res* 31:237–263, 2001.

43. X Ai, M Caffrey, *Biophys J* 79:394–405, 2000; V Cherezov, J Clogston, Y Misquitta, W Abdel-Gawad, M Caffrey, *Biophys J* 83:3393–3407, 2002.

44. E Pebay-Peyroula, R Neutze, EM Landau. *Biochim Biophys Acta* 1460:119–132, 2000.

45. EM Landau, JP Rosenbusch. *Proc Natl Acad Sci* 93:14,532–14, 535, 1996.

46. J Borné, T Nylander, A Khan. *Langmuir* 18:8972–8981, 2002.

47. P Pitzalis, M Monduzzi, N Krog, H Larsson, H Ljusberg-Wahren, T Nylander. *Langmuir* 16:6358–6365, 2000.

48. G Rummel, A Hardmeyer, C Widmer, ML Chiu, P Nollert, KP Locher, I Pedruzzi, EM Landau, JP Rosenbusch. *J Struct Bio* 121:82–91, 1998.

49. J Barauskas, V Razumas, Z Talaikyte, A Bulovas, T Nylander, D Tauraite, E Butkus. *Chem Phys Lipids*, 2003.

50. CM Chang, R Bodmeier. *Int J Pharmaceutics* 173:51–60, 1998.

51. EM Landau. In PI Haris, D Chapman, Eds. *Biomembrane Structures.* Amsterdam: IOS Press, 1998, pp. 23–38.

52. M Loewen, ML Chiu, C Widmer, EM Landau, JP Rosenbusch, P Nollert. In T Haga, G Berstein, Eds. *CRC Methods in Signal Transduction Series.* Boca Raton: CRC Press, 1999, pp. 365–388.

53. Y Qutub, I Reviakine, C Maxwell, J Navarro, EM Landau, PG Vekilov. *J Mol Biol* 343: 1243–1254, 2004.

54. P Nollert, J Navarro, EM Landau. In R Iyengar, JD Hildebrandt, Eds. *Protein Pathways, Methods in Enzymology.* San Diego, Academic Press, 2001.

55. J Borné, T Nylander, A. Khan. *J Phys Chem B* 106:10,492–10, 500, 2002.

56. J Borné, T Nylander, A. Khan. *Langmuir* 17:7742–7751, 2001.

57. E Pebay-Peyroula, G Rummel, JP Rosenbusch, EM Landau. *Science* 277:1676–1681, 1997.

58. N Garti. *Curr Opin Colloid Interface Sci*, 8:197–211, 2003.

59. S Vauthey, C Milo, P Frossard, N Garti, ME Leser, HJ Watzke. *J Agric Food Chem* 48:4808–4816, 2000.

60. T Hofmann, P Schieberle. In AJ Taylor, DS Mottram, Eds. *Flavor Sciences Recent Developments.* Cambridge: The Royal Society of Chemistry, 1996, pp. 182–187.

61. H Bader, K Dorn, B Hupfer, H Ringsdorf. *Adv Polymer Sci* 64:1, 1985.

62. D Pawlowski, A Haibel, B Tieke. *Ber Bunsenges Phys Chem* 102:1865–1869, 1998.

63. W Srisiri, TM Sisson, DF O'Brien, KM McGrath, Y Han, SM Gruner. *J Am Chem Soc* 119:4866–4873, 1997.

64. YS Lee, JZ Yang, TM Sisson, DA Frankel, JT Gleeson, E Aksay, SL Keller, SM Gruner, DF O'Brien. *J Am Chem Soc* 117:5573–5578, 1995.

65. W Srisiri, A Benedicto, DF O'Brien, TP Trouard, G Orädd, S Persson, G Lindblom. *Langmuir* 14:1921–1926, 1998.

66. DH Gray, DL Gin. *Chem. Mater* 10:1827–1832,1998.

67. KM McGrath, CJ Drummond. *Colloid Polym Sci* 274:612–621, 1996.

68. P Rowinski, R Bilewicz. *Mater Sci Eng C* 18:177–183, 2001.

69. SW Jeong, DF O'Brien, G Oradd, G Lindblom. *Langmuir* 18:1073–1076, 2002.

70. F Caboi, J Borné, T Nylander, A Khan, A Svendsen, S Patkar. *Colloids Surf B: Biointerfaces* 26:159–171, 2002.

71. R Wallin, T Arnebrant. *J Colloid Interface Sci* 164:16–20, 1994.

15

Applications of Lipidic Cubic Phases in Structural Biology

EHUD M. LANDAU

CONTENTS

425

15.1 INTRODUCTION

The last decades have witnessed remarkable progress in the methods for elucidating structures of biological macromolecules, including x-ray crystallography, NMR spectroscopy, and electron microscopy. Powerful tools such as third-generation synchrotron facilities and high field magnets, and novel experimental methodologies in combination with software development and advanced computing systems, have provided biologists the necessary technological foundation to elucidate novel structures of ever-larger systems at very high resolution. This development is manifested in the exponential growth of the number of structures deposited in the protein data bank (http://www.rcsb.org/pdb), currently exceeding 21,000, which is about ten times the number that was available one decade ago. With the current rate of thousands of new structures per year, and in light of the promise of an even higher rate of structure solution due to "structural genomics" initiatives, it is expected that this exponential growth will continue in the years to come, reinforcing the important role of structural biology in modern life science.

Of the various classes of biological macromolecules, membrane proteins represent the only exception to this trend, as the number of structures of this class of proteins is still less than 50. Given that approximately 25% of the proteins encoded by the genome are predicted to be membrane proteins and that many of them perform vital processes in cellular systems, alleviating this impasse represents a major challenge for structural biologists. The key reasons for the paucity of membrane protein structures are twofold: First, overexpression and purification of membrane proteins is a very challenging task. Second, crystallization of membrane proteins has proved much more difficult than that of soluble proteins. This chapter describes the relatively new applications of lipidic cubic phase materials in structural biology, with special emphasis on membrane proteins.

15.2 BICONTINUOUS LIPIDIC CUBIC PHASES: COMPATIBILITY WITH MEMBRANE PROTEINS

Conventionally, biochemical and biophysical investigations of membrane proteins were performed in detergent solution. At and above the critical micellar concentration (CMC), equilibrium is established between detergent molecules that are solubilized as monomers and those that aggregate to form micelles (1,2). Upon solubilization with detergent, membrane proteins form protein–detergent comicelles (also called "mixed micelles" or "protein–detergent complex"). These are useful in stabilizing the membrane protein in aqueous solution by virtue of modifying the hydrophobic transmembrane sectors of the protein, rendering them hydrophilic. Biochemical and biophysical studies of comicelles yielded very important information on membrane proteins, and indeed the first successful crystals of membrane proteins were obtained from such media (3–5). These were followed by the seminal first high-resolution structure of a membrane protein (6). Despite these notable advances, detergents pose a number of problems: in some cases purity is a major issue, in a number of cases membrane proteins were found to be unstable in detergent solution, and in general their colloidal properties are difficult to control.

Searching for alternative means for the solubilization, stabilization, and crystallization of membrane proteins, we proposed the use of bicontinuous cubic phases composed of lipids and water (7). Macroscopically, these are highly viscous, transparent, and stable lipid-rich materials, typically composed of 50 to 80% (w/w) lipid, and 50 to 20% (w/w) water. Their molecular architecture consists of a space-filling, curved lipid bilayer that is interpenetrated by structured aqueous channels (Figure 15.1). The membrane of a bicontinuous

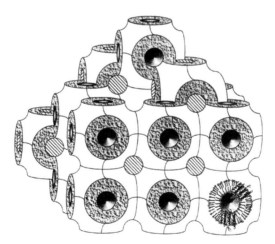

Figure 15.1 Schematic representation of a bicontinuous lipidic cubic phase. The space-filling curved bilayer and the two adjacent aqueous channel systems are shown. Two molecules of bR have been drawn incorporated into the membrane.

lipidic cubic phase comprises a three-dimensional periodic array of saddles, in which each leaflet of the membrane is curved towards the adjacent aqueous compartments. Because of the connectivity in both the lipidic and aqueous compartments, long-range diffusion is possible in both. Indeed, the measured diffusion coefficients of lipids in these materials are of the same order of magnitude as those of lipids in planar bilayers (8,9) and in biological membranes, and water diffuses in the aqueous compartments with a diffusion coefficient of bulk water.

Hydrated monoacylglycerols form the best-studied class of bicontinuous lipidic cubic phases. Specifically, hydrated monoolein (1-monooleoyl-*rac*-glycerol, MO) forms two cubic phases (Ia3d and Pn3m) that are adjacent in the temperature–composition phase diagram (Figure 15.2). As can be expected, the geometric parameters,

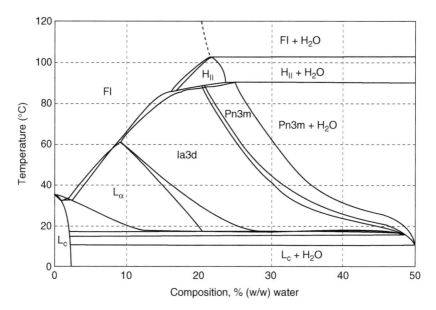

Figure 15.2 Lyotropic and thermotropic phase diagram of the binary MO-water system (from G. Rummel et al. *J Struct Biol* 121:82–91, 1998, with permission). Planar liquid-crystalline (Lc) and lamellar (Lα) phases form at low temperature and hydration. Curved lipidic cubic phases (Ia3d and Pn3m) form at increased hydration levels and at higher temperatures. Unlike detergent micelles, which can be monomerized in solution, lipidic cubic phases are stable in excess water. FI: fluid isotropic phase; H$_{II}$: inverted hexagonal phase.

i.e., bilayer curvature, aqueous channel radius, and the resulting cubic unit cell size, depend strongly on hydration, temperature, and pressure. This plasticity is facilitated by the extraordinary flexibility of the curved bilayers and the "spongelike" nature of the aqueous compartments: the unit cell size increases as a function of hydration and is inversely proportional to the temperature. For the pure monoolein–water system the cubic unit cell size ranges from approximately 97 to 185 Å for the cubic-Ia3d, and approximately 69 to 113 Å for the adjacent cubic-Pn3m phase. For the fully hydrated cubic-Pn3m phase at 25°C, the thickness of membrane's hydrophobic core was found to be ca. 34 Å, and the water channel diameter approximately 46 Å (10).

Based on these remarkable properties, it is possible to solubilize membrane as well as soluble proteins in these materials and perform direct spectroscopic investigations thereon (11–13). Moreover, given the coexistence of hydrophobic and hydrophilic domains, it is possible to form complexes of soluble and membrane proteins in these materials (14). Finally, we suggested that membrane proteins would have a higher propensity to form ordered, three-dimensional crystals in a lipid bilayer than in a nonbilayer environment if they can retain their native properties and diffuse in three dimensions. Because diffusion of hydrophilic and hydrophobic solutes was shown to take place in lipidic cubic phases, we hypothesized that labile membrane proteins might be stabilized in such matrices, and might diffuse along the bilayer to form nuclei and subsequently grow to mature crystals (7). The following sections will describe applications of lipidic cubic phases in spectroscopy, crystallization, x-ray crystallography, and structural dynamics of soluble and membrane proteins.

15.3 SPECTROSCOPY ON PROTEINS SOLUBILIZED IN LIPIDIC CUBIC PHASES

Because lipidic cubic phases are optically isotropic and transparent, they constitute ideal matrices for direct spectroscopic analysis of incorporated macromolecules. The first evidence that an enzyme is functional when immobilized in a lipidic cubic phase was provided in 1991 (11). In this work α-chymotrypsin was immobilized in two cubic phases: one composed of 1-palmitoyl-2-hydroxy-*sn*-glycero-3-phosphocholine (PLPC) and water, the other composed of monoolein and water. Using absorption spectroscopy, it was shown that α-chymotrypsin can be incorporated in the cubic phase, that its spectroscopic properties are essentially the same as in aqueous solution, and, significantly, that the Lambert–Beer law is obeyed in the cubic phase.

Comparison of circular dichrosim (CD) spectra on the immobilized enzyme with CD spectra in solution demonstrated that the protein's conformation is virtually unmodified by the lipidic entrapment. Using a water-soluble chromophoric substrate, the time course of the enzyme-catalyzed hydrolysis was investigated. Immobilized α-chymotrypsin in the gel was shown to retain enzymatic activity, albeit with slower kinetics than in aqueous solution. It was also shown that addition of α-chymotrypsin to the cubic phase does not alter the rheological properties of the latter, suggesting that the gel's global structure is unchanged upon incorporation of the macromolecule.

As mentioned above, lipidic cubic phases are composed of structured aqueous and lipidic compartments. The latter were used to host membrane proteins in order to conduct spectroscopic investigation thereon (12). Bacteriorhodopsin (bR), a seven-transmembrane retinal-binding protein that functions as a light-induced proton pump in the membrane of halophilic archaebacteria, and melittin, the main component of the honey bee venom that binds spontaneously to membranes, were incorporated in a PLPC-based cubic phase. Absorption and CD spectroscopy were used to establish conditions under which membrane proteins are stable at a range of temperatures. In the case of bR, addition of salt stabilized the protein, and the content of α-helicity, computed from CD spectroscopy, was 72 +/– 10%, in good agreement with the values from crystal structures. Melittin in aqueous solution is unfolded at micromolar concentrations, low ionic strength, and neutral pH. It undergoes a transition to a tetrameric α-helical conformation at higher melittin concentration, acidic or basic pH, and high ionic strength. Upon interaction with a membrane, melittin adopts an α-helical conformation as well. In the PLPC cubic phase, melittin is α-helical, with a content of α-helicity of 98 +/–5%, in accord with the crystallographic values.

Immobilization of the photosynthetic reaction center from *Chloroflexus aurantiacus* in a PLPC cubic phase also resulted in a native-like protein, as indicated by visible and CD spectroscopy (13). The reaction center was found to be much more stable in the lipidic environment than in aqueous solution. The photochemical activity of the reaction center was retained, as reversible photooxidation of the primary electron donor could be detected upon continuous illumination. Interestingly, entrapment of the reaction center in the lipidic gel did not affect the kinetics of charge recombination between the primary donor and the primary quinone acceptor. Finally, reconstitution of reaction centers that are devoid of the secondary quinone acceptor together with 1,4-naphthoquinone was accomplished in the cubic phase, but the kinetics of charge recombination in the lipidic cubic phase was dramatically slower than the corresponding reaction in the unreconstituted protein.

To directly probe the interaction of an agonist-activated membrane receptor with intracellular G proteins, bovine rhodopsin, the only G protein–coupled receptor whose x-ray structure has been elucidated to date (15), was reconstituted into the lipidic compartment of a cubic phase (14). Applying ultraviolet (UV)-visible absorption and attenuated total reflectance (ATR)-FTIR difference spectroscopy, it was demonstrated that reconstituted rhodopsin can be activated by light to form metarhodopsin intermediates. Moreover its cognate G protein, transducin, was solubilized in the complementary aqueous channels of the cubic phase and was shown by nucleotide-dependent increase in intrinsic transducin fluorescence to diffuse and interact with light-activated rhodopsin. These experiments show the applicability of lipidic cubic phases as matrices to probe the function and structure of membrane-soluble protein-signaling complexes. Taken together, the spectroscopic evidence on five different proteins—α-chymotrypsin, melittin, bR, a photosynthetic reaction center, and the rhodopsin-transducin complex—clearly establishes the lipidic cubic phase as a structured matrix capable of hosting soluble and membrane proteins in their native conformation and active state and sets the stage for the crystallographic work that will be described in the following two sections.

15.4 CRYSTALLIZATION OF SOLUBLE AND MEMBRANE PROTEINS IN LIPIDIC CUBIC PHASES

Crystallization from solution is a thermodynamically well-understood phase transition. Starting with a homogeneous and monodisperse solution at a relatively high concentration, gradual change in one or more of the parameters may result in formation of nuclei, which, upon growth, may form crystals. Soluble macromolecules possess hydrophilic surfaces. Their solubility properties are well established, and they usually form crystals relatively readily. The situation is more complex with transmembrane proteins, whose surfaces display hydrophilic as well as hydrophobic sectors. Once the protein is released from its native membrane, it is inherently unstable in aqueous solution and tends to aggregate. Conditions must therefore be established prior to crystallization such that these opposing surface characteristics are satisfied. The concept of stabilizing and crystallizing membrane proteins in lipidic cubic phases aims at achieving such conditions.

Detailed and comprehensive phase information of the complex systems (lipid, water, protein, detergent, salt, and possibly endogenous lipids) that are encountered in cubic phase crystallization experiments is lacking. Therefore, when designing crystallization screens one has to resort to the available, albeit simpler, binary phase diagrams for practical guidelines on the lipid and water composition. In a typical experiment, crystallization is brought about by the addition of appropriate precipitating agents to the protein-containing lipidic cubic phase. Microscopic inspection of crystallization setups is easily accomplished due to the transparency of the lipidic cubic phase. Since its introduction (7) and the crystallization of the light-induced proton pump bacteriorhodopsin (Figure 15.3), the concept of lipidic cubic-phase–mediated crystallization of membrane proteins yielded crystals of five additional membrane proteins with varying size, molecular weight, subunit composition, origin, and function (16–20). These systems are: photosynthetic reaction centers from *Rhodopseudomonas viridis* (RCvir) and *Rhodobacter sphaeroides* (RCsph), light harvesting complex 2 from *Rhodopseudomonas acidophila* (LH2), halorhodopsin from *Halobacterium salinarum* (hR), and

Figure 15.3 Photograph of a glass vial (inner diameter 2 mm) containing hundreds of bR microcrystals within a MO-based lipidic cubic phase. The crystals appear as thin hexagonal plates, with typical dimensions of ca. 35 to 70 × 5 to 10 μm. Crystals grow in the bulk cubic phase, which retains its transparency and solidlike texture throughout crystal growth.

sensory rhodopsin II from *Natronobacterium pharaonis* (pSRII). Interestingly, these systems were crystallized from a 60% (w/w) lipid-containing cubic phase. Very recently, well diffracting crystals of the first complex composed of two membrane proteins, the photoreceptor SRII complexed to its cognate transducer HtrII, were generated using the lipidic cubic phase crystallization (21). Significantly, these crystals diffract to higher resolution than the best crystals of SRII alone. In addition, crystallization of soluble proteins in the complementary aqueous compartments of the cubic phase has been reported (22). These examples demonstrate the usefulness and broad range of the lipidic cubic phase crystallization approach.

All the systems discussed above require detergent solubilization of the membrane protein prior to reconstitution in the lipidic cubic phase. Swift as this may be, for some proteins detergent treatment may be harmful. We have recently demonstrated the direct crystallization of bR from a lipidic cubic phase using native purple membrane as the starting material, i.e., without exposure of the protein to detergent (23). This experiment broadens the scope of lipidic cubic phase applications, as it demonstrates the usefulness of curved bilayers in lipidic cubic phases to facilitate the transformation of two-dimensional to three-dimensional crystals.

Lipidic cubic phases are soft gel-like materials. Such solids can be manipulated and shaped at will, and adopt the macroscopic form and size of the container in which they are formed. In contrast to single macromolecules that can diffuse in the cubic phase lattices, crystals are immobilized in the gel materials. This situation lends itself well for applications such as spectroscopy and crystallography. For x-ray diffraction experiments or for trapping intermediates and conducting microspectrphotometrical analyses (see next section), single crystals have to be recovered from the cubic phase, as they need to be mounted inside a glass capillary or onto nylon cryo-loops. Because the cubic phases are soft solids, the simplest procedure for harvesting crystals is by mechanical manipulation. This may introduce shear forces on the crystals and may result in damage and decrease in the diffraction quality. As an alternative, we have established a procedure to enzymatically hydrolyze the lipidic cubic phase (24), which yields single crystals of excellent quality devoid of cubic phase. In this method, monoacylglycerol lipids that form the cubic phase are treated with a lipase, which hydrolyzes the ester bond, yielding a mixture of fatty acid and glycerol. The resulting liquid–liquid biphasic system harbors individual crystals that can be easily harvested, analogous to crystals grown from isotropic solution.

In the case of bR, these crystals were completely devoid of MO, as evidenced by mass spectroscopy (25).

In addition to the conceptual advantages of the lipidic cubic phase approach, such as stabilizing membrane proteins and providing well-defined pathways for diffusion and sites for nucleation, there is an additional practical benefit in minimizing the laborious search for an appropriate detergent. Nonetheless, the exact conditions necessary for a given membrane protein to crystallize have to be established. As in conventional crystallization, the following parameters may have to be tested: concentrations of protein, lipid, detergent, salt, organic solvent, and polymer, as well as temperature, pH, and pressure. Because the available database is still limited, a generally predictive framework for membrane protein crystallization from lipidic cubic phases is outstanding at the present time.

In the case of the best-studied system, bR, a hypothetical mechanism for the "*in cubo*" crystallization has been proposed (26). Using polarization microscopy and low-angle x-ray diffraction, the changes that occur upon crystallization of bR in MO cubic phase were monitored. The model proposed entails that membrane proteins reside in curved lipid bilayers, diffuse into patches of lower curvature, and are subsequently incorporated into lattices, which associate to form highly ordered three-dimensional crystals. Nucleation and growth of bR crystals in lipidic cubic phases is proposed to take place in four or five consecutive steps (Figure 15.4). Upon reconstitution of detergent solubilized bR into the cubic phase bilayer, the number of degrees of freedom is reduced by three (two rotational, one translational), resulting in a system with a total of three degrees of freedom. This compensates in part for the penalty in crystallization entropy. bR molecules can reside in the cubic phase bilayer in one of two orientations (right side up or down), with the transmembrane region located within the hydrophobic core of the membrane, and the hydrophilic loop regions in the aqueous channels. The curvature of the cubic phase bilayers is incompatible with the planar (type I) packing of bR molecules in the crystallographic *ab* plane of the crystals (or, for that matter, with the planar packing of all membrane proteins that have been crystallized in lipidic cubic phase so far), resulting in strain. Local curvature defects were proposed to alleviate this strain, resulting in aggregation of bR and purple membrane lipids above a certain threshold. Partitioning of bR molecules (or prearranged patches) into adjacent local planar domains is proposed to yield nuclei, and crystal growth is facilitated by lateral merging of protein and of small purple membrane–like units. Because the lateral

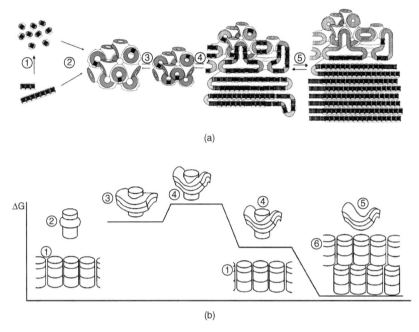

(a)

(b)

Figure 15.4 (a) Detergent-solubilized bR monomer or patches of purple membrane are incorporated into the curved bilayer of the Pn3m cubic phase. Addition of the precipitant salt removes water from the aqueous channels, which results in increased membrane curvature. Lateral diffusion of bR molecules to planar domains adjacent to the cubic phase relieves the strain and facilitates nucleation and crystal growth. Mature crystals coexist with a bR-depleted and highly curved cubic phase. (b) Schematic representation of the changes that occur during crystallization and the respective free energy changes (ΔG). (From P Nollert, H Qiu, M Caffrey, JP Rosenbusch, EM Landau. *FEBS Lett* 504:179–186, 2001. With permission.)

diffusion pathways are defined by the structure of the curved lipid bilayer, the ratio of productive to nonproductive protein encounters necessary for crystal nucleation and growth is larger than the analogous ratio in isotropic detergent solution. Finally, the existence of cubic-to-lamellar phase transition proposed in this mechanism is in accord with the microscopic observation of birefringent lipidic material adjacent to the crystal faces of bR.

Recently, a theoretical analysis of crystallization in lipidic cubic phase has been developed (27). A kinetic barrier-crossing model is presented, which permits determination of the free energy barrier to crystallization from the time-dependent growth of protein clusters. Such an approach, in combination with detailed phase analyses of the complex colloidal systems that may be encountered in the crystallization of membrane proteins from lipidic phases (28–30) will undoubtedly enhance our understanding of these complex phenomena and may result in a predictive framework for the crystallization of membrane proteins.

15.5 HIGH-RESOLUTION STRUCTURES OF MEMBRANE PROTEINS FROM LIPIDIC CUBIC PHASE–GROWN CRYSTALS

Advances in the crystallization of membrane proteins using lipidic cubic phases, better understanding of the underlying complex colloidal chemistry and phase behavior, and conceptual as well as methodological developments in structural dynamics have resulted in a wealth of information on a number of membrane proteins at unprecedented spatial and temporal resolution. The best-characterized system is bacteriorhodopsin, for which high-resolution structures of most of the intermediate states have been solved (31–38), resulting in a coherent sequence of discrete conformational changes that provide a detailed structural picture of this light-induced proton pump in action. The high resolution structure of halorhodopsin (hR), an inwardly directed chloride pump, was also resolved from crystals grown in the lipidic cubic phase (17). Sensory rhodopsin II, a photoreceptor that relays light signals to its cognate transducer protein HtrII, which in turn initiates a phosphorylation cascade that regulates the cell's flagellar motors in order to control phototaxis, has yielded high-resolution structures of the ground (18,19) and first intermediate states (39), as well as the ground state complexed with HtrII (21). Additional structures of SRII's intermediate states in the free and complexed form will undoubtedly facilitate the elucidation of this receptor's mode of action in molecular detail. Very recently, the high-resolution structure of the first non-retinal protein, the photosynthetic reaction center from *Rhodobacter sphaeroides,* grown from a lipidic cubic phase has been determined (20).

15.5.1 Bacteriorhodopsin

Bacteriorhodopsin (bR) is the simplest known light-driven proton pump. This integral membrane protein belongs to the family of archaeal rhodopsins, all of which display a conserved architecture of seven transmembrane α-helices and contain a buried retinal chromophore that is bound to a conserved lysine residue in helix G by a protonated Schiff base linkage. Coupled to its function as an outwardly directed proton pump, bR undergoes a series of light-induced and thermally propagated reactions, called the photocycle (Figure 15.5), during which the molecular environment of the chromophore is altered. A plethora of biochemical, genetic, and spectroscopic data on bR's photocycle has been accumulated over the past three decades, which can now be complemented and contrasted by high-resolution crystallographic data. Taken together, a very detailed picture of this protein's transmembrane proton pumping against an electrochemical gradient has emerged.

The ground state structure of bR, elucidated from cubic phase–grown crystals in the past few years at increasing resolution (25,40) reveals a hydrophilic proton translocation channel in the interior of the membrane, which is orientated perpendicularly to it (Figure 15.6). Starting from the primary proton donor, a protonated Schiff base that links the retinal to Lys216, whose N–H dipole is

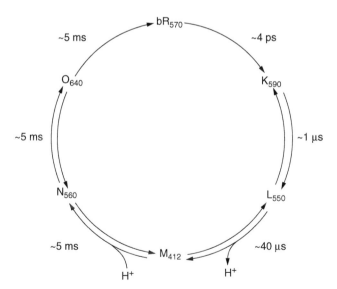

Figure 15.5 Schematic representation of the photocycle of bR. Absorption maxima of the intermediates are given as subscripts. Lifetimes are for bR at room temperature.

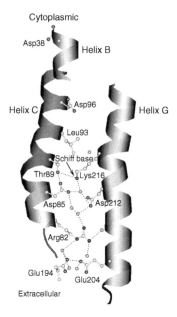

Figure 15.6 Structure of the hydrogen-bonded proton translocation channel in the extracellular half of bR. Starting with the protonated Schiff base and extending to the extracellular surface, this
channel is delineated between helices C and G.

located at the center of the membrane and points towards the
extracellular side, the positions of a number of key charged residues
and water molecules have been determined in the ground state structure. The pKa of the protonated Schiff base is 13.5 (41), whereas that
of Asp85, the primary proton acceptor, is 2.2 (42), ensuring that the
proton resides on the Schiff base rather than Asp85 under physiological conditions. The structure of the Schiff base and its complex
counter ion (Figure 15.7) constitutes a central element in bR's function as a vectorial proton pump that can be switched extremely rapidly by light. A structured hydrogen bonded network that extends
from the Schiff base to the extracellular surface stabilizes this complex counter ion. A key water molecule, Wat402, is located between
and is H-bonded to the Schiff base and the two adjacent aspartates,
Asp85 and Asp212. Together with two additional water molecules
(Wat400 and Wat401), this water molecule stabilizes the unusually
high pKa of the Schiff base and forms a pentagon that links the two
negatively charged aspartates to the positively charged Arg82 further

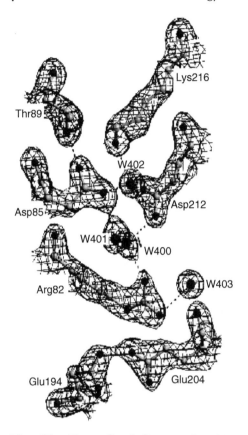

Figure 15.7   The $2F_{obs}$-F_{calc} refined electron density map (contoured at 1.2 σ) shows the structure of the complex counter ion of bR on the extracellular side of the Schiff base. Asp85, Asp212, Arg82, and the three water molecules Wat400, Wat401, and Wat402 stabilize the ground state structure. The external proton release group is formed by Glu194, Glu204, and water molecules. Data from entries 1qhj and r1qhjsf of the PDB. (From R Neutze, E Pebay-Peyroula, K Edman, A Royant, J Navarro, EM Landau. *Biochim Biophys Acta* 1565:144–167, 2002. With permission.)

down the proton translocation pathway towards the extracellular side. The guanidinium moiety of Arg82 is orientated towards the active site in the ground state (Figure 15.6 and Figure 15.7), and as will be discussed in the following, will switch its orientation in preparation for the primary proton transfer event that occurs in the

L-to-M transition. The complex counter ion is completed by H-bonds between Arg82 and the proton release group at the extracellular surface, formed by Glu194, Glu204, and water molecules (43,44). Significantly, an analogous hydrogen-bonded network in the cytoplasmic side does not exist in the ground state, clearly outlining the vectoriality of bR's proton pumping.

The packing arrangement of bR in three-dimensional crystals grown from the lipidic cubic phase (Figure 15.3) consists of homotrimers that are organized in a purple membrane–like arrangement in the *ab* plane and are related by a twofold screw axis along the *c* direction (45). The polar arrangement of bR trimers perpendicular to the plane of the membrane may be favorable for vectorial proton transfer in three-dimensional crystals. Before attempting the structural determination of intermediate states in three-dimensional crystals, conditions must be found such that the protein is functional in the immobilized crystalline state. To this end, an ensemble of bR microcrystals embedded in the lipidic cubic phase was investigated by time-resolved FTIR and resonance Raman spectroscopy. Comparison with analogous spectroscopic data from purple membrane established that the key events in the photocycle — retinal isomerization, conformational changes in the protein backbone, and proton translocation — are almost identical to those in the native membrane, although the kinetics differ somewhat (46). The fact that the ensemble of bR molecules is active when immobilized in three-dimensional crystals set the stage for attempting to trap a sufficient population of intermediate state within crystals.

Because bR's photocycle (Figure 15.5) is light induced and thermally propagated, careful control of both illumination and temperature is necessary when performing intermediate trapping experiments. In particular, two temperature-dependent phase transitions at 150 and 240 K have been observed for bR (47,48). These represent threshold temperatures below which certain motions cannot take place. Trapping the low temperature K-intermediate (K_{LT}) was found to be most efficient by illuminating bR with green light at or below 110 K (49–51). For the L_{LT}-intermediate, illumination with red light at 170 K (52–54), or alternatively, illumination with green light at lower temperature, thereby generating K_{LT}, followed by heating to 170 K (55), were found to be the optimal conditions. Conditions for trapping the M intermediate consist of freezing the crystals in liquid nitrogen, raising the temperature by blocking the cold nitrogen stream for a short period of time, illuminating with either yellow (35,36) or green (33) light, and refreezing.

Correlating spectroscopy of proteins in crystals with x-ray crystallography has not been uncontroversial (56–58) because the specific packing of molecules in three-dimensional crystals, as well as interactions with lipid and detergent molecules, may give rise to differences in the spectra of proteins as compared to spectra from samples in solution or in suspension (59). Compared with purple membrane suspensions, bR in lipidic cubic phase–grown crystals exhibits very similar but not identical FTIR spectra upon photoexcitation (46). Light scattering, orientation of the crystals, and crystal size are factors that have to be controlled in such spectroscopic analyses. In the case of membrane protein crystals grown in lipidic cubic phase, which often tend to form very small microcrystals, these factors are even more pronounced (57). In addition, crystal handling was also found to be an important factor: the combination of absorption, fluorescence, and excitation spectra on lipidic cubic phase–grown bR crystals at room temperature revealed the presence of two protein species in addition to the major bR form of the purple membrane. It was proposed that partial dehydration during crystallization and crystal handling may have caused this heterogeneity (60).

Like all membrane proteins, bR is orientated and functions vectorially within the membrane. Thus proton transfer from the Schiff base to Asp85 further down to the extracellular side, as well as from the external release group to the extracellular medium, cannot be reversed under physiological conditions. Moreover, following proton release, a major switch in accessibility of the Schiff base must occur such that it can only be reprotonated from Asp96 on the cytoplasmic side of the protein. Structural dynamics, i.e., elucidating structures of intermediate states in the photocycle within three-dimensional crystals, should reveal the atomic details of the mechanism of light-induced transmembrane vectorial proton transport mediated by bR.

15.5.1.1 K Intermediate

Because the buildup of the K intermediate under physiological conditions occurs within a few psec after photoisomerization of the retinal (Figure 15.5), one would not expect to observe large structural changes far away from the active site. Using a combination of laser illumination with rapid cooling techniques, a population of 35% of bR was converted to the K intermediate in three-dimensional crystals and its structure elucidated (31). Figure 15.8, left panel, shows the difference Fourier map for K_{LT} overlaid on the ground

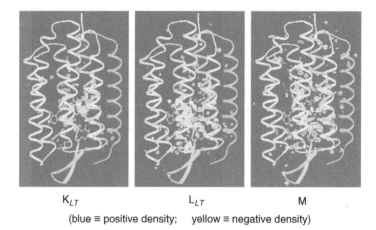

K_{LT} L_{LT} M

(blue ≡ positive density;   yellow ≡ negative density)

Figure 15.8 Difference Fourier maps for the low temperature K, L, and M intermediates of bR, overlayed on the ground-state ribbon structure.

state structure, from which it is evident that most of the structural changes occur near the active site. Figure 15.9 illustrates the detailed changes in K_{LT}: isomerization about the $C_{13}=C_{14}$ double bond is clearly seen as a negative electron density, and the paired positive and negative electron density peaks near the side chain and carbonyl

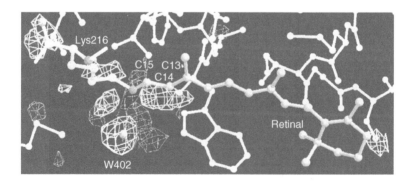

Figure 15.9 Difference Fourier maps for K_{LT} overlayed on the ground-state structure of bR, showing the changes that occur immediately following isomerization of the retinal. (From K Edman, P Nollert, A Royant, H Belrhali, E Pebay-Peyroula, J Hajdu, R Neutze, EM Landau. *Nature* 401:822–826, 1999. With permission.)

oxygen of Lys216 illustrate how its side chain and main chain move in response to retinal isomerization. Interestingly, retinal's polyene chain and β-ionone ring undergo no movement, as they are affixed to their ground state position by the tight binding pocket. This strain in the retinal may play a role in energy storage in K_{LT}, which would be released at a later point of the photocycle. A key water molecule (Wat402), which in the ground state forms hydrogen bonds to the Schiff base, Asp85, and Asp212 (Figure 15.6) becomes disordered in the K state (Figure 15.9). An early movement of Asp85 toward the position originally occupied by the Schiff base is also apparent as paired negative and positive difference electron density peaks at this residue.

15.5.1.2 L Intermediate

Compared with the structural changes observed in K_{LT} (Figure 15.9), those of L_{LT} (32,57) extend further toward the extracellular side (Figure 15.8, middle panel). The most pronounced positive and negative electron densities are observed along the backbone of helix C from Ala81 to Phe88 (Figure 15.10). This novel feature was interpreted

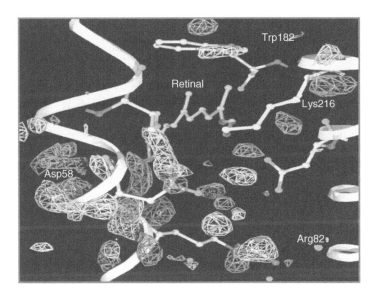

Figure 15.10 Difference Fourier maps for L_{LT} overlayed on the ground state structure of bR. (From A Royant, K Edman, T Ursby, E Pebay-Peyroula, EM Landau, R Neutze. *Nature* 406:645–648, 2000. With permission.)

as a transient flux toward the proton translocating channel, which follows the earlier movement of Asp85 towards the Schiff base observed in K_{LT}. Other features of the structure of L_{LT} are the dislocation of Wat400 and Wat401, reordering of one water molecule in their vicinity (visible as a positive electron density peak), and a flip in the orientation of the guanidinium group of Arg82 towards the extracellular medium, seen as paired negative and positive electron density peaks (Figure 15.10). These extended changes, and disruption of several H-bonds on the extracellular side of the protein at 170 K (32), which were not observed at 110 K (31), are related to the increased thermal energy of the system and coincide with the phase transition at 150 K (47).

15.5.1.3 M Intermediate

A number of x-ray structures of the M intermediate of bR have been presented (33–36,38). On the extracellular side, these structures show rearrangements of water molecules and a change in the orientation of Arg82 as seen in L_{LT} (32). The local flex in helix C, observed for L_{LT}, is not apparent in these structures, indicating that it may have relaxed back to its original position following proton transfer and abolition of the electrostatic attraction of the two groups. The cytoplasmic side of the proton translocation channel acts as a hydrophobic plug in the first stages of the photocycle, preventing the back diffusion of protons through the membrane. In the M intermediate, the structures (33–36,38) reveal a local unwinding of helix G in the immediate vicinity of a π-bulge around Lys216, which provides sites for new water molecules to be positioned. These now define a pathway for the reprotonation reaction from Asp96 almost as far as the Schiff base nitrogen. Interestingly, lower resolution electron crystallographic studies on two-dimensional crystals of bR in projection have demonstrated that in the late M-state, large structural rearrangements of the cytoplasmic portions of helices E, F, and G occur (61,62), but these have only been partially detected in the three-dimensional crystals (33).

Very recently, new structures of an M intermediate of wild-type bR and of an N intermediate of the V49A mutant were presented (38). As expected, the Schiff base in the M structure is unprotonated, while it is reprotonated in the N structure. Because bR undergoes a spectroscopically silent transition in M, this structure, denoted as M_1, exhibits a cluster of three hydrogen-bonded water molecules around Asp96 on the cytoplasmic side, which were not visible in the ground state structure. An earlier structure of M_2 (63) shows that one of these water molecules intercalates between Asp96 and Thr46. The transition from M_2 to N shifts this cluster towards the Schiff base, seen as a hydrogen-bonded chain of four water molecules that connects Asp96 to the Schiff base (38).

In summary, well-diffracting lipidic cubic phase–grown crystals of bR were instrumental in obtaining a series of high-resolution structures of the ground state and most intermediate states of its photocycle. The mechanism of vectorial proton translocation, depicted schematically in Figure 15.11, is now elucidated at unprecedented temporal and spatial resolution. In the ground state, the extracellular half of bR is stabilized by a hydrogen-bonded network of polar residues and water molecules, which connects the protonated Schiff base with the extracellular surface. Light-induced retinal isomerization about the $C_{13}=C_{14}$ bond reverses the orientation of the N–H dipole of the Schiff base and dislocates a strategic water molecule (W402), located in the ground state between the primary proton donor and acceptor. This is followed by gradual disruption of the hydrogen-bonded network on the extracellular side, reorientation of the guanidinium group of Arg82 towards the extracellular medium, and a transient local flex of helix C. As a result, key residues change their pKa values, and proton transfer from the protonated Schiff base to Asp85 is facilitated. Following proton transfer and cancellation of the electrostatic attraction, strain on the retinal and helix C is released. The reorientation of Arg82 aids proton release to the extracellular surface. Later events on the cytoplasmic side of the protein include local unwinding of helix G around Lys216, major conformational changes of helix F that result in opening of the hydrophobic plug, and entry and ordering of water molecules that define a reprotonation pathway between Asp96 and the Schiff base. The outwards movement of helix F results in reprotonation of Asp96 from the cytoplasmic side, followed by thermal reisomerization of the retinal to the all-*trans* configuration and relaxation to the ground state conformation, which is now ready for another cycle.

15.5.2   Halorhodopsin

The light-induced, inwardly directed chloride pump halorhodopsin (hR) was crystallized in space group $P6_322$ from the lipidic cubic phase (17) in a similar fashion to bR. The packing arrangement of hR exhibits a layered array of trimers, with alternate directionality of the stacked bilayers. Ten lipids reside between the trimers, and

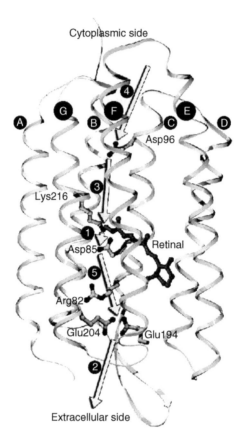

Figure 15.11 Schematic representation of the proton translocation in the photocycle of bR. The ground state structure is shown
as a ribbon representation and the strategic residues are highlighted. (1): the primary proton transfer from the Schiff base to
Asp85 is followed by a proton release to the extracellular side of the
membrane (2). The next step is reprotonation of the Schiff base from
Asp96 on the cytoplasmic side (3), followed by reprotonation of
Asp96 from the cytoplasmic medium (4). The photocycle is completed
by reprotonation of the proton release group on the extracellular side
from Asp85 (5). (From R Neutze, E Pebay-Peyroula, K Edman,
A Royant, J Navarro, EM Landau. *Biochim Biophys Acta* 1565:144–
167, 2002. With permission.)

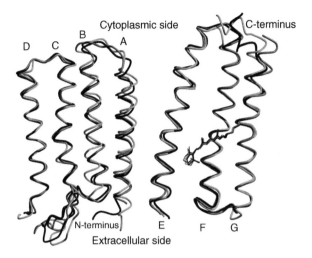

Figure 15.12 Superposition of the backbone structures of bR, hR, and SRII, illustrating the striking similarity in the structure of three proteins with different functions. The retinal is also shown.

three palmitates were found to reside in the middle of the hR trimer. Interestingly, despite differences in ion specificity and vectoriality, the backbone of hR superimposes well on those of bR and SRII (Figure 15.12). The retinal is in the all-*trans*, 15-*anti* configuration, and one chloride ion, located 3.75 Å away from the protonated Schiff base nitrogen, is ion paired thereto and resides 18 Å beneath the membrane surface. The solvation sphere of the anion is highly irregular. The structure of the ground state does not allow for a chloride-conducting pathway between the protonated Schiff base and the cytosolic surface, and therefore such a putative pathway must open up upon light activation. Finally, the charge distribution of the complex counter ion around the Schiff base is almost identical to that of bR because the chloride ion replaces one of the negatively charged oxygen atoms of Asp85, the primary proton acceptor. It is noteworthy that this single replacement switches the kinetic preference and hence the ion specificity from protons to chlorides. In the absence of structural intermediates, this structure serves as a basis for a proposed mechanistic view of this chloride pump. Kolbe et al. (17) have proposed a flipping of the N–H dipole following photoisomerization of the retinal and electrostatic dragging of the chloride across the protonated Schiff base to the cytoplasmic side of the protein.

15.5.3 Sensory Rhodopsin II

The x-ray structure of SRII from *Natronobacterium pharaonis* (pSRII) obtained from lipidic cubic phase–grown crystals was published independently by two groups, with resolution of 2.1 and 2.4 Å, respectively (18,19). As mentioned above, the backbone structure is very similar to those of bR and hR: helices C-G display the highest structural similarity, with root mean square deviation (r.m.s.d.) on the main chain atoms of 0.77Å between SRII and BR and 0.89 Å between SRII and hR (Figure 15.12). The protein forms crystallographic dimers, with the A and G helices at the interface, and like all other membrane proteins grown from cubic phases, is arranged in a layered packing. For SRII cubic phase–grown crystals, which are in space group $C222_1$, the molecules exhibit alternate directionality within the plane. The retinal Schiff base divides the protein into a hydrophobic cytoplasmic half and a hydrophilic extracellular half. As with bR and hR, the retinal is in the all-*trans* configuration, with the β-ionone ring in the 6-s-*trans* configuration. The retinal binding pocket constrains the chromophore into an almost planar conformation, whose bent is different from that in bR. The guanidinium group of Arg72 points toward the extracellular side of the membrane in the ground state structure of SRII, in contradistinction to the orientation of the corresponding Arg82 in bR's ground state structure. A chloride binding site was identified near the extracellular side of helix C. Significantly, a hydrogen bonded network of polar residues and the chloride ion exists in the extracellular channel, in which the protonated Schiff base is linked to Asp75 and Asp201 via H-bonds to Wat402. The cytoplasmic ends of helices F and G reveal a unique surface patch of charged and polar residues, which is not seen in bR or hR. Electrostatic interaction between this patch and the negatively charged cytoplasmic domain of helix 2 of the cognate transducer HtrII was suggested as a possible mechanism of signal relay (18). A different proposal for the interaction between pSRII and HtrII consists of Tyr199, an exposed polar residue on the surface of SRII in the middle of the membrane bilayer, which was suggested to interact with the adjacent helices of the transducer (19).

pSRII's absorption maximum of 497 nm is unique among archaeal rhodopsins, as bR, hR, and SRI have an absorption maximum at $\lambda > 560$ nm. Factors that affect this opsin shift include the polarity near the β-ionone ring, interactions between the Schiff base and its counter ion, and retinal bending (18,19). The structure of the retinal binding pocket of pSRII is largely conserved, as compared

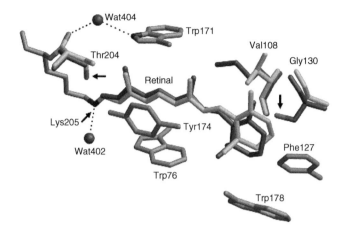

Figure 15.13 The retinal binding pockets in SRII and bR. The retinal is shown as light (SRII) and dark (br). Conserved residues are depicted in gray. Non-conserved SRII residues (Val108, Gly130, and Thr204) and the respective bR residues (Met118, Ser141, and Ala215) are depicted. (From A Royant, P Nollert, K Edman, R Neutze, EM Landau, E Pebay-Peyroula, J Navarro. *Proc Natl Acad Sci USA* 98:10,131–10,136, 2001. With permission.)

to those of the other archaeal rhodospins (Figure 15.13). Variations occur only on three residues: Val108, Gly130, and Thr204 in pSRII, which flank the retinal on both sides and modify its polarity, are Met, Ser, and Ala in the other archaeal rhodospins. The bent of the retinal, the orientation of Arg72, and its distance from the Schiff base were all proposed as factors that regulate the color tuning of SRII.

15.5.3.1 K Intermediate

Illumination of SRII crystals with blue light at 100 K resulted in buildup of the low-temperature K_{LT} intermediate. The structure of this intermediate (39) shows that, as in the case of the K intermediate of bR, all significant changes are clustered in the vicinity of the chromophore. Negative difference electron density peaks appear on two conserved waters (Wat401 and Wat402) in K_{LT} of both pSRII and bR. In SRII, the β-ionone ring, which is exposed to the protein's surface between helices E and F, undergoes a prominent movement following photoisomerization, and a significant displacement of the side chain of the conserved Lys that binds the retinal (205 in SRII) is observed. Thus the structural rearrangements of the retinal observed in K_{LT} of pSRII are considerably more extensive than those observed for bR. This structure illustrates that very small variations have been optimized to achieve either signal transduction, or proton pumping following an identical primary photoisomerization event.

15.5.4 Sensory Rhodopsin II–Transducer II Complex

Very recently, a major advance in lipidic cubic phase crystallization was reported with the co-crystallization of the complex composed of SRII with its cognate halobacterial transducer HtrII (21). This was accomplished using a shortened transducer (residues 1 to 114) comprising the two transmembrane helices (TM1 and TM2) and a small cytoplasmic fragment. The crystals diffracted to very high resolution (1.8 Å). There is one complex per asymmetric unit, and the dimer of the complex is defined by the crystallographic twofold rotation axis. In the structure, transmembrane helices F and G of SRII are in contact with the helices of the transducer. The structure of the complexed receptor is very similar to that of free SRII, with one clear difference at Tyr199, whose aromatic plane has turned by about 90°, forming a hydrogen bond to Asn74 of TM2. The main contacts between the receptor and transducer are van der Waals, with a few hydrogen bonds. Importantly, a crevice formed by helices A, G, TM2, and TM1 is accessible to the cytoplasmic surface. It is expected that structural dynamics studies on the complex would yield important information about the mechanism of light-induced signal transduction between the receptor and its transducer.

15.5.5 Photosynthetic Reaction Center

In addition to the aforementioned, relatively compact retinal proteins, and the SRII-HtrII complex, three larger systems have been successfully crystallized in lipidic cubic phases, demonstrating the generality of the approach: the photosynthetic reaction centers from *Rhodopseudomonas viridis* (RCvir) and *Rhodobacter sphaeroides* (RCsph), as well as the light-harvesting complex 2 from *Rhodopseudomonas acidophila* (16). Very recently, the structure of RCsph grown in lipidic cubic phase was elucidated to 2.35 Å resolution (20). This is the first type I crystal packing reported for a reaction center from a purple bacterium. Comparison with other structures of reaction centers elucidated from detergent-grown crystals reveals that the transmembrane regions are slightly compressed, possibly because of the lipidic scaffold. A cardiolipin molecule that is involved in crystal contacts within the membrane, as well as a Cl⁻ binding site, are observed. The cardiolipin mediates contact between the membrane-exposed surfaces of the H and M subunits, as well as interaction with a loop of the L subunit of a symmetry-related molecule, thereby forming strong crystal contacts at the membrane. In these crystals, molecules are arranged in stacked alternating antiparallel two-dimensional layers. A layered, type-I packing arrangement has been observed in all membrane proteins crystallized from lipidic cubic phases to date, but interestingly in each system specific interactions result in slight variations in the packing motif.

15.6 CONCLUSIONS

Elucidating the atomic structures of membrane proteins and correlating them with the mechanisms of function in health and disease is a central goal in modern membrane biology. This chapter deals with a relatively new application of lipidic cubic phases in structural biology, placing special emphasis on membrane proteins. Starting with a brief description of the current stumbling blocks in the field of structural biology of membrane proteins, a concise discussion of the molecular properties of lipidic cubic phases and their correlation with the properties of membrane proteins ensues. Sections on the advantages and problems of spectroscopic investigations in lipidic cubic phases and crystallization of soluble and membrane proteins follow. These set the stage for a detailed description of structures of the membrane proteins that have been crystallized in lipidic cubic phases and a number of their reaction intermediates, elaborating on the promise and difficulties in the field of structural dynamics. Membrane proteins are badly underrepresented in the protein data bank, and novel concepts and approaches are therefore dearly needed to alleviate this state of affairs.

ACKNOWLEDGMENTS

I would like to thank my colleagues S. Haacke, R. Neutze, P. Nollert, E. Pebay-Peyroula, and J.P. Rosenbusch for many helpful and insightful discussions over the past years. I am grateful to H. Belrhali, M.L. Chiu, M. Dolder, K. Edman, K. Fahmy, A. Hardmeyer, G. Katona,

E. Portuondo, A. Royant, G. Rummel, S. Schenkl, W. Suske, and T. Ursby for experimental contributions. Support from the EU-Biotech, the Swiss National Science Foundation, and the Howard Hughes Medical Institute is gratefully acknowledged.

REFERENCES

1. A Helenius, K Simons. *Biochim Biophys Acta* 415:29–79, 1975.

2. C Tanford, JA Reynolds. *Biochim Biophys Acta* 457:133–170, 1976.

3. RM Garavito, JP Rosenbusch. *J Cell Biol* 86:327–329, 1980.

4. R Henderson, D Shotton. *J Mol Biol* 139:99–102, 1980.

5. H Michel, D Oesterhelt. *Proc Natl Acad Sci USA* 77:1283–1285, 1980.

6. J Deisenhofer, O Epp, K Miki, R Huber, H Michel. *J Mol Biol* 180:385–398, 1984.

7. EM Landau, JP Rosenbusch. *Proc Natl Acad Sci USA* 93:14, 532–14,535, 1996.

8. S Cribier, A Gulik, P Fellmann, R Vargas, PF Devaux, V Luzzati. *J Mol Biol* 229:517–525, 1993.

9. G Lindblom. *Curr Opin Colloid Interf Struct* 1:287–295, 1996.

10. J Briggs, H Chung, M Caffrey. *J Phys II France* 6:723–751, 1996.

11. M Portmann, EM Landau, PL Luisi. *J Phys Chem* 95:8437–8440, 1991.

12. EM Landau, PL Luisi. *J Am Chem Soc* 115:2102–2106, 1993.

13. A Hochkoeppler, EM Landau, G Venturoli, D Zannoni, R Feick, PL Luisi. *Biotechnol Bioeng* 46:93–98, 1995.

14. J Navarro, EM Landau, K Fahmy. *Biopolymers* 67:167–177, 2002.

15. K Palczewski, T Kumasaka, T Hori, CA Behnke, H Motoshima, BA Fox, I Le Trong, DC Teller, T Okada, RE Stenkamp, M Yamamoto, M Miyano. *Science* 289:739–745, 2000.

16. ML Chiu, P Nollert, MC Loewen, H Belrhali, E Pebay-Peyroula, JP Rosenbusch, EM Landau. *Acta Cryst D* 56:781–784, 2000.

17. M Kolbe, H Besir, L-O Essen, D Oesterhelt. *Science* 288:1390–1396, 2000.

18. A Royant, P Nollert, K Edman, R Neutze, EM Landau, E Pebay-Peyroula, J Navarro. *Proc Natl Acad Sci USA* 98:10,131–10,136, 2001.

19. H Luecke, B Schobert, JK Lanyi, EN Spudich, JL Spudich. *Science* 293:1499–1503, 2001.

20. G Katona, U Andreasson, EM Landau, L-E Andreasson, R Neutze. *J Mol Biol* 331:681–692, 2003.

21. VI Gordeliy, J Labahn, R Moukhametzianov, R Efremov, J Granzin, R Schlesinger, G Buldt, T Savopol, AJ Scheidig, JP Klare, M Engelhard. *Nature* 419:484–487, 2002.

22. EM Landau, G Rummel, SW Cowan-Jacob, JP Rosenbusch. *J Phys Chem B* 101:1935–1937, 1997.

23. P Nollert, A Royant, E Pebay-Peyroula, EM Landau. *FEBS Lett* 457:205–208, 1999.

24. P Nollert, EM Landau. *Biochem Soc Trans* 26:709–713, 1998.

25. H Belrhali, P Nollert, A Royant, C Menzel, JP Rosenbusch, EM Landau, E Pebay-Peyroula. *Structure* 7:909–917, 1999.

26. P Nollert, H Qiu, M Caffrey, JP Rosenbusch, EM Landau. *FEBS Lett* 504:179–186, 2001.

27. M Grabe, J Neu, G Oster, P Nollert. *Biophys J* 84:854–868, 2003.

28. C Sennoga, A Heron, JM Seddon, RH Templer, B Hankamer. *Acta Cryst D* 59:239–246, 2003.

29. V Cherezov, J Clogston, Y Misquitta, W Abdel-Gawad, M Caffrey. *Biophys J* 83:3393–3407, 2002.

30. V Chupin, JA Killian, B de Kruijff. *Biophys J* 84:2373–2381, 2003.

31. K Edman, P Nollert, A Royant, H Belrhali, E Pebay-Peyroula, J Hajdu, R Neutze, EM Landau. *Nature* 401:822–826, 1999.

32. A Royant, K Edman, T Ursby, E Pebay-Peyroula, EM Landau, R Neutze. *Nature* 406:645–648, 2000.

33. HJ Sass, G Buldt, R Gessenich, D Hehn, D Neff, R Schlesinger, J Berendzen, P Ormos. *Nature* 406: 649–653, 2000.

34. MT Facciotti, S Rouhani, FT Burkard, FM Betancourt, KH Downing, RB Rose, G McDermott, RM Glaeser. *Biophys J* 81: 3442–3455, 2001.

35. H Luecke, B Schobert, HT Richter, J-P Cartailler, JK Lanyi. *Science* 286:255–260, 1999.

36. H Luecke, B Schobert, J-P Cartailler, H-T Richter, A Rosengarth, R Needleman, JK Lanyi. *J Mol Biol* 300:1237–1255, 2000.

37. S Rouhani, J-P Cartailler, MT Facciotti, P Walian, R Needleman, JK Lanyi, RM Glaeser, H Luecke, *J Mol Biol* 313:615–628, 2001.

38. B Schobert, LS Brown, JK Lanyi. *J Mol Biol* 330:553–570, 2003.

39. K Edman, A Royant, P Nollert, CA Maxwell, E Pebay-Peyroula, J Navarro, R Neutze, EM Landau. *Structure* 10:473–482, 2002.

40. H Luecke, B Schobert, HT Richter, J-P Cartailler, JK Lanyi. *J Mol Biol* 291:899–911, 1999.

41. M Sheves, A Albeck, N Friedman, M Ottolenghi. *Proc Natl Acad Sci USA* 83:3262–3266, 1986.

42. CH Chang, R Jonas, R Govindjee, TG Ebrey. *Photochem Photobiol* 47:261–265, 1988.

43. LS Brown, J Sasaki, H Kandori, A Maeda, R Needleman, JK Lanyi. *J Biol Chem* 270:27,122–27,126, 1995.

44. SP Balashov, ES Imasheva, TG Ebrey, N Chen, DR Menick, RK Crouch. *Biochemistry* 36:8671–8676, 1997.

45. E Pebay-Peyroula, G Rummel, JP Rosenbusch, EM Landau. *Science* 277:1676–1681, 1997.

46. J Heberle, G Buldt, E Koglin, JP Rosenbusch, EM Landau. *J Mol Biol* 281:587–592, 1998.

47. V Reat, H Patzelt, M Ferrand, C Pfister, D Oesterhelt, G Zaccai. *Proc Natl Acad Sci USA* 95:4970–4975, 1998.

48. P Ormos. *Proc Natl Acad Sci USA* 88:473–477, 1991.

49. A Xie. *Biophys J* 58:1127–1132, 1990.

50. SP Balashov, ES Imasheva, R Govindjee, TG Ebrey. *Photochem Photobiol* 54:955–961, 1991.

51. PA Bullough, R Henderson. *J Mol Biol* 286:1663–1671, 1999.

52. B Becher, F Tokunaga, TG Ebrey. *Biochemistry* 17:2293–2300, 1978.

53. A Maeda, J Sasaki, Y Yamazaki, R Needleman, JK Lanyi. *Biochemistry* 33:1713–1717, 1994.

54. MS Braiman, T Mogi, T Marti, LJ Stern, HG Khorana, KJ Rothschild. *Biochemistry* 27:8516–8520, 1998.

55. SP Balashov, FF Litvin, VA Sineshchekov. pp. 1–61 in *Physicochemical Biology Reviews,* Vol. 8, Edited by VP Skulachev (Harwood Academic Publishers GmbH, UK, 1988).

56. SP Balashov, TG Ebrey. *Photochem Photobiol* 73:453–462, 2001.

57. A Royant, K Edman, T Ursby, E Pebay-Peyroula, EM Landau, R Neutze. *Photochem Photobiol* 74:794–804, 2001.

58. R Neutze, E Pebay-Peyroula, K Edman, A Royant, J Navarro, EM Landau. *Biochim Biophys Acta* 1565:144–167, 2002.

59. A Mozzarelli, GL Rossi. *Annu Rev Biophys Biomol Struct* 25:343–365, 1996.

60. S Schenkl, E Portuondo, G Zgrablic, M Chergui, W Suske, M Dolder, EM Landau, S Haacke. *J Mol Biol* 329:711–719, 2003.

61. S Subramaniam, M Lindahl, P Bullough, AR Faruqi, J Tittor, D Oesterhelt, L Brown, J Lanyi, R Henderson. *J Mol Biol* 287:145–161, 1999.

62. S Subramaniam, R Henderson. *Nature* 406:653–657, 2000.

63. JK Lanyi, B Schobert. *J Mol Biol* 328:439–450, 2003.

16

The Controlled Release of Drugs from Cubic Phases of Glyceryl Monooleate

JAEHWI LEE AND IAN W. KELLAWAY

CONTENTS

16.1 INTRODUCTION

Controlled-release technology in pharmaceutical sciences has continuously evolved due to a need to optimize drug therapy. Growing attention has been paid to biocompatible materials in designing controlled-release drug delivery systems. Although synthetic polymeric materials have demonstrated a wide range of applicability, the use of polymeric drug delivery carriers is frequently limited by their intrinsic incompatibility and potential toxic reactions with biological tissues.

Glyceryl monooleate (GMO or monoolein) is commonly used as a food additive or an emulsifier in the cosmetic industry and is generally regarded as safe, biocompatible, and biodegradable (1–3). GMO is commercially available in the form of distilled GMO with high purity (>90 % GMO content) or a mixture of monoglycerides composed of a glycerol with an acyl chain of mainly oleic acid and other fatty acids such as palmitic, stearic, linoleic, linolenic, and arachidonic acids (2). The unique property of GMO from the pharmaceutical point of view is that it can form several types of liquid-crystalline phases such as cubic, hexagonal, and lamellar in the presence of water. Since the first appearance of GMO in 1984 as a pharmaceutical excipient for the preparation of controlled-release formulations, the pharmaceutical uses of GMO as an emulsifier, solubilizer, and absorption enhancer in drug products have steadily expanded (3). In addition, GMO has been proposed in the formulation of oral, parenteral, periodontal, and vaginal drug delivery systems as a promising candidate for matrices or barriers to control or sustain drug release. The information in this chapter is presented to elaborate on GMO and its cubic liquid-crystalline phase with respect to the controlled release of drugs, together with physicochemical properties influencing the *in vivo* performance of GMO–water systems such as muco/bioadhesive properties.

16.2 LIQUID-CRYSTALLINE PHASES OF GMO

16.2.1 Formation and Phase Behavior

GMO is a polar lipid and thus an amphiphilic molecule, with a waxy texture at room temperature and a characteristic odor. It swells to form several types of liquid-crystalline phases; a temperature-dependent phase diagram of the GMO–water system is described in several studies (4–6). When a small amount of water (5% w/w) is added to GMO at 37°C, it causes the formation of reversed micelles (L_2) with very low viscosity. The introduction of additional water leads to phase changes to a lamellar phase (L, ~ 8 to 15 % w/w water content at 37°C) and cubic phase (Q, ~ 20 to 35% w/w water content at 37°C). The reversed hexagonal phase (H_{II}) is observed only at temperatures exceeding 90°C.

The structure of the cubic phase is called bicontinuous because it consists of a curved three-dimensional network of lipid bilayers, separating two congruent water channel networks (7). Two different cubic phases have been identified in the binary system of GMO and water. The cubic phase with lower water content (20 to 34% w/w at 25°C) produces a body-centered lattice (8), while the cubic phase formed with higher water content (33 to 40% w/w at 25°C) causes the formation of a primitive lattice (9). These structures were described as infinite periodic minimal surfaces (IPMS) with space groups of *Ia3d* and *Pn3m* and for the body-centered lattice and space group of *Im3m* for the primitive lattice (10). These space groups correspond to the three fundamental cubic IPMS, the gyroid structure (G), diamond structure (D), and Schwarz's octahedral periodic surface (P), respectively. After reaching an equilibrium with respect to water content, the cubic phase can coexist with excess water without disruption of its intact structure (4).

The cubic phase is an optically isotropic, highly viscous, rigid gel, while the lamellar phase is a visually opaque, relatively less viscous, mucouslike semifluid exhibiting birefringence. Under a polarized light, the cubic phase is characterized by a dark background because of its optically isotropic nature. In contrast, the lamellar phase, which is anisotropic, exhibits multiple textures such as oily streaks and planar bilayers of the lipid with spherical and positive Maltese cross units on a dark background (11). Thus, the cubic and lamellar phases are easily identified by visual inspection or polarized light microscopy. One of the interesting phase properties of the GMO–water system is a lamellar-to-cubic phase transition induced by aqueous media or temperature (12). The lamellar phase is spontaneously transformed into the cubic phase upon contact with an aqueous environment such as biological fluids (e.g., saliva, gastric/intestinal juice, blood). The cubic phase is also formed from the lamellar phase by increasing temperature (e.g., from ambient to body temperature).

16.2.2 Pharmaceutical Characteristics of the Cubic Phase of GMO

16.2.2.1 Amphiphilic Nature of the Cubic Phase

Since the cubic phase of GMO possesses lipophilic and hydrophilic domains in its structure, it can solubilize, in principle, both lipophilic

Table 16.1 Examples of Drugs with Different Molecular Weight and Polarity That Have Been Incorporated in the Cubic Phase of GMO

Compound	MW	Conc. (% w/w)	Ref.
Sodium chloride	58	0.9	4
Melatonin	232	20	15
Pindolol	248	20	15
Pyrimethamine	249	20	15
Propranolol	260	20	15
Atenolol	266	20	15
Lidocaine	270	5	13
Hydroxychloroquine sulfate	434	2	14
Ubiquinone-10	863	0.5	16
Desmopressin	1069	4	4
Gramicidine	1141	6	4
Vitamin E polyethylene glycol succinate	1513	7	14
Insulin	6000	4	4
Bovine serum albumin	67,000	18	17

and hydrophilic drugs as well as drugs with pronounced amphiphilic characters. Table 16.1 lists drugs with varying molecular weights and polarities that have been incorporated in the cubic lattice structure. It also shows that, because of a large interfacial area (~300 to 400 m^2/g cubic phase) (7), polypeptides and proteins with high molecular weights such as insulin and bovine serum albumin, as listed in the table, could be loaded into the cubic liquid-crystalline structure without disrupting the intact liquid-crystalline structure.

If the water solubility of a drug is low, so that the dose exceeds the drug solubility in a certain volume of water used to prepare the cubic phase, then the drug crystals can be uniformly incorporated into the cubic phase by suspending them in the molten GMO and then adding water (14).

16.2.2.2 Bio/Muco-Adhesive Properties of the Cubic Phase

Bio/muco-adhesive materials have attracted significant interest in attempts to retain a drug or drug delivery system at a biological tissue surface to increase drug absorption and improve drug bioavailability (18,19). The cubic and lamellar liquid-crystalline phases of

GMO have been demonstrated to be bio/muco-adhesive (2,7). In these studies, *in vitro* methodologies revealed that both the cubic and lamellar phases possess a mucoadhesive property and that the lamellar phase showed a greater mucoadhesive force than did the cubic phase on mucous tissues including porcine lingual, sublingual, and intestinal mucosas and rabbit jejunum. Further, it was shown that solvents such as water, excipients that affect the formation of liquid-crystalline phases, and incorporated drug concentration influence the mucoadhesive behavior of liquid-crystalline phases of GMO (2).

Recently Lee et al. (20) have reported the impact of experimental conditions on the mucoadhesion properties exhibited by GMO and its lamellar and cubic liquid-crystalline phases, together with a quantitative relationship between the initial amount of water in the phases and the mucoadhesive properties. Two different substrates, mucus gel (20% w/w porcine stomach mucin) and porcine buccal mucosa, were utilized for *in vitro* mucoadhesion studies, where tensile strength was determined as a parameter for mucoadhesive force. Overall, the mucoadhesive force significantly decreased when measured using porcine buccal mucosa, which has less moisture available on the surface. The mucoadhesion force increased greatly when the contact time between a substrate and liquid-crystalline phases increased. This occurs because as the contact time increases, the water uptake by GMO and its liquid-crystalline phases increases. GMO and liquid-crystalline phases of GMO are known to swell upon contact with water or aqueous media until the equilibrium amount of water is contained in the phases (21). Since the cubic phase was found to be an equilibrium state with regard to the amount of water, the water uptake of the cubic phase is minimal. On the other hand, GMO itself, having no initial water, rapidly takes up water to reach equilibrium. Thus, the mucoadhesive force of GMO increased very rapidly with some lag period, during which intimate contact was made between GMO and the substrate. The mucoadhesive force of the liquid-crystalline phases of GMO containing 16, 20, 26, and 35% w/w initial water in the phases was measured (Figure 16.1).

Plotting the peak detachment forces versus initial water content of the liquid-crystalline phases demonstrated an inverse relationship. This quantitative relationship suggested that the amount of water taken up by the liquid-crystalline phases determines the mucoadhesive force since, among the liquid-crystalline phases tested, the phase containing the least amount of water showed the greatest mucoadhesion. From the dependence of the mucoadhesion on water

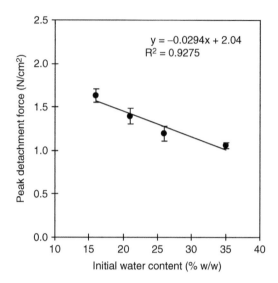

Figure 16.1 Plot of the relationship between peak detachment force and initial water concentration of glyceryl monooleate–water liquid-crystalline phases measured with mucus substrate. The peak detachment force was considered as a mucoadhesive force. The force per square centimeter required to separate the liquid-crystalline phases from the biological substrates was calculated gravimetrically. Mean ± s.d., n = 4. (From J Lee, SA Young, IW Kellaway. *J Pharm Pharmacol* 53: 629–636, 2001.)

uptake, it could be suggested that the mucoadhesion occurs through dehydration of the substrate.

16.2.2.3 Physicochemical Stabilization of Incorporated Drugs

The physical and chemical stabilities of drugs have been improved when incorporated into the cubic phase of GMO. The cubic phase protected cefazolin and cefuroxime from chemical degradation reactions such as hydrolysis and oxidation (22). The cubic phase could also protect insulin from physical instabilities such as agitation-induced aggregation and subsequent precipitation (23). This protective action against physical and chemical instabilities is also considered one of the advantageous features of the cubic phase as a drug delivery system.

16.3 THE CUBIC PHASE OF GMO AS A CONTROLLED-RELEASE DRUG DELIVERY SYSTEM

16.3.1 Oral Delivery

Although a drug-containing cubic phase can be easily prepared by simply dispersing drugs either in molten GMO or in water depending on solubility and subsequently mixing GMO and water, it is, in practice, difficult to produce oral dosage forms from highly ordered bulk cubic phases due to the highly viscous nature of the cubic phase. Alternatively, the lamellar phase and GMO that spontaneously swell to form the cubic phase *in vivo* can be administered. Drugs are dispersed in the semifluid lamellar phase, which is then dispensed into, for example, soft gelatin capsules for oral administration. If drugs are blended with a molten GMO, this molten GMO can be placed in a mold to obtain solid oral dosage forms of GMO, such as small discs or tablets.

For hydrophilic drugs, the drug release from the cubic phase is sustained by water pores, present in the cubic phase, with a diameter of about 5 nm (4). Indeed, the water channels in the liquid-crystalline phases of GMO play an important role in the drug release process. Drug release from liquid-crystalline phases increases with increasing initial water content of the phases (21). Thus, drug release from the cubic phase is greater than from the lamellar phase. In most cases, the drug release follows the typical square root of time dependence as generally observed in a matrix system, indicating that the release of a drug from the cubic phase is a diffusion-controlled process (1,21,24,25). However, if the drug solubility in water is too low, the controlling factor for drug release is dissolution rate, rather than diffusion rate. For pyrimetha-mine, which has limited water solubility (0.0562 mg/ml), an increase in drug concentration from 1 to 20% w/w, which also corresponds to the increase in surface area of the drug exposed to the release medium, resulted in an increased rate and amount of drug dissolved. This increase reached a point where the rate of dissolution was equal to the rate of diffusion through the liquid-crystalline phase. At this point, the drug located in the lipid domain of the liquid-crystalline phase acted as a depot for drug release (26).

When administered orally, the drug release kinetics and phase transition profile of the liquid-crystalline phases are affected by gastrointestinal tract contents, especially surface-active molecules such as bile salts. In fact, sodium taurocholate could dissolve the cubic phase by formation of mixed micelles with GMO (27). The destruction extent was largely dependent on the concentration of sodium taurocholate. This destruction profile of the cubic phase could be predicted by the construction of the phase diagram for GMO, water, and sodium taurocholate (28). Because GMO is formed during fat digestion, and the cubic phase can be formed *in vivo* during this process (29), the presence of lipase (e.g., pancreatic lipases) can therefore have an impact on the integrity of the cubic phase, resulting subsequently in a change in drug release kinetics from the phase (30,31).

16.3.2 Oral Mucosal Delivery

The oral cavity has been selected extensively as a potential site for local and systemic delivery of therapeutic agents. Local therapy has treated conditions such as gingivitis, oral candidiasis, and dental caries (32). Since the successful delivery of glyceryl trinitrate across the sublingual mucosa, the oral cavity has extensively been investigated for systemic delivery (33). Representative delivery systems for this route include bioadhesive tablets, gels, patches, and chewing gums. Of these, gel-type delivery systems are primarily used for a GMO–water oral mucosal delivery system. Tetracycline-containing GMO gel has been formulated for periodontal therapy (34,35). This gel could easily be applied to the oral cavity (e.g., gingival crevice), removed when needed, and cleared by lipolysis. Furthermore, tetracycline demonstrated a long-term prolonged release from GMO gel, ensuring a constant concentration of the drug at the application site. The prolonged effect with regard to drug release from the periodontal pocket is principally attributed to the fact that the liquid-crystalline gel is most likely surrounded by a stagnant layer, causing a slow diffusion of the drug as occurs in nonsink conditions (34).

The prolongation of drug release from intraorally administered liquid-crystalline phases, however, is often limited by mechanical abrasion or erosion caused by the movement of the mouth, especially when speaking and masticating, and the loss or continuous dilution of the drug by continuous oral secretions such as salivary flow. This washed-out fraction of drug will certainly result in a decreased bioavailability. The rate and extent of the washout of a model peptide, [D-Ala^2, D-Leu^5]enkephalin (DADLE), from cubic and lamellar liquid-crystalline buccal delivery systems have been examined using a donor compartment flow-through diffusion cell (36). GMO, a major liquid-crystalline gel matrix component, was liberated from the cubic and lamellar liquid-crystalline matrices and permeated the porcine buccal mucosa. This liberated GMO is believed to act as a permeation enhancer during the *ex vivo* DADLE buccal permeation process. The release rate of GMO was greater from the cubic phase than from the lamellar phase, resulting in a greater permeation-enhancing effect by the cubic phase (37). The cubic phase demonstrated greater resistance than the lamellar phase to the flow of simulated saliva (phosphate-buffered saline, PBS), which may mechanically erode the liquid-crystalline matrix. The difference in the measurement of DADLE and GMO washed out from the cubic and lamellar phases was statistically nonsignificant (Figure 16.2). This is probably because the aqueous release medium (PBS) that flowed through the donor compartment (i.e., the application site of the liquid-crystalline phases) caused a rapid transformation of the lamellar phase to the cubic phase at the interface between the lamellar phase and PBS stream. Although the washout of GMO from the cubic and lamellar phases indicates the erosion of the liquid-crystalline phases, the amount of GMO washed out was very small, implying that there was no substantial breakdown of the liquid-crystalline matrices by the flow of PBS.

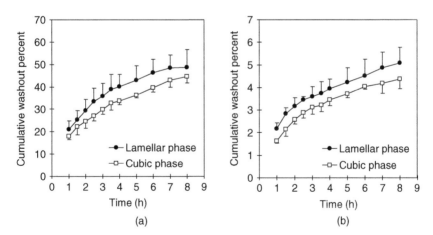

Figure 16.2 Percentage of [D-Ala^2, D-Leu^5]enkephalin (a) and glyceryl monooleate (b) washed out from the cubic and lamellar liquid-crystalline phases of glyceryl monooleate in the donor compartment flow-through diffusion cell. Mean ± s.d., n = 4. Note the scale of the y-axis. (From J Lee, IW Kellaway. *Drug Dev Ind Pharm* 28: 1155–1162, 2002.)

16.3.3 Topical Delivery

The cubic phase of GMO has been employed as a topical delivery system for acyclovir (38). This formulation takes advantage of the bioadhesive property of the cubic phase in order to reduce the number of daily applications. The solubility of acyclovir in the cubic phase was poor (0.1% w/w); thus, the drug had to be suspended in the cubic phase. The diffusion-controlled release of acyclovir from the cubic phase was assumed to follow a three-stage process (38) involving

1. Dissolution of suspended drug in the#l system,
2. Diffusion of dissolved drug through the system
3. Transfer of drug across the interface formed by the cubic phase and release medium

The release rate of acyclovir was faster from the cubic phase than from a cream formulation despite its relatively low solubility in the cubic phase. This study showed that the cubic phase might be a promising topical delivery system.

16.4 FUTURE DIRECTIONS FOR THE CUBIC PHASE DELIVERY SYSTEM

The potential of the cubic liquid-crystalline phase of GMO coupled with its inherent advantages, such as amphiphilicity, biodegradability, biocompatibility, high internal surface area, bioadhesiveness, and stability-enhancing property, will inevitably lead to its utilization in delivering challenging molecules including polypeptides and vaccines as well as low-molecular-weight hydrophilic and hydrophobic compounds. Preliminary reports of such approaches are already appearing in the literature. For instance, the cubic and L_2 phases of GMO have been investigated as parenteral or mucosal immunological adjuvants (39). Since the cubic phase is mainly formulated into a matrix, the kinetics and mechanism of drug release from the cubic phase is observed to follow a square root of time release kinetics. However, little attention has been paid to systematic investigation into drug release mechanisms based on the interaction between the physicochemical characteristics of the drug and the unique structure of the cubic phase gel (40). Furthermore, it will be necessary to consider *in vivo* interactions between the drug, cubic phase delivery system, and the environment, where particular physiological and biochemical events influence drug release and digestion or hydrolysis of the cubic phase. To make the cubic phase suitable for parenteral administration (e.g. intravenous injection), the cubic phase can be dispersed in water. This particulate system will be useful in preparing intravenous formulations for peptide and protein drugs. Currently, many studies focus on the preparation and characterization of cubic liquid-crystalline nano- and microparticles (41–43).

16.5 CONCLUSIONS

This chapter has discussed the cubic liquid-crystalline phase as a controlled-release drug delivery system. The cubic phase of GMO provides a unique opportunity in delivering a wide range of biologically active molecules by various mucosal and parenteral routes. The studies undertaken in the past have provided fundamental knowledge for the design of early controlled-release delivery systems. There is scope for improvement with regard to the prolongation and more precise control of drug release from the GMO cubic phase.

REFERENCES

1. A-S Sallam, E Khalil, H Ibrahim, I Freij. *Eur J Pharm Biopharm* 53: 343–352, 2002.

2. LS Nielsen, L Schubert, J Hansen. *Eur J Pham Sci* 6: 231–239, 1998.

3. A Ganem-Quintanar, D Quintanar-Guerrero, P Buri. *Drug Dev Ind Pharm* 26: 809–820, 2000.

4. S Engström. *Lipid Technol* 2: 42–45, 1990.

5. K Larsson. *J Phys Chem* 93: 7304–7314, 1989.

6. E Boyle, JB German. *Crit Rev Food Sci Nutr* 36: 785–805, 1996.

7. S Engström, H Ljusberg-Wahren, A Gustafsson. *Pharm Tech Eur* 7: 14–17, 1995.

8. K Larsson. *Nature* 304: 664, 1983.

9. W Longley, TJ McIntosh. *Nature* 303: 612–614, 1983.

10. ST Hyde, S Andersson, B Ericsson, K Larsson. *Zeitsch Kristallogr* 168: 213–219, 1984.

11. FB Rosevear. *J Am Oil Chem Soc* 31: 628–639, 1954.

12. S Engström, L Lindahl, R Wallin, J Engblom. *Int J Pharm* 86: 137–145, 1992.

13. S Engström, L Engström. *Int J Pharm* 79: 113–122, 1992.

14. DM Wyatt, D Dorschel. *Pharm Tech* 16: 116–130, 1992.

15. R Burrows, JH Collett, D Attwood. *Int J Pharm* 111: 283–293, 1994.

16. J Barauskas, V Razumas, T Nylander. *Chem Phys Lipids* 97: 167–179, 1999.

17. B Ericsson, K Larsson, K Fontell. *Biochim Biophys Acta* 729: 23–27, 1983.

18. SA Mortazavi, JD Smart. *J Control Rel* 31: 207–212, 1994.

19. A Ahuja, RK Khar, J Ali. *Drug Dev Ind Pharm* 23: 489–515, 1997.

20. J Lee, SA Young, IW Kellaway. *J Pharm Pharmacol* 53: 629–636, 2001.

21. J Lee, IW Kellaway. *Int J Pharm* 195: 29–33, 2000.

22. Y Sadhale, JC Shah. *Pharm Dev Tech* 3: 549–556, 1998.

23. Y Sadhale, JC Shah. *Int J Pharm* 191: 51–64, 1999.

24. S Engström, K Larsson, B Lindman. *Proc Intern Symp Control Rel Bioact Mater* 15: 105–106, 1988.

25. PB Geraghty, D Attwood, JH Collett, Y Dandiker. *Pharm Res* 13: 1265–1271, 1996.

26. R Burrows, D Attwood, JH Collett. *J Pharm Pharmacol* 42 (Suppl): 3P, 1990.

27. B Ericsson, PO Eriksson, JE Löfroth, S Engström. *ACS Symp Ser* 469: 251–265, 1991.

28. M Svärd, P Schurtenberger, K Fontell, B Jönsson, B Lindman. *J Phys Chem* 92: 2261–2270, 1988.

29. JS Patton, MC Carey. *Science* 204: 145–148, 1979.

30. R Wallin, T Arnebrant. *J Colloid Interface Sci* 164: 16–20, 1994.

31. L Zhou, T Landh, B Sternby, Å Nilsson. *Progr Colloid Polym Sci* 120: 92–98, 2002.

32. T Nagai, R Konishi. *J Control Rel* 6: 353–360, 1987.

33. ME de Vries, HE Boddé, JC Verhoef, HE Junginger. *Crit Rev Ther Drug Carr Syst* 8: 271–303, 1991.

34. E Esposito, V Carotta, A Scabbia, L Trombelli, P D'Antona, E Menegatti, C Nastruzzi. *Int J Pharm* 142: 9–23, 1996.

35. N Damani. US Patent 5262164, 1993.

36. J Lee, IW Kellaway. *Drug Dev Ind Pharm* 28: 1155–1162, 2002.

37. J Lee, IW Kellaway. *Int J Pharm* 195: 35–38, 2000.

38. LS Helledi, L Schubert. *Drug Dev Ind Pharm* 27: 1073–1081, 2001.

39. U Schröder, E Björk, P Artursson. *Proc Intern Symp Control Rel Bioact Mater* 24: 573–574, 1997.

40. JC Shah, Y Sadhale, DM Chilukuri. *Adv Drug Del Rev* 47: 229–250, 2001.

41. M Nakano, A Sugita, H Matsuoka, T Handa. *Langmuir* 17: 3917–3922, 2001.

42. PT Spicer, KL Hayden, ML Lynch, A Ofori-Boateng, JL Burns. *Langmuir* 17: 5748–5756, 2001.

43. B Siekmann, H Bunjes, MHJ Koch, K Westesen. *Int J Pharm* 244: 33–43, 2002.

Index

T - #0212 - 111024 - C0 - 234/156/24 - PB - 9780367392871 - Gloss Lamination